Henry Hartshorne

A Hand-Book of Human Anatomy and Physiology

Henry Hartshorne

A Hand-Book of Human Anatomy and Physiology

ISBN/EAN: 9783337365608

Printed in Europe, USA, Canada, Australia, Japan

Cover: Foto ©berggeist007 / pixelio.de

More available books at **www.hansebooks.com**

A

HAND-BOOK

OF

HUMAN·ANATOMY AND PHYSIOLOGY.

FOR THE

USE OF STUDENTS.

BY

HENRY HARTSHORNE, A.M., M.D.,

PROFESSOR OF ORGANIC SCIENCE AND PHILOSOPHY IN HAVERFORD COLLEGE; PROFESSOR
OF HYGIENE IN THE UNIVERSITY OF PENNSYLVANIA, AUXILIARY FACULTY OF
MEDICINE; PROFESSOR OF PHYSIOLOGY AND HYGIENE IN PENNSYLVANIA
COLLEGE OF DENTAL SURGERY; ETC. ETC.

WITH ONE HUNDRED AND SIXTY-SIX ILLUSTRATIONS.

PHILADELPHIA:
HENRY C. LEA.
1869.

Entered according to Act of Congress, in the year 1869, by

HENRY C. LEA,

in the Clerk's Office of the District Court of the United States in and for the Eastern District of Pennsylvania.

PHILADELPHIA:
COLLINS, PRINTER, 705 JAYNE STREET.

PREFACE.

The purpose of the author of this Hand-Book has been to present in the briefest manner all that is most essential in Human Anatomy, and all that is most positive and important in Physiology, for the especial use of the student during his attendance upon lectures. Although designed for medical students, it is believed that it will not be too technical for others; particularly if its use be combined with oral instruction and demonstration. The illustrations have been selected with much care from standard authorities.

LIST OF ILLUSTRATIONS.

ANATOMY.

FIG.		PAGE
1.	Human Skeleton	24
2.	Transverse Section of an Old Tibia	25
3.	Lateral View of the Spinal Column	26
4.	The Atlas	27
5.	A Dorsal Vertebra	27
6.	View of a Lumbar Vertebra	28
7.	Left Temporal Bone	30
8.	Anterior and Inferior Surface of the Sphenoid Bone	32
9.	Upper and Posterior View of the Ethmoid Bone	33
10.	Superior Maxilla	35
11.	The Nasal Cavities	36
12.	The Palate Bone	36
13.	Left Nasal Fossa	37
14.	Front View of the Skull	39
15.	External View of the Base of the Cranium	40
16.	Internal Surface of the Base of the Cranium	41
17.	Front View of the Thorax	42
18.	Anterior View of the Male Pelvis	44
19.	Venter of Scapula	46
20.	Anterior View of Humerus of the Right Side	47
21.	Anterior View of Radius of Right Side	48
22.	Anterior View of Ulna of the Left Side	49
23.	Articulation of Bones of the Carpus	50
24.	Posterior View of the Femur	51
25.	Anterior View of the Tibia	52
26.	The Fibula	53
27.	Upper Surface of the Left Foot	54
28.	Eight Teeth of One Side of the Upper Jaw	56
29.	Eight Teeth of One Side of Lower Jaw	56

FIG.		PAGE
30.	Magnified Section of a Tooth	57
31.	Teeth at Five Years	58
32.	Ligaments of Acromio-Clavicular and Scapulo-Humeral Articulations	62
33.	Internal View of the Elbow-Joint	63
34.	Ligaments of the Hip-Joint and Pelvis	64
35.	The Knee-Joint laid open	65
36.	Vertical Section of the Ankle-Joint and Foot	67
37.	Upper Surface of the Tongue	69
38.	Salivary Glands	70
39.	Parietes of the Abdomen	73
40.	The Peritoneum	74
41.	Organs of Digestion	76
42.	Viscera, after removal of the Fat in the Chest and the Omentum Majus of the Abdomen	78
43.	The Large Intestine	80
44.	The Larynx	83
45.	View of the Larynx from Above	85
46.	Bronchi and Bloodvessels	88
47.	Minute Structure of the Testis	95
48.	Uterus and its Appendages	98
49.	Viscera of Female Pelvis	100
50.	Section of Mammary Gland	101
51.	The Left Ventricle	102
52.	Interior of the Right Ventricle	103
53.	Arteries of the Arm and Shoulder	111
54.	One of the Anomalies of the Brachial Artery	112
55.	Arteries of the Pelvis and Thigh	118
56.	Anterior Tibial Artery	119
57.	Superficial Bones of the Upper Extremity	123
58.	Superficial Veins of the Legs	126

vi LIST OF ILLUSTRATIONS.

FIG.		PAGE
59.	Lymphatics of the Jejunum and Mesentery, injected; the Arteries are also injected	128
60.	Femoral Iliac and Aortic Lymphatic Vessels and Glands .	129
61.	Sebaceous Glands and Follicles of Hairs in the Skin of the Axilla .	131
62.	Sudoriferous Organs of the Sole of the Foot	132
63.	Muscles—front view . .	135
64.	Muscles—back view . . .	141
65.	Superior Muscles of the Upper Front of the trunk . . .	143
66.	Lateral View of the Muscles of the Trunk	146
67.	Abdominal Muscles and Inguinal Canal	147
68.	Second Layer of Muscles of the Back	149
69.	Outer Layer of Muscles on the Front of the Forearm . . .	152
70.	Outer Layer of Muscles on the Back of the Forearm . . .	154
71.	Deep-Seated Muscles on the Posterior Part of the Hip-Joint . .	157

FIG.		PAGE
72.	Muscles of the Back of the Thigh .	159
73.	Muscles of the Front of the Leg .	160
74.	Anterior View of the Brain and Spinal Marrow	162
75.	Base of the Cerebrum and Cerebellum	165
76.	Longitudinal Section of the Brain	166
77.	Lateral Ventricles of the Cerebrum	167
78.	The Second Pair, or Optic Nerves	173
79.	Distribution of the Fifth Pair .	174
80.	The Nerves	181
81.	The Brachial Plexus . . .	182
82.	Nerves of the Front of the Forearm	184
83.	Posterior Tibial Nerve . . .	189
84.	Anterior Tibial Nerve . . .	190
85.	Thoracic Ganglia	194
86.	Section of the Eye . . .	196
87.	Transverse Section of the Eye .	198
88.	Crystalline lens divided . .	198
89.	Muscles of the Eyeball . . .	198
90.	Lachrymal Canals	199
91.	Ossicles of the Ear . . .	200
92.	View of the Ear	201
93.	Arteries of the Perineum . .	204

PHYSIOLOGY.

		PAGE			PAGE
94.	Blood Corpuscles	218	116. Follicles of Pig's Stomach . .	236
95.	Blood Coagula	219	117. Villi of Intestine	237
96.	Blood Coagula	219	118. Piece of Ileum	237
97.	Blood Crystals	219	119. Thymus	238
98.	Nucleated Cells	222	120. The Lymphatics	239
99.	Multiplication of Cells . . .	223	121. Diagram of the Circulation . .	241
100.	Pavement Epithelium . . .	225	122. Semilunar Valves . . .	242
101.	Ciliated Epithelium . . .	225	123. Veins of the Base of the Heart .	243
102.	White Fibrous Tissue . . .	227	124. Capillaries	245
103.	Yellow Fibrous Tissue . . .	227	125. Capillaries of a Tooth . . .	245
104.	Fibrous Cartilage	227	126. Diagram of Air-Cells . . .	247
105.	Osseous Tissue	228	127. Brunner's Gland, magnified . .	252
106.	Fat Vesicles	228	128. Lobule of Liver	253
107.	Striped Muscle	229	129. Lobule of Liver	254
108.	Smooth Muscle	229	130. Section of Liver of the Horse .	255
109.	Nerve Cells	230	131. Bile-duct and Cells . . .	256
110.	Nerve Filaments	230	132. Section of Kidney . . .	257
111.	Hand of Man and of Orang . .	232	133. Structure of Kidney . . .	258
112.	Deglutition	234	134. Section of Skin	259
113.	The Stomach	235	135. Unfertilized Ovum . . .	261
114.	Mucous Membrane of the Stomach, magnified	235	136. Graafian Vesicle and Ovum .	262
			137. Spermatozoa	263
115.	Perpendicular Section of the same	235	138. The Biceps Muscle . . .	266

LIST OF ILLUSTRATIONS.

FIG.		PAGE
139.	Diagram of a Ganglion	268
140.	Ganglion-cells	271
141.	Junction of Spinal and Sympathetic Systems	272
142.	Section of Spinal Cord	273
143.	Diagram of Encephalon	275
144.	Medulla Oblongata	276
145.	Cerebellum	277
146.	Longitudinal Section of the Brain	278
147.	Chiasm of Optic Nerves	280
148.	Papillæ of Palm	285
149.	Pacinian Corpuscles	285
150.	Pacinian Corpuscle	285
151.	Fungiform Papillæ of Tongue	286
152.	The Nasal Cavity	287
153.	The Ear	288
154.	Ossicles of the Ear	289
155.	Labyrinth of the Ear	289
156.	The Cochlea	290
157.	Section of Eyeball	293
158.	Stereoscopic Vision	297
159.	Glottis, from above	299
160.	The Larynx opened	300
161.	The Amnion and Chorion	303
162.	Diagram representing a Human Ovum in the Second Month	303
163.	Diagram of the Fœtus and Membranes about the Sixth Week	305
164.	The Fœtal Circulation	308
165.	Fœtal Skeleton	310
166.	Development of Teeth	311

A MANUAL

OF

ANATOMY.

CONTENTS.

	PAGE
DEFINITIONS	23

CHAPTER I.
BONES.

OSSEOUS TISSUE	23
Composition	23
Structure	24
Number of the Bones	26
SPINE	26
Cervical Vertebræ	27
Dorsal Vertebræ	27
Lumbar Vertebræ	28
Sacrum	28
Os Coccygis	28
HEAD	28
Cranium	28
Frontal Bone	29
Parietal Bones	29
Occiput	29
Temporal Bones	30
Sphenoid Bone	31
Ethmoid Bone	33
Bones of the Face	34
Malar Bones	34
Superior Maxillary Bones	34
Nasal Bones	35
Lachrymal Bones	36
Palate Bones	36
Vomer	37
Inferior Turbinated Bones	37
Inferior Maxillary Bone	38

CONTENTS.

	PAGE
General Remarks on the Head	38

Sutures, 38—Diploë, 38—Exterior Regions, 39—Interior Regions of the Cranium, 39—Orbit, 39—Nasal Cavities, 39—Ossa Triquetra, 40—Fontanels, 40—Diameters, 40—Capacity and Form, 41—Facial Angle, 41.

	PAGE
Hyoid Bone	42
THORAX	42
Sternum	42
Ribs	43
PELVIS	43
Ossa Innominata	43
Ilium	43
Pubes	45
Ischium	45
Acetabulum	45
Thyroid Foramen	45
UPPER EXTREMITY	46
Shoulder	46
Scapula	46
Clavicle	47
Arm	47
Humerus	47
Forearm	48
Radius	48
Ulna	49
Carpus	49
Metacarpal Bones	50
Fingers	50
LOWER EXTREMITY	51
Femur	51
Patella	52
Tibia	52
Fibula	53
Tarsus	53
Metatarsal Bones	55
Toes	55
TEETH	56
Structure	57
Development	57

CHAPTER II.
ARTICULATIONS.

	PAGE
Vertebral Articulations	60
Occiput and Atlas	60
Atlas and Axis	60
Pelvic Ligaments	60
Temporo-maxillary Articulations	61
Thoracic Articulations	61
The Shoulder	62
The Elbow	62
The Wrist	63
Articulations of the Hand	64
Hip-joint	65
Knee-joint	65
Tibio-fibular Articulations	66
Ankle-joint	66
Tarsal Articulations	66
Metatarsal and Phalangeal Articulations	67

CHAPTER III.
DIGESTIVE ORGANS.

Mouth	68
Tongue	68
Salivary Glands	69
Palate	70
Pharynx	71
Œsophagus	72
Abdomen	72
Peritoneum	74
Stomach	75
Intestines	77
Duodenum	77
Jejunum	77
Ileum	77
Mucous Membrane of Small Intestine	77
Large Intestine	79
Cæcum	79
Colon	79
Rectum	79
Mucous Membrane of Large Intestine	79

2*

	PAGE
Liver	
Gall-bladder	82
Spleen	82
Pancreas	82

CHAPTER IV.
ORGANS OF RESPIRATION.

Larynx	83
Trachea	86
Thyroid Gland	86
Thymus Gland	86
Lungs	87
Pleuræ	89

CHAPTER V.
URINARY AND GENITAL ORGANS.

Kidneys	89
Supra-renal Capsules	91
Bladder	91
Urethra	92
Prostate Gland	93
Cowper's Glands	93
Penis	94
Testes	95
Vesiculæ Seminales	96
Spermatic Cord	96
Descent of Testes	97
ORGANS OF GENERATION IN THE FEMALE	97
Ovaries	97
Fallopian Tubes	98
Uterus	98
Vagina	99
Urethra in the Female	100
Mammary Glands	101

CHAPTER VI.
ORGANS OF CIRCULATION.

HEART	101
PERICARDIUM	104

	PAGE
ARTERIES	104
Aorta	104
Pulmonary Artery	105
Coronary Arteries	105
Innominata	106
Primitive or Common Carotid	106
External Carotid	106
Internal Carotid	108
Subclavian Artery	108
Vertebral Artery	109
Thyroid Axis	110
Internal Mammary	110
Superior Intercostal	110
Axillary Artery	110
Brachial Artery	111
Radial Artery	112
Ulnar Artery	113
Thoracic Aorta	113
Abdominal Aorta	113
Cœliac Axis	114
Superior Mesenteric	114
Inferior Mesenteric	114
Supra-Renal Arteries	114
Renal or Emulgent Arteries	114
Spermatic	115
Phrenic	115
Lumbar Arteries	115
Middle Sacral Artery	115
Common Iliac	115
Internal Iliac	115
Sciatic Artery	116
Gluteal Artery	116
Ilio-lumbar	116
Lateral Sacral	116
External Iliac Artery	116
Epigastric Artery	117
Circumflex Iliac	117
Femoral Artery	117
Popliteal Artery	119
Anterior Tibial Artery	119
Posterior Tibial	120
Peroneal Artery	120
Plantar Arteries	120

	PAGE
VEINS	120
Exterior Veins of the Head	121
Veins of the Neck	122
Veins of the Interior of the Skull	122
Veins of the Arm and Hand	123
Veins of the Thorax	124
Spinal Veins	125
Veins of the Lower Extremity	125
Portal System	127
Pulmonary Veins	127
Cardiac Veins	128
LYMPHATICS	128
LEFT THORACIC DUCT	128
RIGHT THORACIC DUCT	129

CHAPTER VII.
THE SKIN.

The Nails	130
The Hair	130
Glands of the Skin	131
Connective Tissue	131
Fat	131

CHAPTER VIII.
MUSCLES.

Muscles of the Head and Face	134
Muscles of the Neck	138
Muscles of the Trunk	143
Muscles of the Shoulder	151
Muscles of the Arm	151
Muscles of the Forearm	152
Anterior, Superficial Layer	152
Deep-seated Anterior Layer	153
Posterior Region	153
Superficial Layer	153
Posterior Deep-seated Layer	154
Muscles of the Hand	155
Muscles of the Pelvis and Thigh	156
Muscles of the Leg	159

CHAPTER IX.
NERVOUS SYSTEM.

	PAGE
THE BRAIN	163
Membranes	163
Dura Mater	163
Arachnoid	164
Pia Mater	164
Brain or Encephalon	164
Cerebrum	164
Cerebellum	169
Medulla Oblongata	170
SPINAL CORD	171
Membranes	171
Fissures of the Cord	171
Columns	172
Structure of Cord	172
Origin of Spinal Nerves	172
CRANIAL OR CEPHALIC NERVES	172
Olfactory, 172—Optic, 173—Motor Oculi, 173—Pathetic, 173—Trifacial, 173—Abducens Oculi, 176—Facial, 176—Auditory, 177—Glossopharyngeal, 177—Pneumogastric, 178—Spinal Accessory, 179—Hypoglossal, 179.	
SPINAL NERVES	180
Cervical	180
Median Nerve	183
Ulnar Nerve	184
Musculo-spinal Nerve	185
Radial Nerve	185
Dorsal Nerves	186
Intercostal Nerves	186
Lumbar Nerves	186
Lumbar Plexus	187
Anterior Crural Nerve	187
Sacral and Coccygeal Nerves	187
Sacral Plexus	188
Great Sciatic Nerve	188
SYMPATHETIC NERVE	190
Cephalic Ganglia; Ophthalmic, Otic, and Submaxillary	191
Cervical Ganglia	192
Cardiac Plexuses	193
Thoracic Ganglia	194
Solar Plexus	194

Lumbar Ganglia
Pelvic Ganglia and Hypogastric Plexus

CHAPTER X.
ORGANS OF SPECIAL SENSE.

The Eye
Appendages of the Eye
The Ear
The Nose

CHAPTER XI.
ANATOMY OF HERNIA

CHAPTER XII.
ANATOMY OF THE PERINEUM

ANATOMY

Anatomy is the science of the *structure* of the human, or of any other, organized body; of all the parts or organs of which it is composed, and their relative positions in it.

Comparative Anatomy includes the study of the organization of other animals with that of man.

General Anatomy, *Histology*, and *Microscopic* Anatomy study the *plan* of structure of the body, the *tissues* of which it is composed, and its *minute* elements.

Special or *Descriptive* Anatomy gives account of the particular organs and all their relations.

Regional or *Surgical* Anatomy takes into consideration the local relations of important parts, especially with a view to surgical operations.

In Descriptive Anatomy, the different organs may be studied as follows: 1. Bones; 2. Articulations; 3. Viscera; 4. Vessels; 5. Tegument; 6. Muscles; 7. Nerves; 8. Organs of Sense.

CHAPTER I.

BONES.

OSSEOUS TISSUE.

Composition.

In bone, *earthy* and *animal* constituents are intimately combined. Of the former there are 66.7 parts to 33.3 of the latter. Phosphate of lime is the most abundant mineral material; being about 51 parts in the 100 of bone. Carbonate of lime 11.3 parts; fluoride of calcium 2 parts. The animal matter of bone is gelatinous, allied to cartilage; originally every bone is developed from cartilage, by *ossification*.

The mineral matter of bone increases with age; making the

24 ANATOMY.

bones of the old more brittle. There is more of it also in som
bones, and parts of bones, than in others.

Structure.

Bones are *long*, *thick*, or *flat* in shape. In the long bones espe

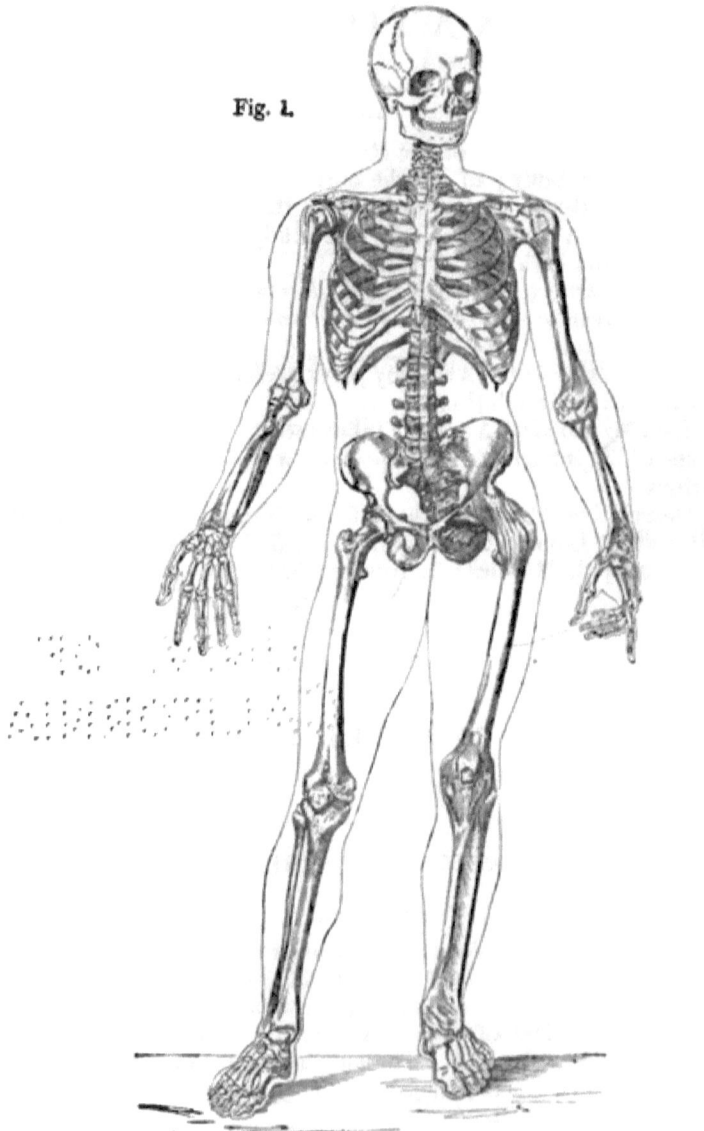

Fig. 1.

HUMAN SKELETON.—Ligaments removed from the right half; remaining on the left

cially, we distinguish a *compact* and a *cellular* structure; the latter at the ends, which are expanded. The shaft of a long bone, which is hollow, is called its diaphysis; each end, an epiphysis; any other projection, a process or apophysis.

When a bone is cut transversely, we see distinctly, with the aid of a microscope, canals (of *Havres*) running in the direction of its length, and communicating laterally, through radiating lesser channels, with minute reservoirs called the *corpuscles of Purkinje*. The nutrition of the bone is secured, by the liquor sanguinis or blood-lymph passing through these canals and cells, from the branches of the artery or arteries, which do not penetrate to nearly all its parts. The foramina or holes upon the surfaces of bones are for their vessels.

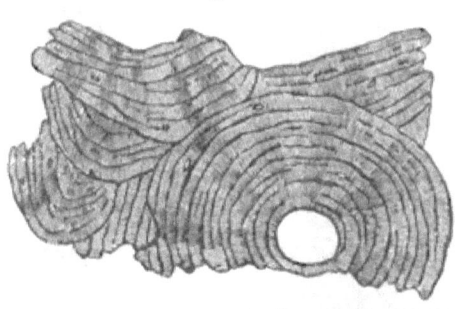

Fig. 2.

TRANSVERSE SECTION OF AN OLD TIBIA. (Magnified.)

By the shaft of the long bones being of compact substance, and hollow, while the ends are cellular and enlarged, the greatest strength is obtained, with economy of material; while the articulations are thus adapted for motion as well as support. The principle of the hollow shaft is illustrated elsewhere, in nature, in the stems of the grasses; and in art, in the tubular bridge.

The *marrow* of bones is a peculiar fatty substance. Bones are covered closely, and lined, by the external and internal *periosteum*.— This is a delicate and yet firm membrane, supplied with blood-vessels, and of great importance to the development, nutrition, and repair of the bone itself.

Development.—In the fœtus, bones commence their formation in *temporary cartilages;* in these, as they grow, osseous matter is deposited at and around the *points or centres of ossification;* which are different in number according to the complexity of the bone. Short bones may have but one such centre. The long bones have one for the shaft or diaphysis, and one for each epiphysis. The junction between the ends and shaft occurs at puberty or during adolescence. Some flat bones, as the frontal, and those of the pelvis, are in early life in separate parts, which afterwards are consolidated together.

Bones continue to undergo absorption and renewal of their particles through life; as interesting experiments upon animals

ANATOMY.

Fig. 3.

Number.

In the adult skeleton there are 206 bones, exclusive of the sesamoid and wormian bones, which are not uniform in number. They are, of the

Cranium	8
Bones of the ear	6
Face	14
Os hyoides, and sternum	2
Ribs	24
Upper extremities	64
Lower extremities	62
	206

The most convenient classification of the parts of the skeleton is into the Head, Trunk, and Extremities. The divisions of the Trunk are, the Spine, Thorax, and Pelvis.

THE SPINE.

In man, the spinal column has four curves; it is convex anteriorly in the neck, concave in the thorax, convex again in the lumbar region, and concave in the pelvis. It is smaller above, the bodies of the vertebræ increasing in size almost regularly from the neck to the sacrum. The vertebræ are separated from each other by the *intervertebral cartilages*, but firmly connected by ligaments. The *spinal cord* is contained within, and protected by, the spinal column; the nerves and bloodvessels of the cord passing through the intervertebral foramina.

There are 24 true, and 4 or 5 *false* vertebræ. Those called false are scarcely movable; viz., the sacrum, and 3 or 4 bones of the coccyx.

Each vertebra has a body and 7

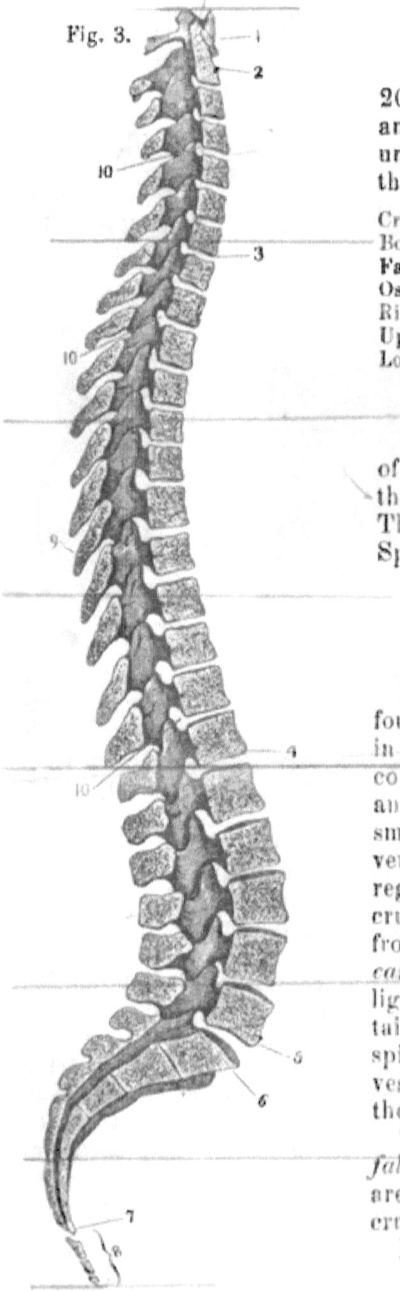

LATERAL VIEW OF THE SPINAL COLUMN.—1. Atlas. 2. Dentata. 3. Seventh cervical vertebra. 4. Twelfth dorsal vertebra. 5. Fifth lumbar vertebra. 6. First piece of sacrum. 7. Last piece of sacrum. 8. Coccyx. 9. A spinous process. 10, 10. Intervertebral foramina.

processes. The processes are, 1 *spinous*, pointing backwards; 2 *transverse;* and 4 *oblique* or articulating processes. Of the last, 2 are above and 2 below. The body of the vertebra is in front of the foramen for the spinal cord.

Seven Cervical Vertebræ.

Small bodies; spinous processes thick, short, straight, and bifid at the end. Transverse processes have each a foramen for the vertebral artery. Oblique processes oval, the superior ones looking upwards and backwards, the inferior, downwards and forwards. The spinal foramen is larger in the cervical than in any other portion of the spine, for free motion of the neck without injury of the cord.

The first vertebra, or *atlas*, is a bony ring merely, with only a tubercle in place of the spinous process; long transverse processes; upper oblique processes, large and concave, to receive the condyles of the occipital bone. The lower oblique processes are round and flat. Its spinal foramen is the largest of all.

The second, *vertebra dentata*, or *axis*, has the odontoid or dentate process rising from its body to enter the ring of the atlas, where it is confined by a ligament. Its spinous process is long and bifid.

The sixth and seventh cervical vertebræ are remarkable for the length of their spinous processes; that of the seventh is the longest. The foramen in its transverse process only gives passage to the vertebral vein; not to the artery.

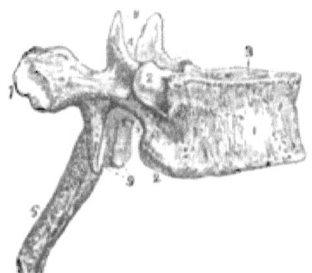

Fig. 4.

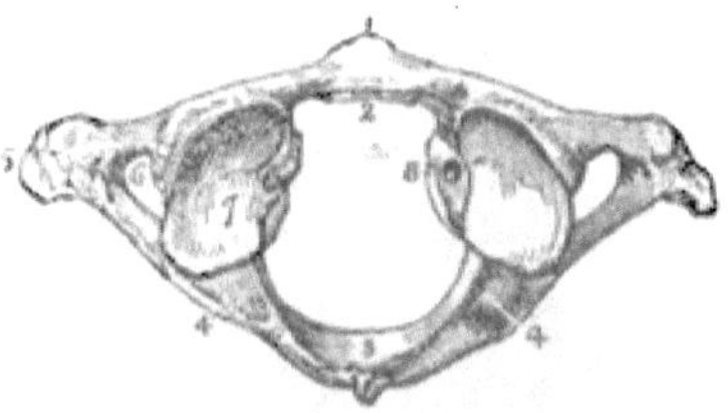

THE ATLAS.—1. Anterior tubercle. 2. Articular face for the dentata. 3. Posterior surface of spinal canal. 4, 4. Intervertebral notch. 5. Transverse process. 6. Foramen for vertebra artery. 7. Superior oblique process. 8. Tubercle for the transverse ligament.

Fig. 5.

A DORSAL VERTEBRA.—1 The body. 2. Face for the head of a rib. 3. Superior face of the body. 4. Superior half of the intervertebral notch. 5. Inferior half of the intervertebral notch. 6. Spinous process. 7. Articular face for the tubercle of a rib. 8. Two superior oblique processes. 9. Two inferior oblique processes.

Twelve Dorsal Vertebræ.

Bodies diminish in diameter from 1st to 3d, then increase to the last.

The 1st, 11th, and 12th have each a complete mark or *fossa* for a rib. Each of the others receives part of the end of one rib above, and part of the end of another below. The oblique processes, so called, are perpendicular; the upper one looking backwards, and the lower ones forwards. Transverse processes long, with marks in front for the junction of the ribs; except the 11th and 12th. Spinous processes long, pointed, and overlapping, as if pressed downwards.

Five Lumbar Vertebræ.

Large bodies, especially so in transverse diameter. Articulating processes vertical, upper ones looking inwards, lower ones outwards, so as to *interlock* the vertebræ. Transverse processes long and straight. Spinous processes short, straight, thick.

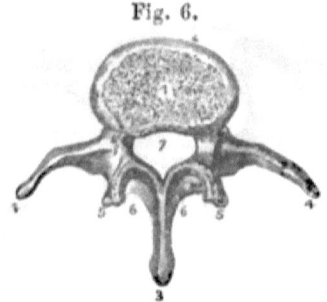

Fig. 6.

VIEW OF A LUMBAR VERTEBRA.—1. Face for the intervertebral substance. 2. Anterior surface of the body. 3. Spinous process. 4. Transverse process. 5. Oblique process. 6. A portion of the bony bridges. 7. The spinal foramen.

Sacrum.

A wedge-shaped bone, with its base above; concave in front, convex and rough behind. Being formed of five vertebræ compacted together, the marks of their union exist; incomplete spinous and oblique processes and foramina, four in front and four behind, on each side, for the nerves. The spinal marrow (*cauda equina*) is continued into a canal in the sacrum. The *base* of the sacrum closely resembles a lumbar vertebra. Its *sides* are rough, for junction with the ilia. Its *apex* has a surface for articulation with the coccyx.

Os Coccygis.

The coccyx consists of either three or four small bones, which frequently in the adult are united into a firm piece with the sacrum. They diminish in size from the first to the last.

THE HEAD.

There are, of the *cranium*, eight bones; of the *face*, fourteen.

CRANIUM.

The eight bones are, the frontal, two parietal, two temporal, occipital, sphenoid, and ethmoid. Each is composed of an *external and internal* table, and a *diploë*, or cellular structure between these.

Frontal Bone.

Convex outside; on each side in front a slight, round protuberance, marking the *puncta ossificationis;* the *orbitar ridge* on either side below, terminating externally in the *external angular process,* and on the nasal side in the *internal angular process.* Below and between the internal angular processes is the *nasal spine.* The *orbitar processes,* passing backwards from the orbitar ridges, roof the orbits of the eyes.

The internal surface of this bone has marks for the convolutions of the brain, and, along its middle, a ridge for the connection of the dura mater, and a fossa for its superior longitudinal sinus; the *foramen cæcum* is the passage for the vein in which this sinus begins.

The *frontal sinuses* are cavities, above the orbitary ridges, between the tables of the bone, and communicating with cavities in the ethmoid bone. They vary in size.

The most notable *foramen* of this bone is the *supra-orbital* foramen, which is sometimes only a notch, over the middle of each brow; it transmits the supra-orbital artery and nerve.

The frontal bone *articulates* with the two parietal, the sphenoid, ethmoid, nasal, upper maxillary, lachrymal, and malar. This bone is double, laterally, in infancy; the two halves uniting into one as life advances.

Two Parietal Bones.

These are four-sided bones, forming the lateral walls of the skull. The *parietal protuberance,* on each, marks the point of ossification. The lower portion, outside, is overlapped by the squamous portion of the *temporal* bone.

Between the upper edges of the two parietal bones is formed the groove for the superior longitudinal sinus of the dura mater.

Behind, the parietal bones connect with the *occiput;* below and in front, by an angular projection, with the *sphenoid;* in front and above, with the *frontal.*

Internally, these bones show marks of convolutions of the brain, and others for the middle artery of the dura mater.

Occiput.

Of rounded form, between and below the parietals, this bone forms the back of the head. It is the thickest of the cranial bones. Outside, is the *external occipital cross;* within, a corresponding *internal cross.*

The *foramen magnum occipitis* is a large opening for the spinal marrow, vertebral artery, and spinal accessory nerves to pass through. Anterior to it is the *basilar process,* going forwards to join with the sphenoid bone.

The *condyloid processes* are on each side of the foramen magnum, resting upon the atlas.

The *anterior condyloid foramen* transmits the ninth or hypoglossal nerve; through the *posterior* condyloid foramen a vein passes to the lateral sinus of the dura mater.

The cross on the internal surface of the occiput is very prominent, giving origin to grooves for sinuses of the dura mater. The posterior lobes of the cerebrum rest upon the superior portions of the concavity of the bone, as marked by the cross; the two halves of the cerebellum are supported by the inferior portions.

The occipital bone articulates with the *two parietal*, the *sphenoid*, and the *two temporal*. The *jugular eminence* is on its lower edge, on each side. Before this is the fossa, which by union with the temporal bone is converted into the *posterior foramen lacerum;* through this pass the eighth nerves and the internal jugular vein.

Two Temporal Bones.

Each of these is composed of a *squamous, mastoid,* and *petrous* portion.

The *squamous* part is thin, and partly overlaps the parietal bone. The temporal artery grooves its external surface; which is covered by the temporal muscle. The zygomatic process projects forward from the middle of the exterior of the bone, to form the *zygomatic arch* with a process of the malar bone.

Fig. 7.

LEFT TEMPORAL BONE.—1. Squamous portion. 2. Mastoid portion. 3. Petrous portion. 4. Zygomatic portion. 5. Tubercle. 6. Temporal ridge. 7. Glenoid fissure. 8. Mastoid foramen. 9. Meatus auditorius externus. 10. Fossa for digastric muscle. 11. Styloid process. 12. Vaginal process. 13. Glenoid foramen. 14. Groove for Eustachian tube.

Beneath the base of this process is the *glenoid cavity* for the articulation with the inferior maxillary bone. Just behind this cavity is the *Glaserian fissure;* back of this, a fossa containing part of the parotid gland. The laxator tympani muscle, and nerve called chorda tympani, pass through the fissure.

The middle artery of the dura mater marks deeply the internal surface of the squamous portion of the temporal bone.

The mastoid (nipple-like) part is behind the ear. It is of a cellular structure, with dentate edge. Outside are the mastoid process and digastric fossa; internally, above, the mastoid foramen; through which passes a vein.

The shape of the *petrous* portion is that of a pyramid, pointing inwards and forwards. The name is given on account of its hardness. It has a *base, apex, inferior, anterior,* and *posterior surfaces.*

In the base is the *external meatus auditorius,* surrounded by a rim for the attachment of the aural cartilages. In the apex is an opening at the end of the canal for the internal carotid artery. The *Eustachian tube* is completed by the junction between the apex and the squamous part of the bone; the tensor tympani muscle passes above it.

On the anterior surface is a groove and a foramen, *hiatus Fallopii,* for the superficial petrous branch of the vidian nerve; also an eminence from the bulging of the labyrinth of the ear. The superior petrosal sinus marks the internal edge; the ganglion of Casser rests upon a depression near the apex.

From the inferior surface goes downwards and forwards the *styloid process.* The *stylo-mastoid foramen* is between it and the mastoid portion. The facial part of the seventh nerve, a small branch of the fifth nerve, and the stylo-mastoid artery pass through the stylo-mastoid foramen. Towards the occiput is the *jugular fossa;* which, when joined to the occipital bone, makes the *posterior foramen lacerum.* Through this pass the eighth nerve and jugular vein. Into it opens the *tympanic canal,* with the *nerve of Jacobson.*

The *carotid canal* winds through the petrous portion, beginning in front of the jugular fossa. In it, besides the carotid artery, is the *ganglion of Laumonier.* The opening for the *aqueduct of the cochlea* is between the carotid canal and the jugular fossa.

In the middle of the posterior surface is the *internal meatus auditorius.* A notch or foramen above this transmits a small artery. The *aqueduct of the vestibule* opens upon this surface.

Sphenoid Bone.

This acts as a *girder* in the architecture of the skull, crossing its base, and holding all of the other bones together. It has a body, four wings, and two descending processes. On the upper surface of the body is the *sella turcica* (turkish saddle), in which rests the pituitary gland. Behind and over this is the posterior clinoid process; at the sides pass the *sulci carotici,* for the carotid arteries. The *olivary process* is a prominence anterior to the sella turcica, and having upon it a mark for the chiasm or junction of the optic nerves.

The nasal lamella of the ethmoid bone articulates with a ridge upon the front of the body of the sphenoid. On each side of this ridge, in the adult skull, is an orifice, belonging to one of the *sphenoidal cells,* which communicate with the posterior ethmoidal cells. The *azygos process* on the lower surface of the body of the sphenoid joins with the vomer.

Behind, the sphenoid articulates, by a rough surface, with the cuneiform process of the occiput.

Fig. 8.

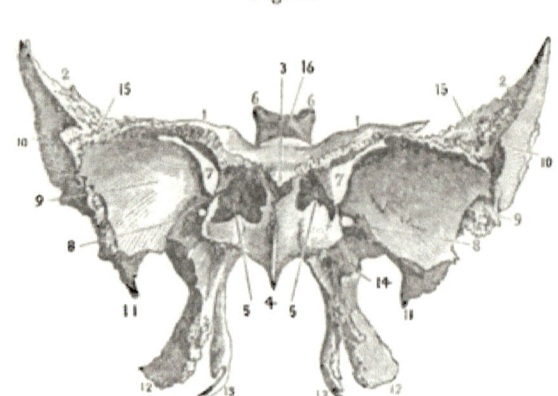

ANTERIOR AND INFERIOR SURFACE OF THE SPHENOID BONE.—1, 1. Apophyses of Ingrassias. 2, 2. The great wings. 3. Ethmoidal spine. 4. Azygos process. 5. Sphenoidal cells, after removal of pyramids of Wistar. 6. Posterior clinoid processes. 7. Sphenoidal fissure.

The *greater wings* of the sphenoid have each three surfaces, the *temporal, orbital,* and *cerebral.* The *temporal* is external; it has a process anteriorly, and is traversed by a ridge. The temporal and external pterygoid muscles lie over it. The *orbital* or anterior surface contributes a large space to the orbit of the eye. Above, by a triangular serrated portion, this connects with the frontal bone. The temporal bone joins it at the side.

On the concave *cerebral* surface are marks for the convolutions of the middle lobe of the cerebrum. Through it, by the *foramen rotundum,* passes the second branch of the fifth pair of nerves. The third branch of the fifth goes through the *foramen ovale.* The *spinous process* terminates this surface behind; and through it, by the *foramen spinale,* passes the middle artery of the dura mater. The lower point of the spinous process is called the styloid process of the sphenoid bone.

The *sphenoidal fissure* lies between the greater and the lesser wings. Through it pass the third, fourth, first branch of the fifth, and sixth pairs of nerves.

The *lesser wings,* or apophyses of Ingrassias, are in front of the greater, and are three-sided. Posteriorly they present the *anterior clinoid processes;* through these pass, by the *optic foramina,* the optic nerves and ophthalmic arteries. The frontal bone articulates with these processes anteriorly.

Downward, on each side, project the *pterygoid processes.* In front, these join with the palate bones. Behind, they present the *pterygoid fossa* and notch, by which each process is divided into two plates, the internal and external. The internal is longer; at its end is the *hamulus*, a hook-like process over which runs the tendon of the circumflexus palati muscle. The internal pterygoid muscle originates upon this internal plate. The external pterygoid muscle has its origin in the external plate of the pterygoid process, which is broader than the internal. Through the base of the process goes the *pterygoid foramen;* which conveys the Vidian or recurrent nerve, a branch of the fifth.

The sphenoid bone articulates with each of the other bones of the palate and vomer of the face, and with the molar.

Ethmoid Bone.

A light, cellular bone, at the base of the skull, below the frontal and in front of the sphenoid bone. Its general shape is cuboid.

Above, it presents the *cribriform plate*, through the orifices of which pass the filaments of the first or olfactory nerve; through the anterior and largest foramen goes the internal nasal branch. The *crista galli* is an upright process or crest from the middle line of the cribriform plate. The falx cerebri is attached to it. In front, it joins the frontal bone. The *foramen cæcum* is here formed, between the two bones; it transmits a small vein.

The *perpendicular plate*, or *nasal lamella*, descends from the middle of the cribriform plate, and constitutes part of the septum of the nose. It articulates with the vomer below.

Each outer surface of the ethmoid is called *os planum*. It forms part of the orbit of the eye, and joins above with the frontal bone, below, with the superior maxillary and palate bones, in front with the lachrymal, and behind with the sphenoid. It has two foramina above, one of which transmits the internal nasal nerve. Below each os planum projects downwards and backwards the irregular *unciform* or hook-like process; which joins with the inferior turbinated bone.

The *superior* and *middle turbinated* bones are curved or scroll-like processes within the ethmoid, seen best from behind. They are attached to the inside of the lateral masses of the bone. The anterior cavities of the nose are thus formed and subdivided. The fissure between the superior and middle turbinated bones is the

Fig. 9.

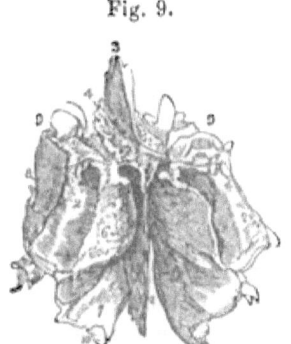

UPPER AND POSTERIOR VIEW OF THE ETHMOID BONE.—1. Nasal lamella. 2. Body or cellular portion. 3. Crista galli. 4. Cribriform plate. 5. Superior meatus. 6. Superior turbinated bone. 7. Middle turbinated bone. 8. Os planum. 9. Surface for the olfactory nerve.

superior meatus of the nose. That beneath the middle, between it and the inferior turbinated bone, is the *middle meatus*. Into the latter empty the anterior ethmoidal cells; the frontal sinuses are connected with the latter. Into the superior meatus empty the sphenoidal sinus and the posterior ethmoidal cells.

The *pyramids of Wistar*, or of Bertin, are hollow, three-sided processes, the bases of which are attached to the superior turbinated bones, the cribriform plate and the nasal lamella, posteriorly. The apices, or points of the pyramids, lie one on each side of the azygos process of the sphenoid. These pyramids are connected with the ethmoid bone only during childhood; afterwards they form the walls of the sphenoidal sinuses.

Besides the frontal and sphenoid, the ethmoid bone articulates with eleven of the facial bones.

BONES OF THE FACE.

These are fourteen in number, all belonging to the superior maxillary region, except the single bone of the lower jaw. They are the two upper jaw bones, two nasal bones, two palate bones, two malar bones, two ossa ungues, or lachrymal bones, two inferior turbinated bones, the lower jaw bone, and the vomer.

The Malar Bones.

Quadrangular and prominent, forming the cheek bones. Each has an outer and inner *surface*, four *processes*, the frontal, orbital, maxillary and zygomatic, and four *edges*, superior, inferior, anterior, and exterior. The outer surface is convex, the inner concave; each has small foramina for vessels and nerves. To the inner, the temporal and masseter muscles are attached; to the outer, the two zygomatic muscles. The *frontal* process joins with the frontal bone. The *orbital* process passes backwards to form part of the outer wall of the orbit; joining the greater wing of the sphenoid. The spheno-maxillary fissure, or foramen lacerum inferius, is bounded by an edge of the orbital process. The *maxillary* process articulates with the malar process of the upper jaw bone. The *temporal* process passes backwards to join the zygomatic process of the temporal bone in forming the zygomatic arch. To the posterior *edge* of the malar bone the temporal fascia is attached.

Two Superior Maxillary Bones.

Each is irregular in shape, with four *surfaces*, four *processes*, and a large cavity within the body. Each forms part of the floor of the orbit, the roof of the mouth, and the floor and outer wall of the nose. The *external* is the *facial* surface. Above the depression called the *canine fossa*, and near the upper edge of the facial surface, is the *infra-orbital foramen;* through which pass the infra-orbital artery, vein, and nerve.

The *posterior* surface is convex; its greatest prominence is below, the *maxillary tuberosity;* within, this joins with the palate bone. The posterior dental artery, vein, and nerve go through its foramina.

The *orbital* surface is superior; it is three-sided. The *nasal* or internal surface presents the opening into the *antrum highmorianum,* or maxillary sinus. The junction of the ethmoid, palate, and inferior turbinated bones narrows this opening very much. The *cavity* is somewhat pyramidal in shape; its base being the outer wall of the nose.

The *malar process* is triangular and rough, for union with the malar bone. The *nasal* process rises to form part of the side of the nose. It joins with the nasal bone in front, the frontal bone above, and, behind, having a groove in common for the lachrymal sac, with the os unguis.

Fig. 10.

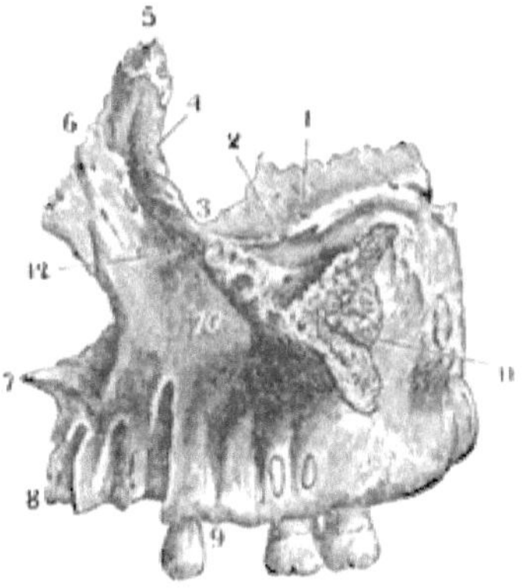

SUPERIOR MAXILLA.—1. Antrum. 2. Ductus ad nasum. 3. Articular surface for os frontis. 4. Articular surface for nasal bone. 5. Surface for nasal cartilage. 6. Floor of the nostril. 7. Surface for the bone of the right side. 8. Foramen incisivum. 9. Palate plate. 10. Surface for palate bone. 11. Ridge for the inferior spongy bone. 12. Articular surface for palate bone behind. 13. Surface for nasal plate of palate bone. 14. Surface for orbitar plate of palate bone. 15. Nasal duct.

The *alveolar* process is thick and spongy; it contains sockets for eight teeth, incisor, canine, and molar.

The *palate* processes make the roof of the mouth. From the median junction of the two arises the *nasal crest;* which articulates with the vomer. The *nasal spine* is at the front of this. Just behind it is the *foramen incisivum;* in which are the ganglion of Cloquet and the naso-palatine nerve.

Posteriorly, the palate processes unite with the horizontal plates of the palate bones.

Two Nasal Bones.

These are long and irregularly four-sided; thick and narrow above, where they join the frontal bone; thin and wide below, for the attachment of the cartilages of the nose. Externally, they are inclosed by the nasal processes of the two superior maxillary bones; internally, they articulate with each other. The posterior surface has a groove for the internal nasal nerve. The inner

36 ANATOMY.

Fig. 11.

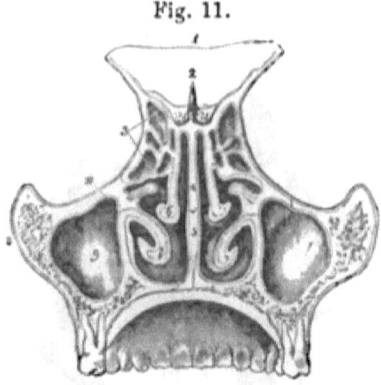

THE NASAL CAVITIES.—1. Part of cranium. 2. Crest of ethmoid. 3. Ethmoid bone. 4. Middle of ethmoid. 5. Vomer. 6. Middle turbinated bone. 7. Inferior turbinated bone. 8. Malar bone. 9. Antrum.

border has a crest for junction with the nasal spine of the frontal above, and with the perpendicular plate of the ethmoid, below.

Two Lachrymal Bones.

Also called *ossa ungues*, from resemblance to finger-nails. They are located at the front part of the inner wall of the orbit. A vertical ridge divides the orbital surface into two portions. In front is the ungual part of the lachrymal groove, completed by the nasal process of the superior maxillary. In the upper part of this groove lies the lachrymal sac; in the lower, the nasal duct. The ridge gives origin to the *tensor tarsi*, or Horner's muscle.

The lachrymal bones articulate with the frontal, ethmoid, superior maxillary, and inferior turbinated bones.

Two Palate Bones.

Fig. 12.

THE PALATE BONE.—1. Palate plate on its nasal surface. 2. Nasal plate. 3. Pterygoid process. 4. Surface for articulating with its fellow. 5. Half of the crescentic edge and spine for the azygos uvulæ muscle. 6. Ridge for the inferior spongy bone. 7. Spheno-palatine foramen. 8. Orbital plate.

Irregular; formed chiefly in two portions. The *horizontal plate* on each side completes the roof of the mouth and palate, by junction with the palate processes of the upper jaw bone. Part of the nasal crest which articulates with the vomer is formed at the junction of the two plates. Behind, in the same line, is the *posterior nasal spine*. From this, the azygos uvulæ muscle arises.

The *vertical* plate articulates internally, by a ridge, with the inferior turbinated bone; externally, with the superior maxillary. In the latter articulation is the *posterior palatine foramen*, for the palatine artery and nerve.

The *pterygoid process* is behind and below. It is triangular, and receives the two plates of the pterygoid process of the sphenoid bone.

The *spheno-palatine foramen* is formed by a notch between the processes at the upper part of the vertical plate being completed by junction with the sphenoid bone. Through this foramen pass the spheno-palatine artery and nerve.

Anteriorly, and above, is the *orbitar* process, which passes between the superior maxillary and the ethmoid, to contribute a small portion to the orbit of the eye. The other process passes backwards, at the top of the vertical plate; it is called the pterygoid apophysis, and joins the sphenoid bone.

Vomer.

The posterior nares are divided by this bone; which is flat, thin, and vertical in position, with its antero-posterior diameter the longest. Its upper border is thickest, and articulates with the sphenoid. Its anterior border joins in front with the triangular cartilage of the nose, and, behind this, with the ethmoid. The inferior border articulates in front with the superior maxillary bones, and further with the palate bones. The posterior border is concave, thin-edged, and free.

Two Inferior Turbinated Bones.

Small, spongy, *scroll-shaped*, one on each side of the nose, within the fossa, attached to its outer wall. Its internal convex

Fig. 13.

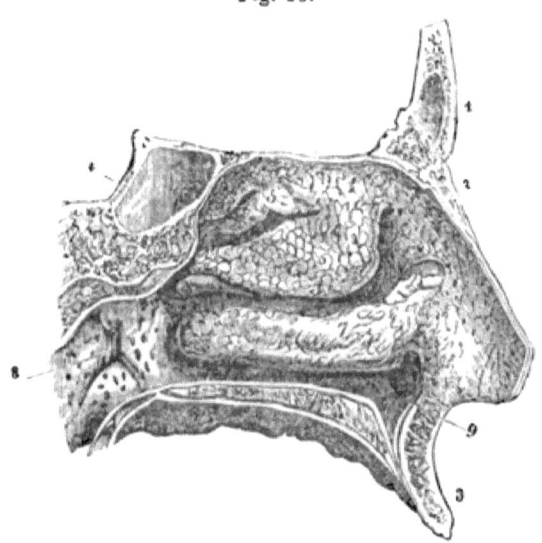

LEFT NASAL FOSSA.—1. Frontal bone. 2. Nasal bone. 3. Superior maxillary. 4. Sphenoid. 5, 6, 7. Superior, middle, and inferior turbinated bones.

surface is covered by the lining membrane of the nose; its external surface is concave, and forms part of the inferior meatus.

Each bone articulates with the ethmoid, superior maxillary, palate, and lachrymal bones.

Inferior Maxillary Bone.

This consists of the *body*, horizontal and convex anteriorly, and two perpendicular *rami*.

In the median line of the body is the ridge of the *symphysis*, where join the two parts in which the bone is developed. At the base of this ridge is the triangular *mental process*. Each side of the symphysis is the *incisive fossa;* outside of this, the *mental foramen*, through which pass the mental artery and nerve. The internal surface is marked by the *four genial tubercles*, for muscular insertions. Near them are fossæ for the sublingual glands.

The *superior border* is hollowed out for the sockets of sixteen teeth, in the adult.

Each *ramus* is four-sided, with two prominent processes above. Its flat external surface is covered by the masseter muscle. Its *internal surface* has the aperture of the *inferior dental canal*, for the nerve and vessels of the same name. This canal passes into the substance of the body of the bone, to distribute nerves and vascular branches.

The *condyloid process* is short and thick, with a condyle for articulating with the temporal bone, and below this, a neck.

The *coronoid process* is flat, thin, and triangular; anterior to the condyloid, with the *sigmoid notch* between them. It is the point of attachment of the temporal muscle.

General Remarks on the Head.

The Sutures.—These are the somewhat irregular and generally serrated lines of articulation of the bones of the head. The principal are the *coronal, sagittal, squamous* and *lambdoidal* sutures.

The *coronal* suture is between the frontal and the two parietal bones, across the head above.

The *sagittal* suture is between the two parietals, passing longitudinally.

The *squamous* suture connects the temporal with the parietal and sphenoid bones. It is rough and undulated rather than serrated.

The *lambdoidal* suture (named for the Greek letter lambda) is between the parietal bones and the occiput. It is formed of two limbs separating at an angle.

Diploë.—This is the cellular bony structure between the outer and the inner tables of the skull. Branching canals through it contain veins. The outer table of the bones of the skull is the strongest and least brittle. In infancy they are closely connected with but little diploë.

Exterior Regions of the Head.
—These are the coronal region, the facial, the two lateral, and the basal region.

The *coronal* region is seen in a vertical view of the cranium in its natural position.

The *facial* region is bounded by the parietal protuberances, the zygomatic arches, the coronal suture, and the margin of the lower jaw.

The *lateral* regions are contained between the temporal ridge above, the margin of the orbit front, and the lambdoidal suture behind.

The *base of the skull* is very irregular and complex; capable of being studied to advantage only with a specimen or plate. Its general outline is oval. Its principal foramina are, the *foramen magnum occipitis, foramen incisivum, posterior palatine foramen, foramen lacerum anterius, foramen lacerum posterius,* and the special foramina of the sphenoid bone.

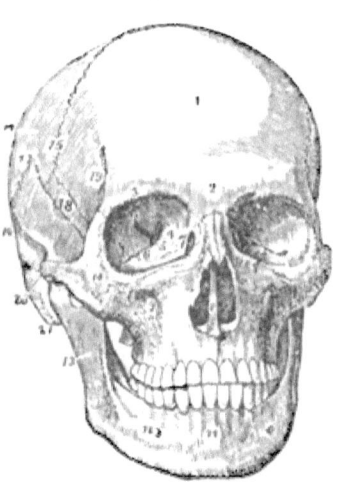

Fig. 14.

FRONT VIEW OF THE SKULL.—1. Os frontis. 2. Nasal tuberosity. 3. Supra-orbital ridge. 4. Optic foramen. 5. Sphenoidal fissure. 6. Spheno-maxillary fissure. 7. Lachrymal fossa. 8. Opening of the anterior nares, and the vomer. 9. Infra-orbital foramen. 10. Malar bone. 11. Symphysis of the lower jaw. 12. Anterior mental foramen. 13. Ramus of the lower jaw-bone. 14. Parietal bone. 15. Coronal suture. 16. Temporal bone. 17. Squamous suture. 18. Great wing of the sphenoid.

Interior Regions of the Cranium.—The *calvaria*, or vaulted roof of skull, is generally smooth within, though marked for the convolutions of the brain, and also by a groove for the longitudinal sinus of the dura mater, and depressions, on each side of the sagittal suture, for the *glands of Pacchioni*.

The *base of the cranium*, within, exhibits three regions, *anterior, middle,* and *posterior.* The anterior fossa lodges the anterior lobes of the cerebrum; the middle fossa the middle lobes of the same; while the pons Varolii, medulla oblongata, and cerebellum are contained in the posterior fossa.

The Orbit.—The cavity for the eye is a somewhat quadrangular pyramid, formed of seven bones; the frontal, superior maxillary, malar, sphenoid, ethmoid, palate, and lachrymal bones.

Nasal Cavities. (See fig. 11.)—These are contained between the cribriform plate of the ethmoid and the sphenoid bones above; in front the ossa nasi and cartilages of the nose; below, the palate processes of the superior maxillary and palate bones; outside, the superior and maxillary, ethmoid, and inferior turbinated bones. Between the two cavities is the *septum narium*, formed by the vomer, nasal lamella of the ethmoid, and the nasal cartilage.

Fig. 15.

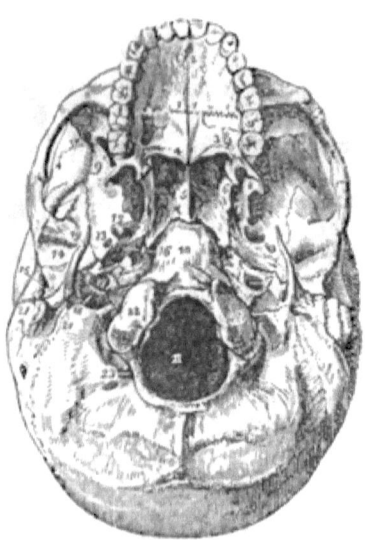

EXTERNAL VIEW OF THE BASE OF THE CRANIUM —1. Hard palate. 2. Foramen Incisivum. 3 Palate plate of palate bone. 4. Crescentic edge. 5. Vomer. 6. Internal pterygoid process of sphenoid bone. 7. Pterygoid fossa. 8. External pterygoid process. 9. Temporal fossa. 10. Basilar process. 11. Foramen magnum. 12. Foramen ovale. 13 Foramen spinale. 14. Glenoid fossa. 15. Meatus auditorius externus. 16. Foramen lacerum anterius. 17. Carotid foramen. 18. Foramen lacerum posterius 19. Styloid process. 20. Stylo-mastoid foramen. 21. Mastoid process. 22. Condyles of occipital bone. 23. Posterior condyloid foramen.

The outlets from these fossæ in front, are the *anterior nares;* behind, the *posterior nares.* The passages, nearly horizontal, through the nasal cavities are, the *superior, middle,* and *inferior meatus.* The superior is the smallest. The middle contains the opening into the *antrum maxillare.* The inferior presents the orifice of the *ductus ad nasum* from the lachrymal sac.

Ossa Triquetra, or Wormiana.—These are small, irregular bones, not always present or of the same shape, included in the sutures, especially the lambdoidal.

Fontanels.—These, most interesting in obstetric anatomy, are places of deficient ossification at the junction of the bones in the head at and for a time after birth. They are the *anterior* and *posterior* fontanels, and two smaller ones on each side.

Diameters of the Head.—In the adult of European race, the

cranium is about six inches and a half in length, five inches in height, and five and a half transversely.

Fig. 16.

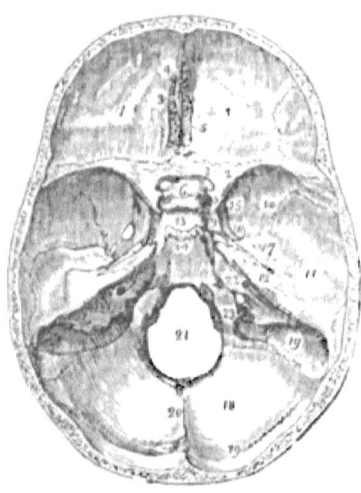

INTERNAL SURFACE OF THE BASE OF THE CRANIUM.—1. Anterior fossa for anterior lobes of the cerebrum. 2. Lesser wing of the sphenoid bone. 3. Crista galli. 4. Foramen cæcum. 5. Cribriform plate. 6. Processus olivaris. 7. Foramen opticum. 8. Anterior clinoid process. 9. Groove for the carotid artery. 10. Greater wing of the sphenoid bone. 11. Middle fossa for middle lobes of the cerebrum. 12. Petrous portion of temporal bone. 13. Sella turcica. 14. Basilar gutter for the medulla oblongata. 15. Foramen rotundum. 16. Foramen ovale. 17. Foramen spinale. 18. Posterior fossa for the cerebellum. 19. Groove for the lateral sinus. 20. Ridge for the falx cerebelli. 21. Foramen magnum. 22. Meatus auditorius internus. 23. Posterior foramen lacerum for the jugular vein.

Internal Capacity.—Morton gives, as the average capacity of the Anglo-Saxon and German cranium, ninety cubic inches; of the native African races, eighty-five inches; other races coming between.

Form of the Head.—Dividing the human species into five *varieties*, the head in the *Caucasian* or *European* may be described as of a rounded oval shape. In the *Mongolian* race it is pyramidal, rising from the prominent cheek bones and almost vertical occiput toward the sagittal suture. In the *Malay* the same shape of the head is seen, with a broader and flatter face. The head of the *American* Indian is also like that of the Mongolian, with somewhat greater prominence of the face. The *Negro* has a skull long antero-posteriorly, with low forehead, full occiput, and prominent maxilla.

Facial Angle.—After Camper, this is obtained by drawing a line from the anterior margin of the upper jaw to the most prominent part of the forehead, and crossing this by a horizontal line from the external meatus of the ear to the lower edge of the nose. Morton asserts the mean facial angle of the Caucasian race to be 80°; of the Mongolian, 77°; of the Malay, American Indian, and Negro, 75°. All that the facial angle determines is the proportion of development between the head and face.

HYOID BONE.

Connected in man with no other bone, the os hyoides is attached to the root of the tongue and the larynx. It is shaped like the letter U, or a horse-shoe, the convexity being forward. Besides the central body, it has two greater and two lesser *cornua*. The *greater* project backwards, each terminating in a tubercle. The *lesser* cornua ascend, to the length of a few lines, from the junction of the body and the greater cornua; they are generally cartilaginous. Several muscles and ligaments are connected with this bone.

THE THORAX.

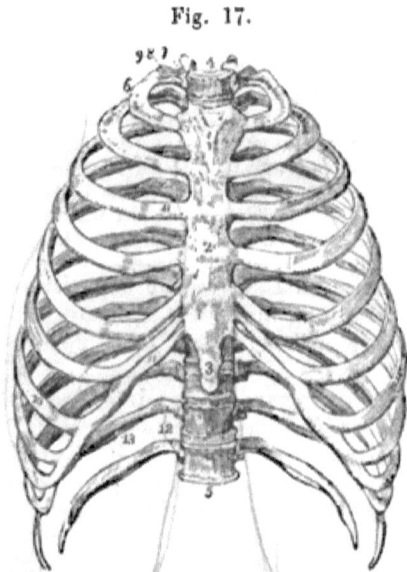

Fig. 17.

FRONT VIEW OF THE THORAX.—1. First bone of the sternum. 2. Second bone of the sternum. 3. Third bone or ensiform cartilage. 4. First dorsal vertebra. 5. Last or twelfth dorsal vertebra. 6. First rib. 7. Its head. 8. Its neck. 9. Its tubercle. 10. Seventh or last true rib. 11. Its cartilage. 12. Angle of eleventh rib. 13. Its body.

The *sternum, ribs*, and *dorsal vertebræ* inclose the chest. Its shape is that of an imperfect cone, notched in front, below, and flat above, concave behind, and open at the top.

Sternum.

Composed of three pieces, this bone is flat and oblong. In old age the pieces are consolidated into one.

The first or upper piece is the thickest. The clavicle on each side articulates with it at the upper corner; lower down are small cavities for the first rib and part of the second.

The *second* or middle piece is the longest, but is narrower than the first, though widening below. With its sides articulate the cartilages of part of the second rib, the whole of the third, fourth, fifth, sixth, and part of the seventh.

The *third* piece is often a cartilage only. It varies much in shape, being sometimes bifurcated. It is often called the xyphoid or ensiform cartilage. Part of the cartilage of the seventh rib is attached to its side.

Ribs.

Twelve on each side; *seven* whose cartilages reach the sternum, called *true* ribs, and *five* others below, the *false* ribs. The last two, with free ends, are called *floating* ribs.

The sternal end of each rib is larger than the vertebral end. The latter is rounded, with a ridge dividing it into two surfaces. Beyond this round head is a neck, and an inch from the head is a tubercle, which articulates with the transverse process of a vertebra. A smaller tubercle receives the *external costo-transverse ligament*. The *internal* costo-transverse ligament is inserted into the upper edge of the neck of the rib.

Besides a twist in its whole shape, each rib has an *angle*, as though it had been bent. Along its rounded upper edge the intercostal muscles are inserted. Within the thin and cutting lower edge, for two-thirds of its length, is a groove for the intercostal nerve and vessels.

The *first* rib is the smallest and most simple in form. The subclavian artery rests in a fossa on its upper surface; to which are attached the scalenus anticus and scalenus medius muscles. This rib has no intercostal groove.

The longest rib is the eighth; after that they decrease to the last; which also is without an intercostal groove.

THE PELVIS.

Two *ossa innominata*, the *sacrum* and the *coccyx*, inclose this cavity.

OSSA INNOMINATA.

Large, irregular, 8-shaped bones, one on each side, known as the hip-bones. Until puberty each innominatum consists of three bones united; the *ilium*, *pubes*, and *ischium*.

Ilium.

A very large, flat or concavo-convex bone, the uppermost of the pelvis. Its outer surface or *dorsum* is smooth in the main, and bounded above by the *crest*, below by the *acetabulum*, before and behind by the anterior and posterior *borders*. These semi-

Fig. 18.

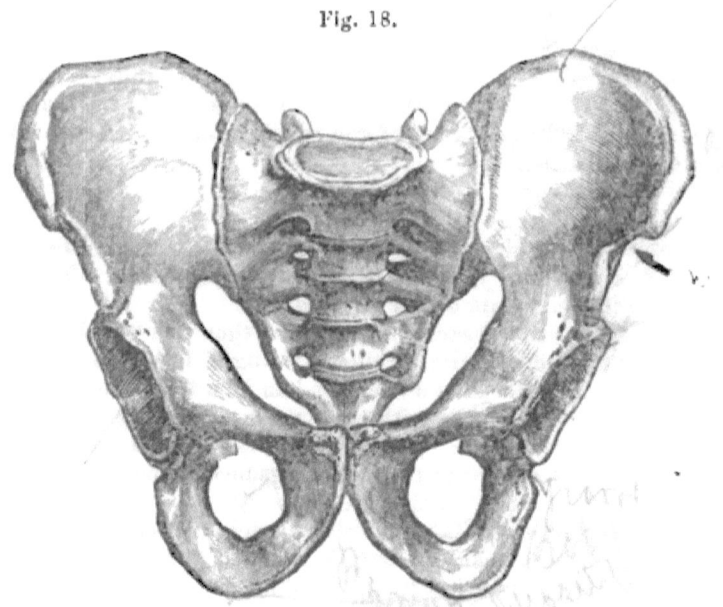

ANTERIOR VIEW OF THE MALE PELVIS.

circular lines cross the dorsum ilii. Between the upper and the crest originates the gluteus maximus muscle. Between the upper and middle curved lines the gluteus medius arises. The space between the middle and lower lines gives origin to the gluteus minimus.

Of the internal surface, the anterior or larger part, the *venter*, or *iliac fossa*, is smooth; the iliacus internus muscle lodges upon it. Behind this fossa is a rough surface; the upper part of which has attached to it the posterior sacro-iliac ligaments, and the lower part articulates with the side of the sacrum.

The *crest* of the ilium is a convex, thick, curved line; thickest behind. In front it ends in the anterior superior spinous process; behind, in the posterior superior spinous process. To the outer edge of the crest are attached the tensor vaginæ femoris, obliquus externus, and latissimus dorsi muscles; to the inner edge the transversalis abdominis, and two other muscles; between the two edges the obliquus internus.

The *anterior superior spinous process* of the ilium gives origin to the tensor vaginæ femoris, sartorius, and iliacus internus muscles, and attachment to the fascia lata and Poupart's ligament. Just beneath the process is a notch, ending in the anterior inferior spinous process. To this is attached the straight tendon of the rectus femoris muscle.

On the posterior border of the ilium, separated by a notch, are the posterior superior and posterior inferior spinous processes. The great *sacro-sciatic notch* is below the inferior process.

Pubes.

This forms the anterior part of the innominatum. It consists of a horizontal *body* and a descending *ramus*.

The outer extremity of the body is the thickest, and constitutes one-fifth of the acetabulum. At the inner extremity is the *symphysis pubis*.

The upper surface of the body of the pubes presents its *spine*, and, going backward from this, a ridge called the *linea ilio-pectinea*. To this line parts of Poupart's and Hey's ligaments are attached. The under surface of the body has a sharp margin for the upper part of the circumference of the thyroid or obturator foramen.

The *ramus* of the pubes goes outwards and downwards to join the ramus of the ischium. Its outer border forms part of the margin of the thyroid foramen. The crus penis in the male is attached to its inner border.

Ischium.

Situated beneath the ilium, and behind and below the pubes, this bone is at the lowest part of the innominatum. It has a *body* and a *ramus*. Of the body, the smooth inner surface is called the *plane* of the ischium. At its posterior part is the *spine* of the ischium; to which is attached the lesser sacro-sciatic ligament. Lower than this is the *tuberosity* of the ischium, on which we sit. It gives origin to the semi-tendinosus and semi-membranosus muscles, and the long head of the biceps flexor cruris. The great sacro-sciatic ligament is attached to a ridge in front of this. The *ramus* of the ischium ascends to join that of the pubes. Below, its inner surface gives a sharp border to form part of the margin of the obturator foramen.

Acetabulum.

The *acetabulum* or *cotyloid cavity*, the socket for the thigh bone, is a deep hemispherical depression, formed in its upper two-fifths by the ilium, internal one-fifth by the pubes, and lower and posterior two-fifths by the ischium. Its rim is prominent but uneven. On the inner side is the deep *cotyloid notch;* at the end of which is a circular depression at the bottom of the cavity, in which lodges a mass of fat, and around which arises the ligamentum teres.

Thyroid Foramen.

The *thyroid* or *obturator foramen*, oval in the male, triangular in the female, is a large opening, bounded by the pubes and ischium. Except a groove above for the obturator nerve and vessels, it is filled by a membranous ligament.

The sacrum and coccyx have been described with the spine.

UPPER EXTREMITY.

SHOULDER.

Two bones in man form the shoulder; the scapula or shoulder-blade, and the clavicle or collar-bone.

Scapula.

A thin, flat, three-sided bone, reaching downward from the second to the seventh rib, behind the thorax on each side.

Fig. 19.

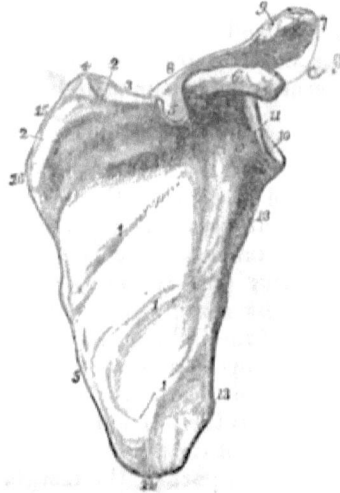

VENTER OF SCAPULA.—1, 1, 1. Oblique ridges. 2, 2. Fossa for subscapularis muscle. 3. Superior border. 4. Superior angle. 5. Suprascapular notch. 6. Coracoid process. 7. Acromion process. 8. Spine of scapula. 9. Articular surface. 10. Glenoid cavity. 11. Head of scapula. 12. Neck. 13. Inferior border. 14. Inferior angle. 15. Posterior border. 16. Origin of spine.

The *venter* or *costa* of the scapula is the concave face which presents anteriorly towards the ribs, and is occupied by the subscapularis muscle. The *dorsum* or posterior surface is unequally divided by the *spine* of the scapula. In the smaller space or fossa above the spine arises the supra-spinatus muscle; in the larger space below it, the infra-spinatus.

The *acromion* process is at the outer extremity of the spine. In front, this process articulates with the clavicle.

Part of the deltoid muscle arises from the edge of the spine of the scapula; and the trapezius muscle is inserted into it.

At the *external angle* of the scapula is a process hollowed out for the *glenoid cavity* of the shoulder joint. This is a shallow socket of an oval shape. At its top is a mark for the origin of the long head of the biceps muscle. The glenoid process has a narrow *neck*; from this, forwards and outwards, extends the *coracoid process* of the scapula. Its name is given (from *corax*, a crow) from its shape resembling a crow's beak. From its end originate the coraco-brachialis muscle and the short head of the biceps. The pectoralis minor muscle is inserted into it.

The *superior* angle of the scapula is almost a right angle; into it is inserted the levator scapulæ muscle. The *superior edge* is thin; the *coracoid notch* divides it; this notch, made a foramen by a ligament, transmits the supra-scapular nerve and artery.

The *inferior angle* is pointed. The teres major muscle arises from its posterior surface.

Clavicle.

A long, transverse, *f*-shaped bone, extending from the upper part of the sternum to the scapula. The two-thirds nearest the sternum are convex in front; the humeral third is concave in front. Near the sternal end, inferiorly, is a roughness for the attachment of the costo-clavicular ligament; next to this the subclavius muscle is inserted; and near the humeral end is attached the coraco-clavicular ligament, to a ridge and tubercle. The pectoralis major muscle arises in part from the anterior edge of the clavicle.

The sternal end of this bone is the thickest, and is elongated behind and below. The humeral end has a face for articulation with the acromion process of the scapula.

ARM.

The *arm* consists of a single bone; the humerus.

Humerus.

This long, cylindrical bone consists of a round *head*, an ill-defined *neck*, a *shaft*, and two *condyles*. The head is hemispherical, and fits into or against the glenoid cavity of the scapula. The *anatomical* neck of the bone is a *groove*, just beyond the articulation. A *greater external*, and a *lesser internal tuberosity* are below this neck, with a perpendicular groove between them for the tendon of the long head of the biceps muscle. Into the greater tuberosity are inserted the supra-spinatus, infra-spinatus, and teres minor muscles; into the lesser tuberosity, the subscapularis. The pectoralis major, teres major, and latissimus dorsi muscles are inserted into the humerus below its head, near the edges of the bicipital groove

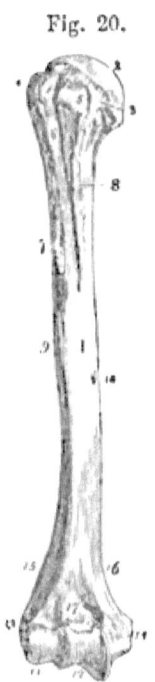

Fig. 20.

ANTERIOR VIEW OF HUMERUS OF THE RIGHT SIDE.—1. Shaft or diaphysis. 2. The head. 3. Anatomical neck. 4. Greater tuberosity. 5. Lesser tuberosity. 6. The bicipital groove. 7. External bicipital ridge for pectoralis major. 8. Internal bicipital ridge. 9. Point of insertion of deltoid muscle. 10. Nutritious foramen. 11. Face for head of the radius. 12. Face for the ulna.

Between the anatomical neck and these insertions is the so-called *surgical* neck of the bone.

The deltoid muscle is inserted on the outer side of the middle of the humerus. The coraco-brachialis is attached on its inner side. The *nutritious foramen* is a little lower; and above this a spiral groove, occupied by the profunda major artery and the muscular spiral nerve.

The humerus flattens below, and widens, having a ridge on each side extending to the *internal* and *external condyles*. Of these the internal is the most prominent. Anteriorly, the lower extremity of the bone presents the *lesser sigmoid cavity*, for the coronoid process of the ulna. Posteriorly is found the *greater sigmoid cavity*, for the olecranon process.

FOREARM.

The *forearm* contains two bones; the *radius*, on the same side with the thumb, and the *ulna*, on the side of the little finger.

Radius.

Small above, where it contributes but little to the elbow-joint; larger below, for firm union with the wrist. The upper end presents a rounded head, having a shallow depression which meets a slight projection of the end of the humerus. Below this is a distinct neck, embraced by a ligament within which it turns. Internally, below this neck, is a tubercle for the biceps tendon.

The outer surface of the radius is curved. The thick inferior end articulates with the scaphoid and lunare, bones of the wrist, as well as, laterally, with the ulna. The *styloid process* of the radius is below, on the outside; to it is attached the external lateral ligament.

Three grooves on the back of the lower part of the radius transmit the tendons of the extensor muscles to the hand. The extensor major pollicis has the deepest mark.

Several muscles arise from the radius, which will be hereafter described.

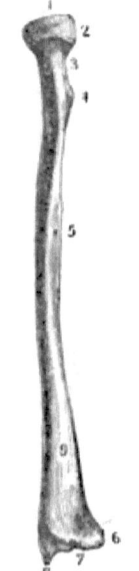

Fig. 21.

ANTERIOR VIEW OF RADIUS OF RIGHT SIDE.—1. Cylindrical head. 2. Surface for lesser sigmoid cavity of the ulna. 3. Neck of the radius. 4. Its tubercle, for insertion of biceps muscle. 5. Interosseous ridge. 6. Concavity for lower end of the ulna. 7. Carpal surface. 8. Styloid process. 9. Surface for pronator quadratus muscle.

CARPUS.

Ulna.

This is longer than the radius, and upon its inner side in the skeleton. It is largest at its upper end. Behind the elbow-joint, it ascends in the *olecranon process.* Into this the tendon of the *triceps extensor cubiti* muscle is inserted. In front is the *coronoid* process; into which is inserted the *brachialis anticus* muscle. The *greater sigmoid cavity* lies between these two processes; it receives the end of the humerus. The *lesser* sigmoid cavity is outside of the coronoid process, and receives the head of the radius. The anconeus muscle is inserted behind it; and the supinator radii brevis arises from a ridge near it.

The upper three-fourths of the anterior face of the ulna give origin to the flexor profundus digitorum muscle. Below this, lies the pronator quadratus muscle. Posteriorly, the ulna gives origin to the extensores pollicis and indicator muscles. To its outer edge the interosseous ligament is attached.

At its lower end the ulna is round, with a smooth surface for the radius, on its outside. From the inner side goes off the styloid process, to which the internal lateral ligament is attached. Behind the process is a groove for the extensor carpi ulnaris tendon.

CARPUS.

The carpus or wrist consists of eight bones, in two rows. The wrist is convex posteriorly, concave anteriorly; through the concavity pass the flexor tendons.

The upper row of bones contains the *scaphoid, lunare, cuneiform,* and *pisiform.* The lower row has the *trapezium, trapezoid, magnum* and *unciform.*

The *scaphoid*, or boat-shaped bone, is on the radial side of the upper row. Above, it is convex, to articulate with the end of the radius; below, concave for the magnum. It also articulates with the lunare, trapezium, and trapezoid.

The *lunare*, or crescent-shaped bone,

Fig. 22.

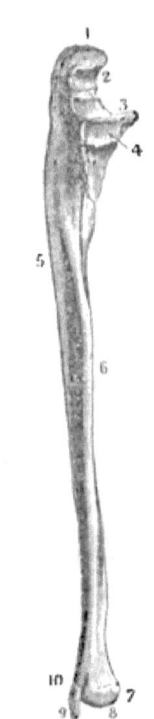

ANTERIOR VIEW OF ULNA OF THE LEFT SIDE —1. Olecranon process. 2. Greater sigmoid cavity. 3. Coronoid process. 4. Lesser sigmoid cavity. 5. External surface. 6. Ridge for interosseous ligament. 7. Small head for the radius. 8. Carpal surface.

Fig. 23.

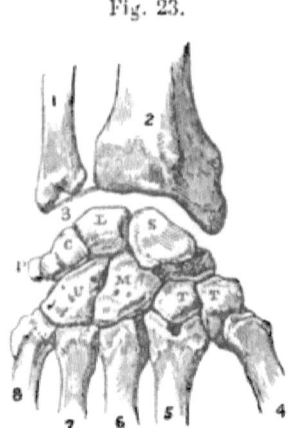

ARTICULATIONS OF BONES OF THE CARPUS.—1. Ulna. 2. Radius. 3. Inter-articular fibro-cartilage. 4. Metacarpal of thumb. 5. Metacarpal of first finger. 6. Metacarpal of second finger. 7. Metacarpal of third finger. 8. Metacarpal of fourth finger. S Scaphoides. L. Lunare. C. Cuneiform. P. Pisiform. T. T. Trapezium and trapezoides. M. Magnum. U. Unciform.

has a convex upper surface, which joins the radius, and articulates below with the magnum, and at the side with the cuneiform.

The *cuneiform*, or wedge-shaped bone, is joined below to the unciform, and at the side to the pisiform. It has a round face to meet the latter bone.

The *pisiform*, or pea-shaped bone, is the smallest of the carpus. It unites only with the cuneiform. It forms a prominence on the palmar face of the wrist, at the ulnar side.

The *trapezium* is many-sided. It articulates with the trapezoid, the scaphoid, and the metacarpal of the thumb.

The *trapezoid* is smallest of the lower row; it has a pyramidal shape, the apex being on the palmar side.

The *magnum* is the largest bone of the carpus. It is four-sided, with a rounded head or tubercle on its dorsal surface.

The *unciform* is remarkable especially for a hooked process on the palmar side.

METACARPAL BONES.

These are five, one for each digit. Each metacarpal has a round *head*, to join with the first phalanx below; a rough four-sided *base* above, for its carpal articulation; and a prismatic *shaft*, on the sides of which the interossei muscles are attached.

The longest metacarpal is that of the index or fore-finger. That of the thumb is short and thick. The smallest is that of the little finger.

FINGERS.

Each finger has a *first, second,* and *third* phalanx; counting from the metacarpal bone. The thumb has but two phalanges. The first phalanx is always the largest, and is convex on its dorsal surface, flat on the palmar side. A concavity at its upper end receives the round head of the metacarpal bone. Below it presents to the end of the second phalanx two small tuberosities with a groove between them.

The second phalanx has at its upper end two concavities with a ridge between them; at its upper end two tubercles with an intermediate groove, to meet the third.

The third phalanx is smallest. Its upper end resembles that of the second; its lower extremity is thin, rough, and flat.

The fore-finger is called the *index*. The middle and longest finger has been sometimes named the *impudicus*; the third, *annularis*; the little finger, *auricularis*.

Sesamoid Bones.—Two of these (named from *sesamum*, oriental barley) are usually in the metacarpo-phalangeal joint of each thumb, connected with the flexor tendon. They are not unfrequently wanting.

LOWER EXTREMITY.

Femur.

This, the thigh bone, is the longest of all the bones. It has a *head, neck, shaft*, and condyles.

The *head* is nearly spherical, to be received into the acetabulum. Near its middle, above, is a depression for the ligamentum teres. The *neck* shortens and approaches a right angle in its direction in old age, when it is more liable to fracture. Outside, at the lower end of the neck, is the large process called the *trochanter major*. The gluteus medius, gluteus minimus, and other muscles are inserted into this.

The *trochanter minor* is lower and more internal; the iliacus internus and psoas magnus muscles are inserted into it. Into a ridge between the trochanters behind is inserted the quadratus femoris.

The *shaft* of the femur is arched, the convexity being in front. Over its anterior surface is the origin of the cruræus muscle. The *linea aspera* is a strong ridge up and down the posterior surface of the bone. It has two edges; to the outer are attached the tendon of the gluteus maximus, and the muscular origin of the vastus externus and the short head of the biceps. Into the inner edge of the linea aspera are inserted the pectineus and adductor magnus; and from this edge arises the vastus internus.

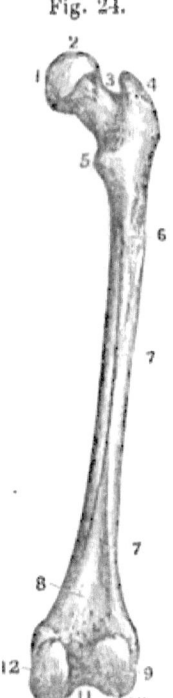

Fig. 24.

POSTERIOR VIEW OF THE FEMUR.—1. Depression for round ligament. 2. The head. 3. Depression for rotary muscles. 4. Trochanter major. 5. Trochanter minor. 6. Roughness for gluteus maximus tendon. 7, 7. Linea aspera. 8. Surface for gastrocnemius muscle. 9. External condyle. 10. Depression for anterior crucial ligament. 11. Depression for posterior crucial ligament. 12. Origin of internal lateral ligament.

The *condyles*, internal and external, are at the lower end of the femur, which widens towards them. The *external* condyle has upon its posterior face the origin of the popliteus, part of the gastrocnemius and plantaris muscles. The *internal* condyle is longest. It also gives part origin to the gastrocnemius. Each condyle has marks for the crucial and lateral ligaments.

The fossa in front of the condyles for the patella receives its largest contribution from the external condyle.

Patella.

A flat bone, of a roundish triangular shape, commonly known as the cap of the knee or knee-pan. It is thickest and widest above, where the tendon of the quadriceps femoris is inserted into it. Below, the continuation of the same tendon, in which (like a sesamoid bone) the patella may be considered as situated, is called the ligament of the patella: this is inserted in the tibia. The knee-pan is covered anteriorly only by the skin.

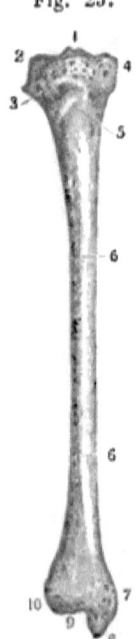

Fig. 25.

ANTERIOR VIEW OF THE TIBIA —1. Spinous process. 2. Surface for condyles of the femur. 3. Face for head of the fibula. 4. The head. 5. The tubercle. 6, 6. Spine and shaft of the bone. 7. Internal malleolus. 8. Process for internal lateral ligament of the ankle. 9. Tarsal surface. 10. Face for lower end of fibula.

Tibia.

The inner and larger of the two bones of the leg. Its *head*, or large upper end, presents an oval surface, divided by the upright *spinous process* into two parts, each oval, for the condyles of the femur. Projections on each side are called the external and internal condyles of the tibia. At the back part of the external condyle is a small face for the articulation of the fibula.

In front, below the head, is a tubercle for the tendon or ligament of the patella. Posteriorly, is the insertion of the popliteus muscle, and the origin of the soleus.

Of the *body*, the inner surface is covered only by skin. The outer surface is occupied by the tibialis anticus and extensor digitorum. The tibialis posticus and flexor digitorum arise from its posterior surface.

The outer *edge* of the tibia has attached to it the interosseous ligament. To the upper portion of the rounded inner edge are attached the tendons of the sartorius, semi-tendinosus and gracilis muscles.

The lower end of the tibia is smaller

than the upper. Internally, it presents the *internal malleolus* or large process of the ankle. Outside, it articulates, by a fossa, with the lower end of the fibula. Between these parts, at the end of the bone, is a concavity which rests and moves, by a hinge-like joint, upon the astragalus.

The extensor tendons lie over the anterior surface of the lower end of the tibia; the tendon of the flexor longus pollicis marks its posterior surface.

Fibula.

Much more slender than the tibia, this bone is external to it, and, above, somewhat behind it; not reaching to the articulation of the knee. It supports the head of the tibia, having an enlargement or head to articulate with its outer condyle. The biceps flexor cruris is inserted into its *styloid process*.

Fig. 26.

The body of the fibula has three surfaces; internal, external, and posterior. The internal or tibial surface presents a long ridge for the interosseous ligament. The extensor communis and extensor proprius pollicis arise in front of this ridge; behind it, the tibialis posticus partly arises.

The wide external surface is a long spiral, anterior above and becoming posterior below. The peroneus longus and peroneus brevis muscles arise from the upper portion of this surface, and their tendons pass through a groove on its lower part. On the posterior surface, which is somewhat spiral, arise the soleus and flexor longus pollicis muscles.

The *external malleolus* is a long descending process of the lower end of the fibula. It is somewhat triangular, with a pointed termination, the coronoid process, to which the external lateral ligament is attached. Within, the malleolar process articulates with the astragalus.

THE FIBULA.—1. Head. 2 Articular face 3. Insertion of external lateral ligament. 4. Shaft. 5, 5. External face. 6. Interosseous ridge. 7. Face for lower end of tibia. 8. Malleolus externus. 9. Tarsal surface.

TARSUS.

Seven bones form the tarsus; the *astragalus, os calcis, scaphoid, cuboid,* and *internal, middle, and external cuneiform*.

Astragalus.—Composed of a *head* and a *body* The *body* is rounded above for the tibial articulation; nearly flat at the sides, for the two malleoli; underneath, the body is concave. The anterior part or head is convex in front, and widest transversely. A constriction or neck intervenes between the head and the body.

Os Calcis.—This, the heel-bone, is the largest of the tarsus. It articulates above with the astragalus by two surfaces, having between them a groove for the interosseous ligament.

Externally the os calcis is covered by skin. The tendons of the peroneus longus and peroneus brevis mark the surface with grooves. On its internal surface is the *sinuosity*, occupied by tendons, nerves, and bloodvessels going to the plantar region of the foot.

Behind, the bone is rough below for the insertion of the tendo Achillis; above, it is smooth. *Underneath,* the os calcis has three tuberosities, for ligaments and tendons.

In *front,* are the greater and lesser *apophyses.* The greater joins the cuboid bone. The lesser is hook-shaped, passing forward and ascending to meet the astragalus. The tendon of the flexor longus pollicis passes through a groove on its surface.

Scaphoid.—Thickest above, convex in front, concave behind. The anterior surface has three facets for the cuneiform bones. The astragalus fits into the posterior deep concavity. The tibialis posticus muscle is inserted into a tubercle on the inner face of this bone.

Cuboid.—Convex and rough above. Underneath, at the end of a ligamentous ridge, is a groove for the peroneus longus tendon. In front, the cuboid articulates with the last two metatarsals; behind, with the great apophysis of the heel bone; internally, with the external cuneiform.

Internal Cuneiform.—Largest of the three. The point of the wedge is uppermost. It is concave externally, convex internally, where it is covered by tegument only. Behind, it joins the scaphoid by a triangular surface; in front, it meets the metatarsal of the great toe. The tibialis anticus muscle is inserted into the inner side of the rounded base or lower part of the bone.

Fig. 27.

UPPER SURFACE OF THE LEFT FOOT.—1. Astragalus. 2. Its anterior face. 3. Os calcis. 4. Naviculare, or scaphoides. 5. Internal cuneiform. 6. Middle cuneiform. 7. External cuneiform. 8. Cuboid bone. 9, 9. Metatarsal bones. 10. First phalanx of the big toe. 11. Second phalanx of the big toe. 12, 12, the first; 13, 13, second; and 14, 14, third phalanges of the other toes.

Middle Cuneiform.—Smallest of the three. The apex of the wedge is downwards. In front, it joins the second metatarsal, internally, with the internal cuneiform; externally, with the external cuneiform, and behind with the scaphoid.

External Cuneiform.—Size intermediate between the two others. The narrow part of the wedge is below. It joins, in front, the third metatarsal; on its inner side, by two surfaces, the internal cuneiform and second metatarsal; behind, the scaphoid. Outside, it has a process which joins anteriorly with the fourth metatarsal bone, and posteriorly with the cuboid.

METATARSAL BONES.

Five in number, one for each toe. At their bases they articulate with the cuboid and cuneiform bones. By rounded heads they join the first phalanges of the toes.

The *first* metatarsal, on the inside, is the thickest and shortest. It articulates by a large base with the internal cuneiform. The peroneus longus tendon is inserted into it. Sesamoid bones are met with in its phalangeal articulation.

The *second* is the longest of all. It articulates with the middle cuneiform behind, internally with the internal cuneiform, externally with the external cuneiform and third metatarsal.

The *third* joins by its base with the external cuneiform, and, outside, by two facets, with the fourth metatarsal.

The *fourth* articulates with the cuboid, and, at its sides, with the third and fifth.

The *fifth* and smallest articulates with the cuboid and fourth metatarsal. From its base projects outwards and backwards a *tubercle*, which receives the tendons of the peroneus tertius and peroneus brevis muscles.

TOES.

Each has a *first, second*, and *third phalanx*, counting from the metatarsus, except the great toe, which has but two.

The *first* phalanx is always smaller than the first of the corresponding finger. Their bodies are narrow, their bases concave, their distal ends have two small convexities with a groove between them.

The *second* phalanges have small bodies; their bases have two concavities with a ridge between; the anterior ends, two convexities with an intermediate groove.

The *third* phalanges are still smaller. The articulating base has two concavities and a ridge. The extremity is rough and flattened.

THE TEETH.

Man has two sets of teeth; first, twenty small *deciduous* or *milk* teeth; afterwards thirty-two permanent teeth. Of the former,

Fig. 28.

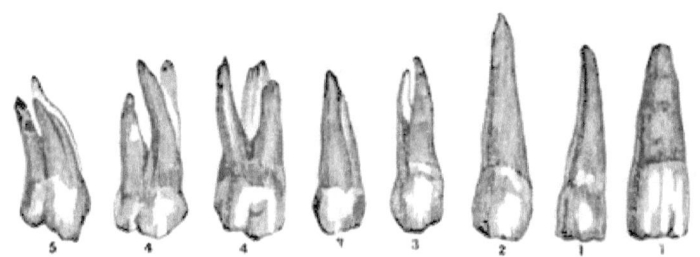

Eight Teeth of One Side of the Upper Jaw.—1. Incisors. 2. Cuspids or canine teeth. 3. Bicuspids. 4. First two molars. 5. Dens sapientiæ.

Fig. 29.

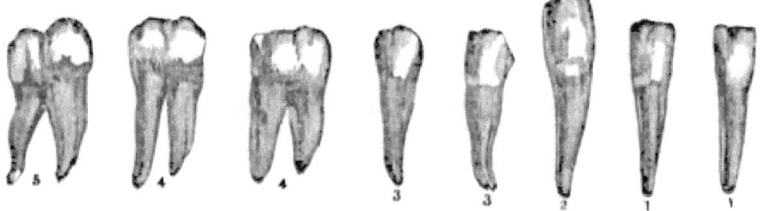

Eight Teeth of One Side of Lower Jaw.—1. Incisors. 2. Cuspids or canine teeth. 3. Bicuspids. 4. First two molars. 5. Dens sapientiæ.

there are in each jaw four *incisors*, two *canines*, and four back teeth or *molars*. Of the permanent set, in each jaw, *four* in front are incisors, *two* (one on each side) are canines, *four* (two on each side) are bicuspids or premolars, and *six* (three on each side) molars.

Each tooth has a projecting *crown*, a *neck*, and a concealed *fang* or root. The root is implanted in the *alveolus* of the jaw; which is lined with periosteum.

The edge of the *incisor* tooth is cutting, as the name implies; the crown is wedge-shaped; the fang is long and single. The upper incisors are largest.

The *canines* have a conical, somewhat pointed crown, and a single fang, longer than that of the incisors. The upper ones are largest.

The *premolar* or *bicuspid* teeth have the surface of the crown divided, not deeply, into two cusps or prominences. Their fangs are single, but with a partial division, especially at the apex. The upper ones are larger than the lower.

The *molars* have broad crowns, surmounted by four or five tubercles. The *first* and *second* molars have three fangs in the upper jaw, two in the lower; the *third* (wisdom tooth) has one fang, grooved as though becoming divided into three or two.

Structure of the Teeth.

Each tooth, in vertical section, shows an interior cavity containing the *pulp*, with the bloodvessels and nerves belonging to it. The solid tooth consists of *dentine*, *enamel*, and *cementum* or *crusta petrosa*.

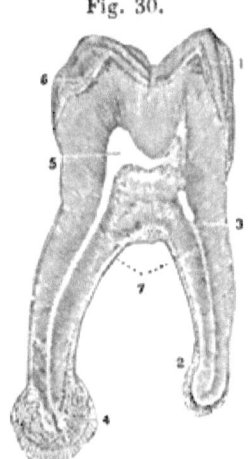

Fig. 30.

Dentine differs in structure from bone; being shown by the microscope to consist of minute wavy and branching *tubuli*, about $\frac{1}{1500}$ of an inch in diameter, imbedded in the dense *intertubular tissue*. The tubuli are vertical in the summit of the crown, oblique in the neck, and inclined downwards in the lower part of the neck.

Enamel is the hardest part of a tooth. It forms a thin crust over the exposed part of the crown, thinning down towards the neck. Microscopic examination shows it to be composed of parallel six-sided columns, directed vertically at the summit of the crown, and horizontally at the sides.

Cementum or *crusta petrosa* is intermediate in compactness between enamel and dentine. It covers the fangs of the teeth. Its structure is like that of bone, having the Haversian canals and lamellæ.

MAGNIFIED SECTION OF A TOOTH.—1. Enamel. 2, 7. Cementum. 3. Ivory. 4. Foramen. 5. Dental cavity. 6. Osseous corpuscles.

Chemically, teeth are composed of phosphate and carbonate of lime, traces of fluoride of calcium, and other salts, and a little gelatinoid animal matter. Dentine has seventy-two parts of mineral matter and twenty-eight parts of gelatin.

Development of the Teeth.

In the sixth or seventh week of fœtal life the germs of the milk-teeth begin to form in a groove of the maxillary mucous membrane. Calcification of the permanent teeth commences a little before birth; both sets of teeth being thus in the jaws together, long before their eruption.

Early in fœtal life, the dental groove becomes closed over and subdivided by septa into *follicles*, within each of which a *papilla* arises. The follicles then change into *dental sacs*, and the papillæ into *tooth-pulps*. Within the enlarging sac, but at the expense of the pulp, the dentine forms. From the lining of the dental sac is developed (at first quite soft) the *enamel organ*, composed of fibres, united to the dentine of the pulp-surface. In the place of this contact, called the *enamel membrane*, the mineral deposition which gives the enamel hardness, occurs.

Eruption of the teeth takes place when their size and hardness induce absorption of the gum by pressure.

Of the *milk teeth*, the *central incisors* come through the gum at about the seventh month of infancy; those of the lower jaw usually first. *Lateral incisors*, eighth to tenth month; *anterior molars*, twelfth to fourteenth month; *canines* (stomach and eye teeth), fourteenth to twentieth month; *posterior molars*, eighteenth to thirty-sixth month.

Fig. 31.

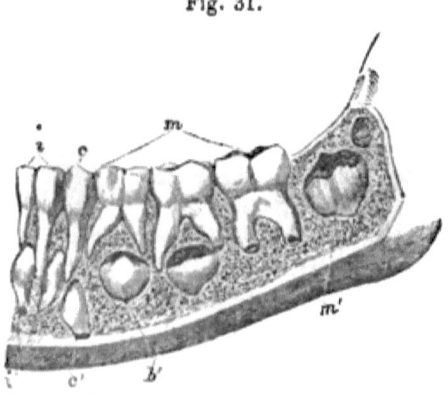

TEETH AT FIVE YEARS.—*i*. Temporary incisors. *c*. Temporary canine. *m* Temporary first and second molar, and first permanent molar. *i'*. Permanent incisors. *c'*. Permanent canine. *b'*. Permanent bicuspids. *m'*. Permanent second molar.

Permanent teeth come out as follows: Between six and a half and seven years of age, the *first molars*; seventh year, *middle incisors*; eighth year, *lateral incisors*; ninth year, *first premolar*; tenth year, *second premolar*; eleventh to twelfth year, *canine*; twelfth to thirteenth year, *second molars*; seventeenth to twenty-first year, *last molars*.

CHAPTER II.
THE ARTICULATIONS.

Ligaments, cartilages, and *synovial membranes* constitute the apparatus of the joints between the bones.

Ligaments are either of *white fibrous* or *yellow elastic* tissue. Of the latter the *ligamentum nuchæ* and *ligamenta subflava* of the spine are the principal examples.

Cartilages are either *temporary* (becoming ossified) or *permanent*. The latter are numerous in the body; being, 1, *articular* cartilages; 2, *costal* cartilages; and 3, various *lamellar* cartilages, as those of the ear, the nose, eyelids, Eustachian tube, larynx, and other parts of the air-passages. The tissue of cartilage is, when minutely examined, found to consist of cells or corpuscles in an intercellular fibro-granular substance. It is, in mass, firm, but elastic and flexible; either pearly white or yellow in color.

Synovial membrane resembles serous membrane in structure, but secretes a peculiar fluid, *synovia;* which is glairy like the white of an egg.

Bursæ are membranous cavities between surfaces which move upon each other; as between the patella and the skin, over the olecranon, outside of the malleoli, between the trochanter major of the femur and the gluteal muscles, &c.

Articulations are of three kinds: *immovable, synarthrosis; movable, diarthrosis;* and *mixed, amphiarthrosis.*

Synarthroses are either *sutura*, with a series of interdentations; *schindylesis*, or dove-tailing, by a thin plate of one being received into a fissure between two laminæ of another; or *gomphosis*, where a conical process is fastened into a socket, as the tooth-fangs are in the alveoli.

Diarthroses are of four kinds: *arthrodia*, which admits of gliding movement, as in the temporo-maxillary articulation; *enarthrosis*, or the ball and socket joint, as at the hip; *ginglymus* or hinge, as at the elbow; and *diarthrosis rotatorius*, as between the atlas and axis vertebræ, and between the upper ends of the radius and ulna.

Amphiarthrosis is an articulation with but limited motion; as, for example, that between the ossa pubis—the *symphysis pubis*.

VERTEBRAL ARTICULATIONS.

In front of the bodies of the vertebræ, from the second cervical to the first sacral, lies the *anterior vertebral ligament*. It widens as it descends. Behind, upon the bodies of the vertebræ, from the occiput to the coccyx, is the *posterior vertebral ligament*. Between each two spinous processes is an *inter-spinal ligament;* almost wanting, however, in the cervical region. Those in the dorsal region are three-sided; the lumbar ones quadrangular.

The *ligamentum nuchæ* takes the place of the interspinal ligaments in the back of the neck, extending from the occiput to the last cervical spine. It is strong and elastic.

The *ligamenta subflava*, also of elastic fibrous tissue, join the laminæ or bridges of the vertebræ, below the second cervical.

Capsular ligaments surround the oblique or articulating processes; and vertical fibres, somewhat corded in the dorsal region, but scanty elsewhere, join the transverse processes.

The *intervertebral fibro-cartilages*, twenty-three in number, are disks, formed of concentric laminæ at the circumference, and an elastic pulp at the centre; those in the lumbar region are the largest.

Occiput and Atlas.

The articulation here consists of an *anterior* and a *posterior* ligament, and a *capsular* ligament, including together the condyle of the occiput on each side and the oblique process of the atlas.

Atlas and Axis.

Across the ring of the atlas, behind the odontoid or dentate process, stretches the *transverse* or *cruciform* ligament; sending also a fasciculus up to be attached to the occiput, and one down to connect with the odontoid process.

There are also two *anterior atlo-axoid* ligaments, one *posterior atlo-axoid*, and two *capsular* ligaments; the latter connecting the oblique processes.

PELVIC LIGAMENTS.

The sacrum and ilium form together a *symphysis*, on each side, with a cartilage over each articular surface; and, during infancy and pregnancy, a synovial membrane partially developed. Around the symphysis are the short and strong fibres of the *sacro-iliac ligament*, sometimes divided into anterior and posterior. The *sacro-spinal* ligament extends from the transverse processes of the lower sacral vertebræ to the posterior inferior spinous process.

The *ilio-lumbar* ligament connects the crest of the ilium, behind, with the last lumbar transverse process. The *lumbo-sacral* ligament joins the transverse process of the last lumbar vertebra with the upper part of the sacrum on each side. Between the sacrum and the ischium extend the *greater* and *lesser sacro-sciatic* ligaments. The sacro-sciatic notch is by them divided into two foramina for the passage of vessels and nerves, &c.

The *obturator* ligament occupies the obturator or thyroid foramen; it is perforated near its upper margin by vessels.

The triangular *sub-pubic* ligament is immediately under the pubic arch. *Anterior, superior* and *posterior* ligamentous bands surround this arch.

The *symphysis pubis* is an amphiarthrosis, with two oval articular cartilages, with an interspace lined with epithelium; this is more distinct at the time of pregnancy, when a very slight movement of the bones upon each other may be possible.

TEMPORO-MAXILLARY ARTICULATION.

An *external* and an *internal lateral* ligament and a *capsular* ligament constitute the periphery of this joint. The first is broad, the second forms a sheath for vessels and nerves, the third envelops the condyle of the jaw and the margin of the glenoid cavity of the temporal bone. An inter-articular cartilage and two synovial membranes are contained within the articulation.

The *stylo-maxillary* ligament passes from the styloid process of the temporal bone to the *angle* of the lower jaw.

THORACIC ARTICULATIONS.

Ribs and Vertebræ.—Around the *head* of each rib is a *capsular* ligament. Another capsular ligament joins the *tubercle* of the rib to the *transverse process* of a vertebra.

The *anterior radiated* ligament extends from the head of the rib to the two vertebræ with which it is connected, and to their intervertebral cartilage. The *inter-articular* ligament (except with the first and two last ribs) extends from the head of the rib to the intervertebral cartilage; a distinct synovial membrane is on each side of it. There are, further, the *external, internal,* and *middle costo-transverse* ligaments, whose names locate them.

Sternum and Ribs.—The *anterior* and *posterior radiated* or *costosternal* ligaments, at the anterior ends of the true ribs, pass from the cartilages of the ribs to the sternum. The anterior is most fully developed. A thin *capsular* ligament completes the connection.

The *costo-xiphoid* ligament joins the cartilages of the sixth and seventh ribs to the sternum.

Sternum and Clavicle.—A *capsular* ligament of considerable thickness surrounds the end of the clavicle and connects it with the sternum. There is an articular cartilage between the bones. Between the sternal ends of the two clavicles passes the *inter-clavicular* ligament. The *rhomboid* ligament extends from the sternal end of clavicle downwards and inwards to the cartilage of the first rib.

THE SHOULDER.

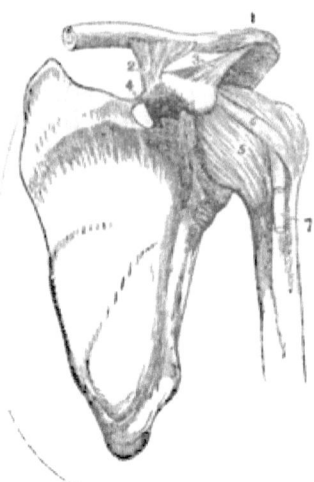

Fig. 32.

LIGAMENTS OF ACROMIO-CLAVICULAR AND SCAPULO-HUMERAL ARTICULATIONS. —1. Superior acromio-clavicular ligament. 2. Coraco-clavicular ligament. 3. Coraco-acromial ligament. 4. Coracoid ligament. 5. Capsular ligament of the shoulder-joint. 6. Ligamentum adscititium, or coraco-humeral ligament. 7. Tendon of long head of the biceps muscle.

Clavicle and Scapula.—This articulation has the *superior* and *inferior acromio-clavicular* ligaments, the *coraco-clavicular*, which *divides* into the *conoid* and the *trapezoid*, and, lastly, the small *coracoid* ligament, which bridges the coracoid notch; through the foramen thus made of this notch pass the suprascapular nerve and artery. There is a fibro-cartilage in the scapuloclavicular articulation.

Scapula and Humerus.—This is a ball-and-socket joint, with a shallow glenoid cavity; the ligaments being powerfully supported by the surrounding *tendons* and *muscles*.

The ligaments of the shoulder joint are the *capsular*, the *glenoid*, and the *coraco-humeral* ligaments.

The *capsular* ligament is penetrated by the long tendinous head of the biceps. The *coraco-humeral* is sometimes called the *ligamentum adscititium*.

The *glenoid* ligament is a firm marginal band surrounding the upper part of the glenoid cavity, and deepening it.

The *synovial membrane* of this joint communicates with a bursal sac for the tendon of the subscapularis muscle, and sometimes with one for that of the infra-spinatus. The tendon of the biceps has its own synovial sheath, not communicating with that of the joint.

The muscles related to this articulation are, the *supra-spinatus*, long head of the *triceps, subscapularis, infra-spinatus, teres minor*, and long head of the *biceps;* with the *deltoid* more superficially.

THE ELBOW.

A *ginglymoid* articulation, chiefly between the humerus and the ulna. The radius supports it by its head, receiving a tuberosity of the humerus into a shallow cup; while the radius also rotates upon the ulna at its upper end.

The ligaments at the elbow are, the *anterior, posterior*, and *external* and *internal lateral* ligaments. The synovial membrane of the joint is an extensive one.

Around the neck of the radius, suspending it to the ulna, passes the *orbicular* or *annular* ligament, allowing of rotary motion.

Between the elbow and wrist the radius and ulna are connected, nearly the whole distance, by the *interosseous* ligament. The *round* (*teres*) or *oblique* ligament (sometimes wanting) is a fibrous cord reaching downwards and outwards from the ulna at the base of the coronoid process to the radius below the insertion of the biceps tendon.

THE WRIST.

Lower Radio-ulnar Articulation.—This consists of an *anterior* and a *posterior* ligament, and an articular fibro-cartilage. The synovial membrane of this connection is called *sacciform* from its looseness. Pronation and supination of the forearm and hand depend upon the rotary movement of the radius upon the ulna at their lower junction.

Radio-carpal or *Wrist-joint.*—The end of the radius and the *inter-articular cartilage* join with the *scaphoid, lunare*, and *cuneiform* bones of the carpus. The surfaces are all covered by cartilage. The ligaments are the *anterior, posterior*, and *external* and *internal lateral*. The synovial membrane is a simple one.

Fig. 33.

INTERNAL VIEW OF THE ELBOW JOINT.—1. Capsular ligament. 2, 2. Internal lateral ligament. 3. Coronary ligament. 4. Ligamentum teres. 5. Interosseous ligament. 6. Internal condyle.

Articulations of the Carpal Bones.—These are, those of the *superior row with each other*, those of the *second row together*, and of the *two rows with each other*. For the first, there are *two palmar, two dorsal*, and two interosseous ligaments; for the second, *three palmar, three dorsal*, and *two interosseous;* between the two rows, an *anterior* or *palmar*, a *posterior* or *dorsal*, an *external lateral*, and an *internal lateral*. There are *two* distinct *synovial membranes* in the articulations of the carpal bones. The more extended lines the *scaphoid, lunar*, and *cuneiform* bones below, and separates them by its prolongations; covers also, and separates, the bones of the lower row; and covers the articular ends of the metacarpal bones. The other synovial membrane lies between the cuneiform and pisiform bones.

Carpo-metacarpal Articulations.—The metacarpal of the *thumb* has a *capsular* ligament connecting it with the *trapezium*; the joint is lined by a distinct synovial membrane.

Between the metacarpal bones of the *four fingers* and the carpus pass the *palmar, dorsal,* and *interosseous* ligaments. Similar ligaments, also, unite the carpal extremities of the metacarpals to each other. A *transverse* ligamentous band unites their digital ends underneath.

ARTICULATIONS OF THE HAND.

The *metacarpo-phalangeal* joints are ginglymoid, the round heads of the metacarpals being received into cavities of the upper ends of the first phalanges. They have an *anterior* and *two lateral* ligaments.

The *inter-phalangeal* articulations are, like the above, ginglymoid; and their ligaments are also *one anterior* and *two lateral*.

HIP-JOINT.

The strongest ball-and-socket joint in the body, formed by the head of the femur with the acetabulum. The ligaments are, the *capsular, cotyloid, teres, ilio-femoral,* and *transverse*.

Fig. 34.

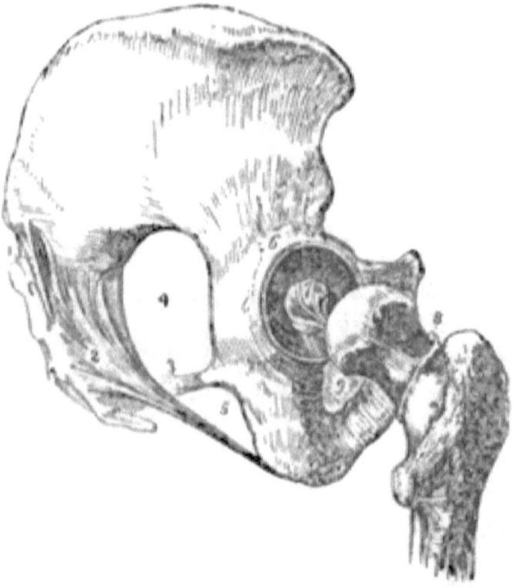

The *capsular* ligament extends from the margin of the acetabulum to the neck of the femur, surrounding the whole joint. It is dense and strong, especially above and in front. A synovial bursa separates it from the iliacus and psoas muscles.

The *cotyloid* ligament is an almost cartilaginous ring which deepens the margin of the cavity of the acetabulum. It is thickest above and behind.

LIGAMENTS OF THE HIP-JOINT AND PELVIS.—1. Posterior sacro-iliac ligament. 2. Greater sacro-sciatic ligament. 3. Lesser sacro-sciatic ligament. 4. Greater sacro-sciatic notch. 5. Lesser sacro-sciatic notch. 6. Cotyloid ligament around the acetabulum. 7. Ligamentum teres. 8. Line of attachment of the capsular ligament of the hip-joint, posteriorly. 9. Obturator ligament.

The *ligamentum teres* or round ligament is a triangular band, whose base is attached to the bottom of the acetabulum, while its apex is connected with the head of femur, below and behind its centre. It is thus quite within the joint.

The *ilio-femoral* ligament passes obliquely across from the anterior inferior spine of the ilium to the anterior inter-trochanteric line.

The *transverse* ligament crosses the notch at the lower portion of the acetabulum, and converts it into a foramen. It is continuous with the cotyloid.

The synovial membrane of this joint is extensive. There is also a mass of fat contained in a fossa of the acetabulum.

KNEE-JOINT.

A hinge between the condyles of the femur and the head of the tibia, with the patella in front. *Outside* of the joint are the *anterior* ligament (ligamentum patellæ); the *posterior* (ligamentum Winslowii); the *internal lateral;* two *external lateral;* and the capsular ligament. *Within* the articulation are the two *crucial* ligaments (anterior external, and posterior internal); two *semilunar cartilages;* the *transverse* and *coronary* ligaments; *ligamentum mucosum,* and *ligamenta alaria*.

The *ligament of the patella* connects the lower point of that bone with the tubercle of the tibia. It is about three inches long. A synovial bursa is between the patella and its ligament and the skin; and a smaller one between the ligament and the tuberosity of the tibia.

The *crucial* ligaments are so called because they *cross* each other, X-like. They are respectively called anterior and posterior, according to the place of their tibial insertion; each passing from a condyle of the femur across to the other side of the tibia.

The *semilunar fibro-cartilages* are crescent-shaped, and deepen the shallow surfaces (for the condyles) on the head of the tibia. The circumference of each is thicker than its inner concave margin.

The *transverse* ligament connects the anterior parts of the two semilunar cartilages.

The *coronary* ligaments are numerous short fibrous bands joining the margin of the semilunar cartilages with the head of the tibia and surrounding ligaments.

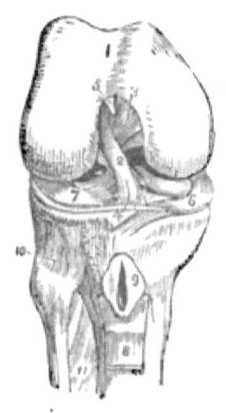

Fig. 35.

THE KNEE-JOINT LAID OPEN.
—1. Lower end of the femur. 2. Anterior crucial ligament. 3. Posterior crucial ligament. 4. Transverse fasciculus. 5. Attachment of ligamentum mucosum. 6. Internal semilunar cartilage. 7. External semilunar cartilage. 8. Ligamentum patellæ. 9. Its bursa laid open. 10. Superior peroneo-tibial articulation. 11. Interosseous ligament.

The *synovial membrane* of the knee is the most extensive in the body, projecting above and below the joint beneath tendons and aponeuroses. The *ligamentum mucosum* is a fold of it, of a triangular shape, under the patella. The *ligamenta alaria* are fringe-like folds passing on each side from the ligamentum mucosum to the sides of the patella.

TIBIO-FIBULAR ARTICULATIONS.

Superior Junction.—This has an *anterior* and a *posterior* ligament, and a synovial membrane.

Inferior.—Of this the ligaments are, the *anterior, posterior, transverse,* and *interosseous.* The synovial membrane is connected with that of the ankle-joint.

ANKLE-JOINT.

A perfect hinge, between the tibia and fibula above, and the astragalus below. The tibia rests upon the astragalus; the fibular malleolus supports the side of the articulation. The ligaments are, the *anterior*, the *internal lateral*, and the *external lateral*.

The *anterior* ligament is simple, the *internal lateral* is in two layers, the superficial and the deep. The *external lateral* has three *fasciculi*. The anterior and posterior of these fascicles connect the internal malleolus with the astragalus; the middle one passes from the malleolus to the os calcis.

The synovial membrane of the ankle-joint invests the lining of the ligaments and goes for a short distance between the tibia and fibula.

TARSAL ARTICULATIONS.

The *calcaneo-astragaloid* ligaments are, the *external*, the *posterior*, and the *interosseous*. The *last* is the principal connection between the bones. It is composed of many fibres, vertical and oblique. This articulation has two synovial membranes.

Between the *scaphoid*, the *cuboid*, and the *three cuneiform* bones the union is maintained by the *dorsal, plantar,* and *interosseous* ligaments.

The *os calcis* is connected with the *cuboid* by two *dorsal* ligaments—the *superior* and the *internal calcaneo-cuboid*, and by two *plantar*, the *long* and the *short calcaneo-cuboid* ligaments. The *os calcis,* and *scaphoid* are united by two ligaments, the *superior* and the *inferior calcaneo-scaphoid.*

The *astragalus* forms with the *scaphoid* a limited ball-and-socket joint; the posterior concavity of the scaphoid receiving the round head of the astragalus. Dislocation sometimes occurs in this articulation.

PHALANGEAL ARTICULATIONS.

Four synovial membranes exist in the tarsus: one, *posterior calcaneo-astragaloid;* one, *anterior calcaneo-astragaloid* and *astragalo-*

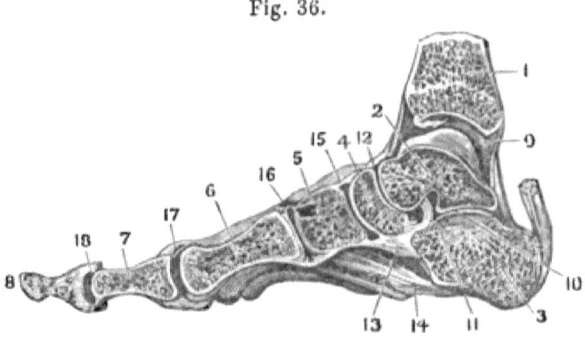

Fig. 36.

VERTICAL SECTION OF THE ANKLE-JOINT AND FOOT.—1. Tibia. 2. Astragalus 3. Os calcis. 4. Scaphoides. 5. Cuneiforme internum. 6. Metatarsal bone of the great toe. 7. First phalanx of the great toe. 8. Second phalanx of the great toe. 9. Articular cavity between the tibia and astragalus. 10. Synovial capsule between astragalus and calcis. 11. Calcaneo-astragaloid interosseous ligament. 12. Synovial capsule between astragalus and scapholdes. 13. Calcaneo-scaphoid ligament. 14. Calcaneo-cuboid ligament. 15. Synovial capsule between scaphoides and cuneiform internum. 16. Synovial capsule between cuneiforme internum and first metatarsal bone. 17. Metatarso-phalangeal articulation of the great toe, with the sesamoid bones below. 18. Phalangeal articulation of the great toe.

scaphoid; a third, *calcaneo-cuboid;* a fourth, between the *scaphoid* and *three cuneiform, between* the *cuneiform,* between the *cuboid* and the *external cuneiform,* and between the *middle* and *external cuneiform* and the *second* and *third metatarsal* bones.

METATARSAL AND PHALANGEAL ARTICULATIONS.

Tarso-metatarsal.—The *three cuneiform* bones and the *cuboid* join with the five metatarsals. The *internal cuneiform* receives that of the *great toe.* The *second* metatarsal goes against the *middle* cuneiform, between the internal and external ones. The *third* metatarsal is connected with the *external cuneiform;* the *fourth* with the same bone and also the *cuboid;* and the *fifth* with the *cuboid.* Inter-articular *cartilages* cover the surfaces, between which there are *three* synovial membranes; the strength of the union being also maintained by *dorsal, plantar,* and *interosseous* ligaments.

Inter-metatarsal ligaments are, the *dorsal, plantar,* and *interosseous.*

Metatarso-phalangeal.—The round heads of the metatarsal bones are received into concavities of the first phalanges; the connecting ligaments are, the *anterior* or *plantar,* and two *lateral.*

Phalangeal Articulations.—Like those of the hand, the phalanges of each toe are united together by (at each joint) one *anterior plantar*, and two *lateral* ligaments. Synovial membranes line these articulations.

CHAPTER III.
DIGESTIVE ORGANS.

Mouth.

THE *roof* of the mouth is formed by the hard *palate* in front, and the soft palate behind; its *floor* by the *mylo-hyoid muscles*. It opens posteriorly into the *fauces*. The mouth is lined by a mucous membrane, continuous with the lining of the pharynx, larynx, and nares, and, upon the lips, with the skin. Under the tongue is a doubling of this membrane, the *frænum linguæ*. A frænum also exists within each lip at its middle, and one in front of the epiglottis.

The *lips* are chiefly composed of the fibres of the orbicularis muscle, covered externally by fat and skin.

The *gums* are formed of a dense fibrous tissue, connected with the alveolar periosteum around the necks of the teeth; their covering mucous membrane is vascular, but slightly sensitive. Papillæ of capillaries and nerves are numerous upon both the lips and gums.

Tongue.

This, the organ of taste and, in part, of mastication, as well as of articulation, is made up principally of muscular fibres, covered by a mucous membrane supplied with papillæ. The mucous membrane is much thickest on the dorsum or upper side of the tongue. It consists of a *corium* or basement membrane covered with *epithelium*.

The *papillæ* of the tongue are described as *maximæ* (circumvallate), *mediæ* (fungiform), and *minimæ* (conical and filiform).

Of the *maximæ* there are eight or ten, all at the posterior part of the dorsum of the tongue, arranged in a V-shape, the point behind.

The *mediæ* or fungiform papillæ are numerous, and scattered over the dorsum, chiefly at the sides and tip; they are deep red in color, and rounded.

The *minimæ*, conical and filiform, cover two-thirds of the tongue anteriorly. They are minute, and are arrayed in lines nearly parallel with the rows of the *maximæ;* only more transverse near the apex of the tongue. The *filiform* ones have a very thick epithelium, which gives them a whitish appearance; they are also covered by *secondary papillæ*.

In structure the lingual papillæ, like those of the true skin, consist essentially of capillary loops, invested by nervous terminations, and enveloped by epithelial cells. Over the tongue, as well as the lining membrane of the mouth, are many *mucous glands* and *follicles*. The glands abound especially upon the posterior third of the tongue.

The two halves of the tongue are distinctly separated by a fibrous septum. The muscles on each side are, the *hyo-glossus*, *genio-hyo-glossus*, *stylo-glossus*, *palato-glossus*, and in its substance the *superior* (or *superficialis*), and *inferior longitudinal* (or *lingualis*) and the *transverse*.

The *arteries* of the tongue are branches of the *lingual*, *facial*, and *ascending pharyngeal*.

Its *nerves* are three: the lingual branch of the *fifth* pair, the *hypoglossal*, and the lingual branch of the *glosso-pharyngeal*.

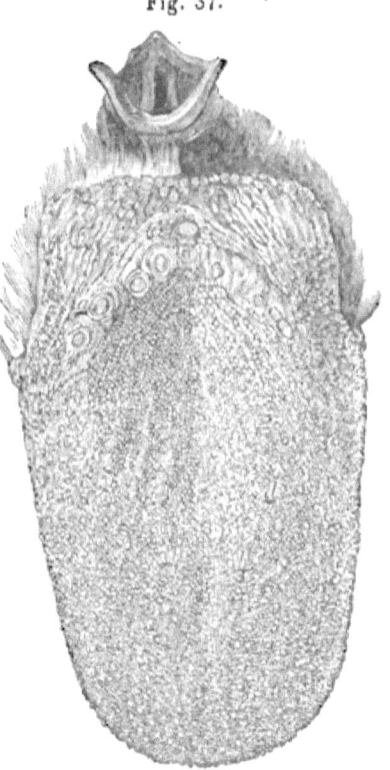

Upper Surface of the Tongue.—*a*. One of the circumvallate papillæ. *b*. One of the fungiform papillæ. *d*. Conical papillæ. *e*. Glottis and epiglottis.

Salivary Glands.

These are the *parotid*, *submaxillary*, and *sublingual*.

The *parotid*, the largest, is placed just below and in front of the ear, extending from the zygoma above, to the level of the angle of the jaw below; anteriorly, it stretches a short distance over the masseter muscle; posteriorly, it reaches as far as the external meatus, and, below it, to the mastoid process.

The inner surface of the parotid has two processes, one in front of the styloid process of the temporal bone, and one behind it. The external carotid artery passes through the substance of the parotid gland; and, outside of this, also, the common trunk of the temporal and internal maxillary veins. The *socia parotidis* is a small lobe of the gland, occasionally detached from it.

Fig. 38.

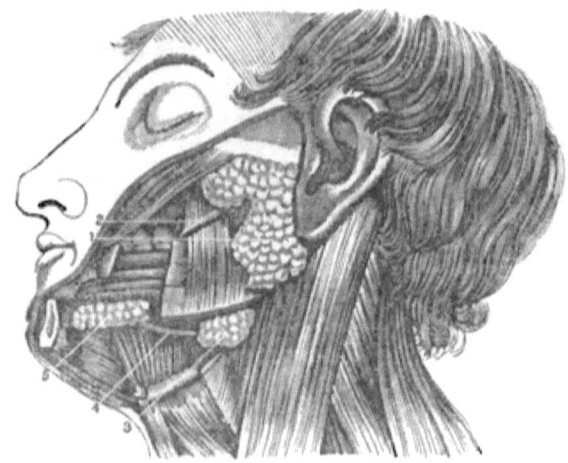

SALIVARY GLANDS.—1. Parotid gland. 2. Duct of Steno. 3. Sub-maxillary gland. 4. Its duct. 5. Sub-lingual gland.

The *duct* of the parotid (duct of *Steno*) opens inside of the cheek opposite to the second molar tooth of the upper jaw. It is about two inches and a half in length.

The *submaxillary* gland is of considerably smaller size. It lies in a fossa of the inner face of the lower jaw-bone, near its angle. The platysma myoides muscle covers it. Its duct (of *Wharton*) opens under the tongue, near its frenum.

The *sublingual* is the smallest of the three glands. It is almond shaped, and lies under the tongue, on each side, imbedded between the mucous membrane and the mylo-hyoid muscle. It has from eight to twenty ducts (ducts of *Rivinus*), which open at the side of the frenum; some of them connect with the duct of the submaxillary.

The salivary glands are all conglomerate in structure, made of lobes subdivided into lobules; each of the latter consisting of many closed cells, connected with a common duct.

Palate.

The *hard* palate reaches from the alveoli in front of the upper jaw to the line of junction of the soft palate behind. A ridge or *raphe* runs along its middle line, continuous with a similar line upon the soft palate. The small palatal mucous glands are numerous between the mucous membrane and the bone.

The *soft* palate (velum pendulum) is a thick flexible fold of mucous membrane, embracing muscular fibres, bloodvessels, &c.

It is convex behind, where it is continuous with the floor of the posterior nares. At the sides it passes into the walls of the pharynx; below, its border is free.

The *uvula* is a conical projection, of similar structure, downward from the soft palate. On each side, from its base, pass the *anterior* and *posterior half arches* of the palate; the anterior, to the base of the tongue, the posterior to the pharynx. The space from side to side between the opposite arches is the *isthmus* of the *fauces*. Between the two half-arches, anterior and posterior, on each side, lies the *tonsil*. This is a round gland of variable size, often morbidly enlarged. It is about opposite to the angle of the jaw. The internal carotid and ascending pharyngeal arteries pass outside of it. The tonsil has twelve or more small orifices of minute ducts or follicles.

The muscles of the palate are as follows: *Levator Palati*; which originates from the petrous portion of the temporal bone, and the Eustachian tube, and is inserted into the soft palate.

Tensor or *Circumflexus Palati*; arising from the spinous process of the sphenoid and the Eustachian tube. Its tendon passes around the hook of the internal pterygoid process of the sphenoid, to be inserted into the posterior edge of the palate.

Constrictor Isthmi Faucium; originating at the middle of the soft palate, and passing along the anterior half arch to be inserted into the side of the base of the tongue.

Palato-pharyngeus; origin, soft palate; course, through the posterior half arch; insertion, the wall of the pharynx. Its action is to approximate the palate to the pharynx.

Azygos Uvulæ; arising from the posterior nasal spine, it passes through the middle of the soft palate to near the end of the uvula. Its action is to draw up and shorten the latter.

Pharynx.

This, opening downward from the fauces, behind the glottis, is a mucous canal surrounded by connective or cellular tissue and muscles. Below, it is continuous with the œsophagus. Above, the posterior nares and Eustachian tube are in communication with it. It lies against the spinal column, from the occiput to about the fifth vertebra. Its length is five inches in the adult; its width is greater above than below.

The *epithelium* of the mucous membrane of the pharynx is columnar and ciliated above, and squamous below.

The *superior constrictor* muscle of the pharynx is thin and pale. It arises from the internal pterygoid process of the sphenoid and contiguous parts of the palate bone, upper jaw bone, and side of the tongue. Its insertion is described as being into the middle line of the pharynx; some fibres, partly aponeurotic, passing up and back as far as the basilar process of the occiput.

The *middle constrictor* arises from the greater and lesser cornua of the hyoid bone and the stylo-hyoid ligament. It is inserted into the median raphe of the pharynx. This muscle overlaps the superior constrictor and the stylo-pharyngeus and palato-pharyngeus muscles.

The *inferior constrictor* muscle is the thickest of the three. It arises from the cricoid and thyroid cartilages. Part of its fibres are horizontal, the rest ascend obliquely and overlap the the middle constrictor.

The *stylo-pharyngeus* muscle is long and slender; round above, broad and thin below; arising from the styloid process of the temporal bone, to be inserted into the side of the pharynx. Its action is to draw the pharynx upwards and dilate its upper part.

Œsophagus.

This canal begins where the pharynx is narrowest, opposite the fifth cervical vertebra; its length is about nine inches; its width gradually increases below. It passes through a foramen in the diaphragm, and opens by the cardiac orifice into the stomach.

Outside of its mucous and cellular coats, the œsophagus has two layers of muscular fibres; the internal circular and the external longitudinal. The circular fibres are continuous with the inferior constrictor of the pharynx.

The mucous membrane of the œsophagus is reddish above and pale at the lower part. Its epithelium is of the variety called *squamous*. Small compound glands are numerous in the submucous tissue of this canal; each has a single excretory duct.

Abdomen.

This important cavity is, for convenience, divided in description into nine regions. The three upper ones are the *right* and *left hypochondriac*, and, between these, the *epigastric*. The middle regions, the *right* and *left lumbar*, and the *umbilical*. The three lower ones, the *right* and *left iliac*, and the *hypogastric*.

In the *right hypochondriac* region are the *right lobe of the liver*, the *gall-bladder*, *duodenum*, part of the arch of the *colon*, top of the *right kidney*, and *right supra-renal capsule*.

In the *epigastric* region are the left half of the *stomach*, including the *pylorus*, the left lobe of the *liver*, and the *lobulus Spigelii*, the *hepatic artery and vein* and *portal vein*, the *pancreas*, the *semilunar ganglion*, and part of the *aorta*, as well as of the *ascending vena cava*, *vena azygos*, and *thoracic duct*.

The *left hypogastric region* contains the large end of the *stomach*, the *spleen*, the left end of the *pancreas*, part of the *colon*, upper part of the left *kidney*, and left *supra-renal capsule*.

The *right lumbar* region has the *ascending colon*, part of right *kidney*, and part of the *ileum and jejunum* (small intestine).

ABDOMEN.

The *umbilical* region contains the transverse part of the *colon*, part of the *omentum majus* and *mesentery*, part of the *duodenum*, and other portions of the *small intestine*.

Fig. 39.

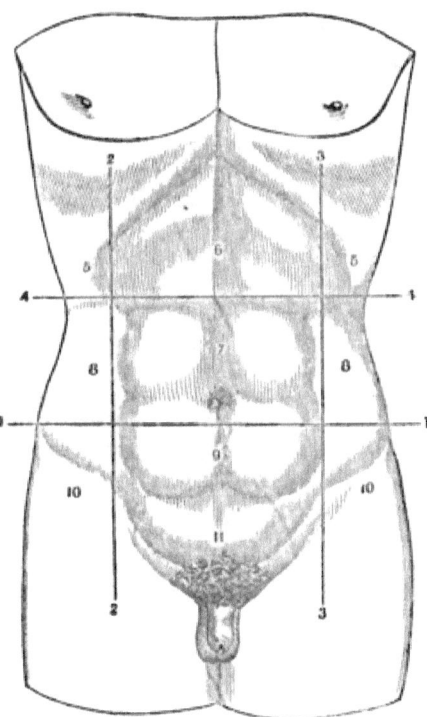

PARIETES OF THE ABDOMEN —1, 1. Line from the highest point of one ilium to the same point of the opposite one. 2, 2. Line from the anterior superior spinous process to the cartilages of the ribs. 3, 3. A similar one for the opposite side. 4, 4. Line drawn perpendicularly to these. 5, 5. Right and left hypochondriac regions. 6. Epigastric region. 7. Umbilical region. 8, 8. Right and left lumbar regions. 9. Hypogastric region. 10, 10. Right and left iliac regions. 11. Lower part of the hypogastric, sometimes called pubic.

In the *left lumbar* region are the *descending colon*, lower part of left *kidney*, and part of the *small intestine*.

In the *right iliac* region lie the *cæcum* or *caput coli*, with the *vermiform appendix*, the *ureter*, and the *spermatic vessels*.

In the *hypogastric* region are portions of the small intestine, the *bladder* in the child, or in the adult when it is distended, and the *uterus* in the *pregnant* female.

The *left iliac* region holds the sigmoid flexure of the *colon*, the left *ureter*, and *spermatic vessels*.

Peritoneum.

This is the most extensive serous sac in the body; thin, transparent, and moistened with serum like other serous membranes. It is duplicated over all the viscera of the abdomen and the inner wall of the cavity itself; while certain folds of it act the part of ligaments to fix or suspend the viscera. *Omentum* is the name applied to the intermediate double folds of the peritoneum.

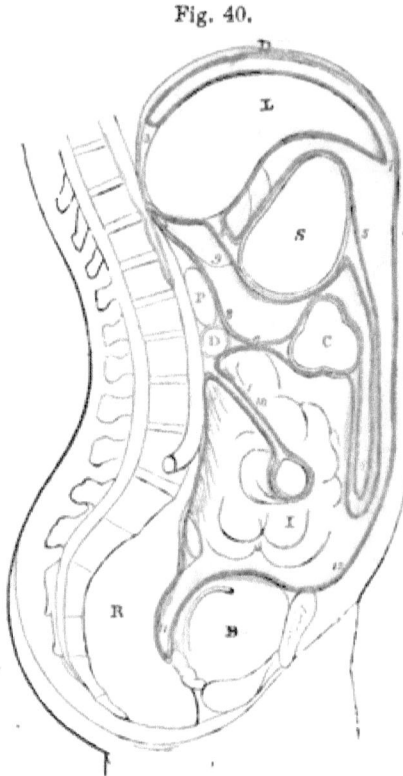

Fig. 40.

THE PERITONEUM.—D Diaphragm L Liver. S. Stomach. C. Transverse colon. D. Transverse duodenum. P. Pancreas. I. Small Intestines. R. Rectum. B. Bladder.

The *omentum minus* connects the *stomach and the liver; omentum majus* passes from the *stomach to the colon;* the *meso-colon* fixes the colon to the vertebral column; the *mesentery* connects the folds of the small intestine with the abdominal walls. The gastro-colic or greater omentum hangs over the intestines, apron-like; it is sometimes named the *caul.*

The reflections of the peritoneum may be successively traced as follows, beginning at the umbilicus: upward, within the wall of the abdomen, to the diaphragm; backwards under that; forwards over the liver, and back under it most of its width; then forward over the stomach, and down, apron-like (part of *omentum majus*) in front of the colon, to ascend again to its under surface, pass under it and back to the spine; thence obliquely forward and downward around the small intestine, and returning to the spinal column (*mesentery*), to descend in front of the rectum to near the lower posterior part of the bladder; forward and upward over the upper surface of the bladder, and thence upward within the abdominal wall to the starting point at the umbilicus.

The *foramen of Winslow* is a communication between the *cavity of the greater omentum* and the general peritoneal cavity, where the gastric and hepatic arteries pass forward, from the arterial trunk called the cæliac axis, to the stomach and liver. This foramen is bounded above by the lobulus Spigelii, in front by the lesser omentum, behind by the ascending vena cava, and below by the hepatic artery.

In the female, the peritoneal reflections deviate from the lower part in front of the rectum, going thence upwards over a small part of the vagina over the body of the uterus, from the sides of which it extends in the form of the *broad ligaments* to the pelvic walls, and then descending in front of the uterus to the bladder; thence upwards and forwards as in the male, it covers the upper part of the bladder and ascends within the abdominal wall.

The lower part of the rectum, the neck, base, and front of the bladder, and the lower part of the vagina, have no covering of peritoneum. It is deficient also at the ends of the Fallopian tubes in the female.

The *appendices epiploicæ* are pouches of the peritoneum holding masses of fat, along the colon and rectum, especially connected with the transverse colon.

Stomach.

The stomach is placed next within the front wall of the abdomen, below the diaphragm and liver; chiefly in the left hypochondriac and epigastric regions. It is irregularly rounded, the left end much the largest. When full, it is about twelve inches in transverse diameter, and four vertically, in the adult.

The left end of the stomach is sometimes called the *splenic* end, being connected by omentum with the spleen. The right end is the *pyloric* portion, it touches the lower surface of the liver.

The œsophagus empties into the *cardiac orifice*, which is nearest the large end, in the upper portion of the stomach.

The *pylorus* is the valvular opening from the stomach into the duodenum.

Between the cardiac and pyloric orifices, on the upper surface, is the *lesser curvature* of the stomach. The *greater curvature* is between the same points around the lower surface.

Four *coats* of the stomach are described: the *serous, muscular, cellular,* and *mucous* coats.

The *serous* coat is an extension of the peritoneum over almost the whole organ.

The *muscular* coat consists of three layers of fibres: *longitudinal, circular,* and *oblique.* The first are most superficial; the second are next within them over the whole stomach, but most developed at the pylorus, where they make a ring-like valve. The oblique fibres are most abundant about the cardiac orifice.

ANATOMY.

Fig. 41.

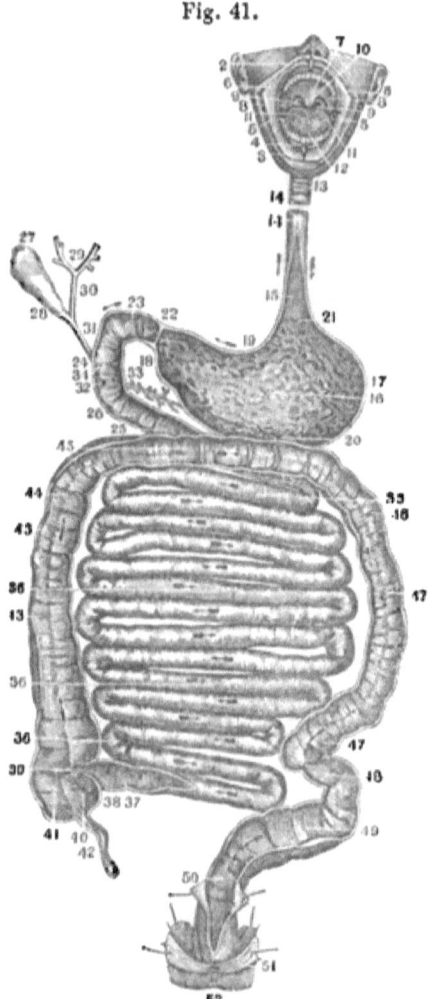

The *cellular* coat is formed of loose connective tissue, containing the bloodvessels.

The *mucous* coat is thick, soft, and velvety; pink in color in the young, pale yellow or gray in the adult. Under the microscope it exhibits a honeycomb structure, being covered with depressions from $\frac{1}{100}$th to $\frac{1}{350}$th of an inch in diameter. The *gastric follicles* are minute tubes at the bottom of these depressions; at the pyloric end the follicles are convoluted; elsewhere simple. They secrete mucus in the pyloric region; in other parts, those called the *peptic glands* secrete the *gastric juice*. Simple follicles also exist numerously over the mucous membrane of the stomach. Its epithelium is *columnar*.

The *arteries* of the stomach are, the *gastric*, branches of the *hepatic* (right gastro-epiploic and pyloric), and branches of the *splenic* (left gastro-epiploic and vasa brevia). Its *veins* terminate in the *splenic* and *portal* veins.

ORGANS OF DIGESTION.—1. Upper lip. 2. Frænum. 3. Lower lip. 4. Frænum. 5. Cheek. 6. Duct of Steno. 7. Roof of mouth. 8. Half arches. 9. Tonsils. 10. Velum pendulum. 11. Tongue. 12. Papillæ. 13. Trachea. 14. Œsophagus. 15. Its interior. 16. Stomach. 17. Its greater end. 18. Its lesser end. 19. Lesser curvature. 20. Greater curvature. 21. Cardiac orifice. 22. Pylorus. 23, 24, 25. Duodenum. 26. Valvulæ conniventes. 27. Gall-bladder. 28. Cystic duct. 29. Hepatic ducts. 31. Ductus communis choledochus. 32. Its opening. 33. Jejunum. 36. Ileum. 37. V. conniventes. 38. Ilium. 39. Ileo-colic valve. 40, 41. Cæcum. 42. Appendix vermiformis. 43–48. Colon. 49, 50. Rectum. 51. Levator ani muscle. 52. Anus.

The *nerves* of the stomach are branches of the right and left pneumogastric, and of the ganglionic or sympathetic.

Intestines.

The small and large intestine together have a length of between thirty and thirty-five feet in all; of which about twenty feet belong to the upper or small intestine.

The *small intestine* is divided in description into the *duodenum*, *jejunum*, and *ileum*. All of these have a *serous, muscular, cellular,* and *mucous* coat. The serous coat is a mesenteric extension. The muscular coat has longitudinal and circular fibres. The cellular coat is merely connective. The mucous coat has some peculiarities in the different parts.

Duodenum.

This is named from its length, which is the breadth of twelve fingers or nine or ten inches. It is curved in position, horseshoe like, first ascending, then descending, and then its longest portion going transversely to end in the jejunum. It is in contact, at different parts, with the liver, gall-bladder, pancreas, colon, diaphragm, aorta, and vena cava. The interior of the duodenum is usually stained with bile.

Jejunum.

This makes two-fifths of the small intestine below the duodenum. It has a somewhat greater diameter than the ileum, with thicker walls and more vascularity and color. There is no boundary whatever between the two—the names being somewhat arbitrary.

Ileum.

Three-fifths of the small intestine, almost, have this name. The ileum ends in the *ileo-cæcal valve*, which is between it and the cæcum, in the right iliac fossa.

Mucous Membrane of Small Intestine.

This is covered by *columnar* epithelium. To the unaided eye it presents numerous transverse foldings, called *valvulæ conniventes*. The depth of these is sometimes two-thirds of an inch, usually less. They first appear an inch or two from the pylorus. Large in the duodenum and upper part of the jejunum, they afterwards diminish, and are almost entirely absent in the lower part of the ileum. Their use is to retard the passage of food during digestion and absorption.

Villi are minute projections from the intestinal mucous membrane, either conical, pyramidal, or cylindrical in shape. From forty to ninety of them have been counted upon the square of a line ($\frac{1}{12}$ of an inch). Each villus contains a minute network of capillaries and lacteal tubes inclosed in basement membrane, on which is a single layer of columnar epithelial cells, perpendicular to the surface. The length of the villi varies from $\frac{1}{36}$ to $\frac{1}{48}$ of an inch.

The *follicles* or *crypts* of *Lieberkühn* are scattered over the lining of the whole of the small intestine. Each is a tubular depression of the mucous membrane, $\frac{1}{500}$ of an inch in diameter, having a circular outlet.

The *glands* of *Brunner* are found only in the duodenum and upper part of the jejunum. They are small, flat, and granular in appearance, with minute ducts—most abundant near the pylorus.

Fig. 42.

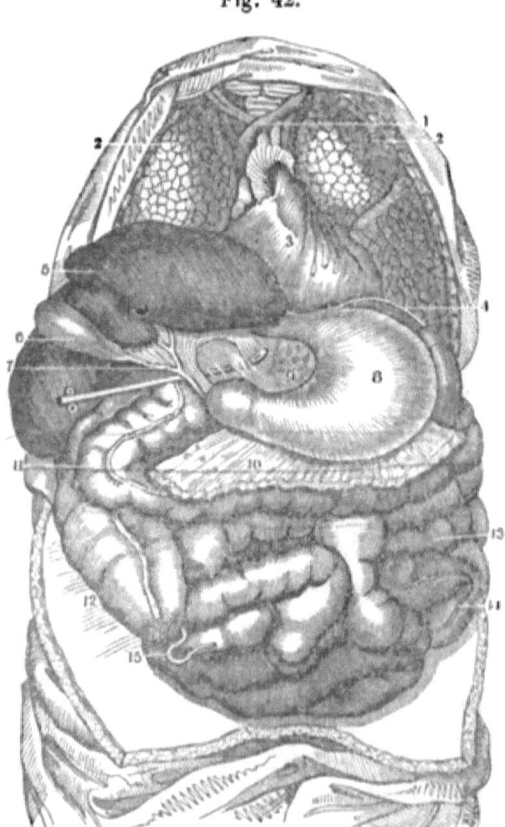

VISCERA, AFTER REMOVAL OF THE FAT IN THE CHEST AND THE OMENTUM MAJUS OF THE ABDOMEN. THE LIVER ALSO TURNED BACK.—1. Great bloodvessels of the heart. 2. Lungs of each side. 3. Heart. 4. Diaphragm. 5. Liver. 6. Gall-bladder. 7. Ductus choledochus. 8. Stomach. 9. The gastro-hepatic, or lesser omentum. 10. Gastro-colic, or greater omentum, cut off. 11. Transverse colon. 12. Its ascending portion. 13. Small intestines. 14. Sigmoid flexure. 15. Appendix vermiformis.

The *solitary glands* are met with in all parts of the small intestine, especially in the lower part of the ileum. They are round and whitish, about $\frac{1}{24}$ of an inch in diameter. Each is a closed sac, with no duct, although around each is a circle of orifices like those of the Lieberkühnian follicles.

Peyer's glands (glandulæ agminatæ) are round or oval *patches* of glands like the solitary glands. The patches vary in length from half an inch to four inches.

Large Intestine.

This comprises the *cæcum*, *colon*, and *rectum*. Its whole length is about five feet. Its diameter is considerably greater than that of the small intestine, and it is more fixed in position. Its division, in description, into three parts is arbitrary, but convenient.

Cæcum.

The cæcum or *caput coli* begins at the ileo-cæcal valve in the right iliac fossa. It has a diameter of about 2¼ inches.

The *vermiform appendix* is attached to this part of the bowel. It is about as thick as a goose-quill, and from three to six inches long; it opens into the cæcum by an incomplete valve.

The *ileo-cæcal valve* (valve of Bauhin) is formed of two folds of mucous membrane of semilunar shape, so disposed that distension of the cæcum forces the margins of the folds together, and closes the valve.

Colon.

The *colon* has an *ascending*, a *transverse*, and a *descending* portion (*arch* of the colon), and a *sigmoid flexure*. The diameter of the colon is less than that of the cæcum.

The *sigmoid flexure* (named from the letter S) ends in the rectum, opposite to the left sacro-iliac symphysis.

Rectum.

The terminal and nearly, though not quite, straight part of the intestine is thus named. It is six or eight inches in length. Its size increases as it descends to the *anus*, its outlet, which is provided with a sphincter muscle.

Mucous Membrane of Large Intestine.

This coat is smooth, not villous, but laid in crescentic folds. It is gray or pale yellow in color, darker in the rectum; where also it is thicker and more vascular. Its epithelium is columnar.

Near the lower part of the rectum there are from two to four semilunar *pouches*, half an inch in width.

Simple *follicles*, or tubular depressions, are more numerous in

Fig. 43.

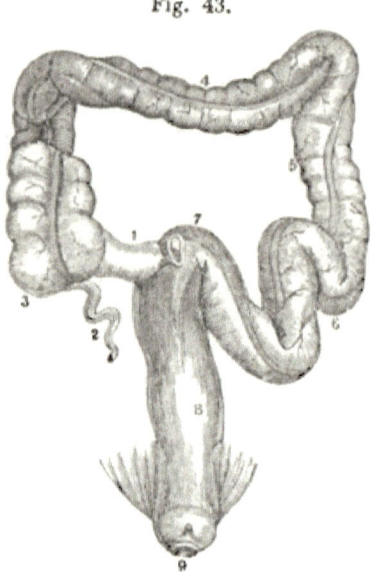

THE LARGE INTESTINE.—1. End of the Ileum. 2. Appendix vermiformis. 3. Cæcum, or caput coli. 4. Transverse colon. 5. Descending colon. 6. Sigmoid flexure. 7. Commencement of rectum. 8, 8. The rectum. 9. Anus.

the large than in the small intestine. They have minute round openings.

Solitary glands also are abundant, especially in the cæcum and appendix. They are small, flask-shaped, whitish, and with a very small central outlet.

Liver.

This is the *largest gland* in the body. It chiefly occupies the right hypochondriac region, immediately under the diaphragm, reaching over, however, through the epigastric into the left hypochondriac. Its transverse diameter is from ten to twelve inches; antero-posterior, six to seven inches. Its greatest vertical thickness, three inches; weight, from three to upwards of four pounds.

The liver is convex on its upper surface, and concave below; posterior border round and wide, anterior border thin and sharp, with a deep notch. This border nearly corresponds with the margin of the ribs. The right half of the liver is much the thickest.

It is divided by a fissure and by the *broad* or *suspensory ligament* (peritoneal) into the *right* and *left lobes*. The *right* lobe is much the largest; it is quadrilateral in shape. On the under surface of this lobe are three fissures: the *transverse* fissure, that for the *gall-bladder*, and for the *vena cava*. The colon, right kidney and suprarenal capsule are in contact with it.

The *left* lobe is convex above, concave over the stomach below. Behind, it reaches nearly to the cardiac orifice of the stomach.

Beneath the right lobe is a portion of the liver called the *lobulus quadratus*, or square lobe.

The *lobulus Spigelii* projects from the back part of the under surface of the right lobe.

The *lobulus caudatus* extends obliquely outwards from the base of the lobulus Spigelii to the under part of the right lobe.

The liver then has, as just described, *five lobes*; it also has *five ligaments* and *five fissures*. The *ligaments* are, all but one, folds of peritoneum. They are called, respectively, the *longitudinal* (broad, suspensory), *two lateral, coronal*, and *round* ligaments.

The *longitudinal* or *broad* ligament is principally attached above to the diaphragm; in front, to the sheath of the rectus abdominis muscle.

The *lateral* ligaments are triangular, and are attached to the diaphragm. The *coronary* ligament connects the posterior border of the liver with the diaphragm.

The *round* ligament (*teres*), is a fibrous cord, the remainder of what was the *umbilical vein*. It may be traced from the navel to the anterior notch of the liver, and along the longitudinal fissure underneath it, as far as to the vena cava.

The *fissures* of the liver are, the *longitudinal*, that of the *ductus venosus* (of the foetus), the *transverse*, that of the *gall-bladder*, and that for the *vena cava*.

The *vessels* of the liver are, the *hepatic artery*, *portal vein*, *hepatic vein*, *hepatic duct*, and the lymphatics.

The *capsule of Glisson* is a coat of loose connective tissue which envelops the vessels of the liver and accompanies them for some distance through the organ.

The liver is supplied with *nerves* from the *hepatic plexus* of the ganglionic or *sympathetic*, from the *pneumogastrics*, and from the *right phrenic*.

Structure of the Liver.—Its substance (seen to be granular, when torn, by the naked eye) is made up of a great number of minute lobules, each called an *acinus*. The whole liver is penetrated by the vessels already named, and is covered by a *fibrous* and a *serous* or *peritoneal* coat.

Each acinus is about $\frac{1}{10}$ to $\frac{1}{20}$ of an inch in diameter. Its shape, transversely, is polygonal. It is suspended, as it were, by its capillaries, from a branch of the hepatic vein; it is interpenetrated by a plexus of capillaries from the portal vein and hepatic artery, and surrounded by a plexus of biliary tubuli or ducts; the mass of each acinus being formed of *cells*.

These cells have a diameter, each, of $\frac{1}{1000}$ to $\frac{1}{2000}$ of an inch. They are *nucleated*, sometimes with two nuclei. They contain yellow biliary matter, which they secrete from the blood.

The origin of the bile-ducts is yet undetermined. Kölliker considers them to commence in a network *outside* of the acini. Kiernan and Leidy believe them to ramify *through* and *within* each acinus, its cells being held in their meshes, as well as in those of the capillaries.

All of the biliary ducts conjoin to form two, one for the right and one for the left lobe, which issue at the transverse fissure, and, uniting, make the hepatic duct. This, after an inch and a half, about, of length, joins at an acute angle the *cystic* duct (of the gall-bladder), thus constituting the *ductus communis choledochus*, or common biliary duct, which empties into the duodenum. This *ductus communis* is three inches long, and about as large as a goose-quill. It passes close to, and sometimes through the pancreas.

Gall-Bladder.

This reservoir for bile lies under the liver. It is pear-shaped; having an anterior rounded *fundus*, and a posterior narrow *neck* or stem. It is about four inches long by one inch broad; and holds eight or ten fluidrachms. It has three coats: serous, fibro-muscular, and mucous. The serous or peritoneal coat covers only its under surface.

Spleen.

A ductless gland, in the left hypochondriac region (opposite the ninth, tenth, and eleventh ribs); oblong, flattened and rounded, about five inches long, three or four inches wide, and an inch and a half in thickness; in color, dark bluish-red. It is covered by the peritoneum, and connected with the stomach by omentum. A *suspensory ligament* joins it above to the diaphragm. A vertical fissure called the *hilus*, divides its inner surface.

Within the *serous* or peritoneal coat is the *elastic fibrous coat* of the spleen. This coat, besides embracing the whole organ, is, at the *hilus*, extended inwards over the vessels. From the sheaths so formed, and from the rest of the coat, many bands (*trabeculæ*) pass in every direction; by their interunion the peculiar areolar structure of the spleen is constituted.

By the presence of this tissue a great degree of *elasticity* is conferred upon the spleen, admitting of great variations of size. It is sometimes remarkably enlarged in ague.

In the interspaces of the trabeculæ is the *proper substance* of the spleen, which is soft and pulpy, and of a dark reddish-brown color.

The *Malpighian corpuscles* are spherical, gelatinous, whitish, semi-transparent bodies, $\frac{1}{2\bar{5}}$ to $\frac{1}{6\bar{0}}$ of an inch in diameter, scattered through the substance of the spleen. They are most distinct in the young subject. They are attached to the sheaths of the smaller arteries, "like moss-rose-buds." Each consists of a capsule formed from the substance of the vessel-sheaths, and containing a soft white pulpy substance, consisting of granules, cells, and nuclei. On the surface of each Malpighian body are ramifications of the arteries, of veins, and a capillary network.

The *splenic artery*, and the *splenic veins*, are very large for the size of the organ. The splenic artery is also tortuous.

The spleen has lymphatic vessels, some of which are deep-seated, and others superficial.

Pancreas.

An oblong, flattened, hammer-shaped conglomerate gland, the *head* being at its right end, embraced by the duodenum; the left end or *tail* reaching to the spleen and left kidney. Its length is six or eight inches; breadth an inch and a half; thickness half an inch to an inch.

The posterior surface of the pancreas is separated from the first lumbar vertebra by the superior mesenteric vessels, vena cava, vena portæ, aorta, left kidney, supra-renal capsule, and vessels.

The *duct* of the pancreas (*ductus Wirsungii*) passes from left to right to open into the duodenum near its middle, by an orifice generally common to it and the ductus communis choledochus. The pancreatic duct is of the size of a goose-quill; it is occasionally double.

In *structure*, the pancreas closely resembles the salivary glands. It is supplied with arteries, veins, lymphatics, and nerves.

CHAPTER IV.
ORGANS OF RESPIRATION.

The *windpipe* is composed of the *larynx, trachea*, and *bronchial tube*, with its branches, communicating with the *lungs*.

Larynx.

This is the special organ of *voice*. It lies just below the root of the tongue, in front of the pharynx and under the skin. It is chiefly composed of cartilages, with a mucous membrane, ligaments, and muscles. In shape, it is triangular above, having a prominent vertical ridge; and cylindrical below.

Nine cartilages enter into the larynx: the *thyroid, cricoid, epiglottis*, two *arytenoid*, two *cornicular*, and two *cuneiform*.

The *thyroid* cartilage is the largest. It consists of two flat sides or wings, meeting in front in the ridge called *pomum Adami* or Adam's apple. This ridge is most strongly marked in the male. It is surmounted by a deep notch. Within, the surface of this cartilage is lined with mucous membrane, to which the *chordæ vocales* are attached.

The lower border of the thyroid cartilage is connected to the cricoid, by a membrane in front, and by muscles at the sides.

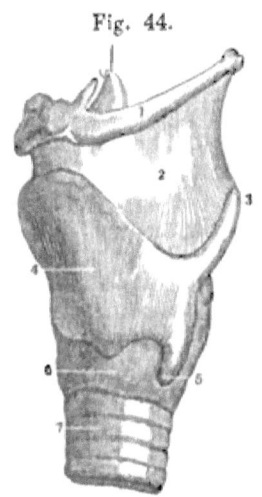

Fig. 44.

The Larynx.—1. Os hyoides. 2. Thyreo-hyoid ligament. 3. Cornu majus of thyroid cartilage. 4. Its angle and side. 5. Cornu minus. 6. Lateral portion of cricoid cartilage. 7. Rings of trachea.

The posterior border ends above in the *superior cornu*, on each side, which is long and narrow. Below, the same border ends in the *inferior cornu* on each side; which is short and thick.

The *cricoid* cartilage is ring-like in shape, smaller but thicker than the thyroid, and situated below and behind it. It may be described as composed of two halves, anterior and posterior. On each side, at the junction of the two halves, there is a small elevation, to which is attached the lower cornu of the thyroid cartilage.

The under border of the cricoid is connected by membrane with the first ring of the trachea.

Its upper border slopes upwards and backwards between the wings of the thyroid. At the highest point, on each side, it supports on an oval surface the arytenoid cartilage.

The cricoid cartilage is lined by mucous membrane continuous with that of the thyroid.

The *arytenoid* cartilages are small and pyramidal; they rest upon the cricoid at the back of the larynx, one on each side. The apex of each is curved backwards and inwards, and surmounted by the small conical *cornicular* cartilage (cartilage of Santorini).

The *cuneiform* cartilages (of Wrisberg) are just in front of the arytenoid; they are quite small, of elongated shape, lodged in the mucous membrane between the arytenoid cartilages and the side of the *epiglottis*.

The *epiglottis* is the lid of the upper aperture of the larynx called the *glottis*, which is considered to extend to the inferior vocal chords. The *rima glottidis* is the fissure between those chords.

The epiglottis is a thin fibro-cartilage, attached to the upper front border of the thyroid by a narrow neck, and having a broad and round free margin, which is vertical during respiration, but closed backwards over the glottis in swallowing.

The *external ligaments* of the larynx are as follows:—

Connecting the thyroid cartilage with the hyoid bone, three—the *thyro-hyoid membrane* and two *lateral thyro-hyoid ligaments*.

Between the thyroid and cricoid cartilages, three—the *crico-thyroid* membrane and two capsular ligaments, each with a synovial membrane. The latter join the cricoid to the lower cornua of the thyroid.

The cricoid and arytenoid cartilages are connected by strong *posterior crico-arytenoid* ligaments and thin and loose *capsular* ligaments.

The ligaments of the *epiglottis* are the *hyo-epiglottic* and *thyro-epiglottic* ligaments. It is connected also with the base and sides of the tongue by three folds of mucous membrane.

The *superior aperture* of the larynx is almost cordiform, widest in front. The laryngeal cavity extends from the epiglottis to the lower edge of the cricoid cartilage. The vocal chords and their connections divide it into two parts.

The *chords* are called *inferior* and *superior*, or *true* and *false* vocal chords. The latter are formed of mucous membrane only; in the inferior or true vocal chords are ligamentous fibres also. The orifice of the glottis varies in shape and size during vocalization and respiration.

The *ventricle* of the larynx (of Galen) is a cavity on each side, between the superior and inferior vocal chords.

Fig. 45.

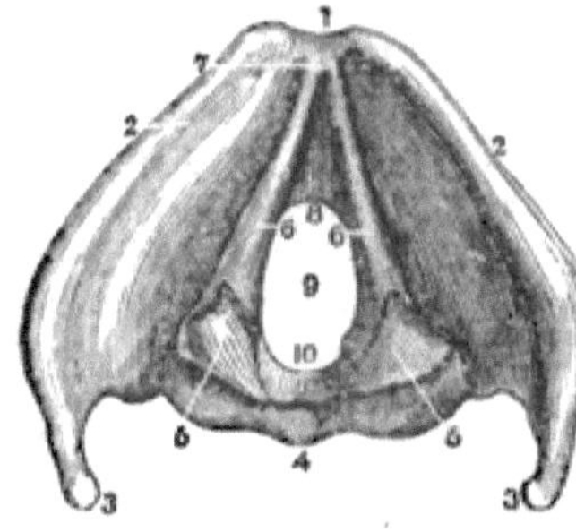

VIEW OF THE LARYNX FROM ABOVE.— 1. Superior edge of the larynx. 2. Its anterior face. 3. Cornua majores of thyroid cartilage. 4. Posterior face of cricoid cartilage. 5, 5 Arytenoid cartilages. 6, 6. Thyreo-arytenoid ligaments. 7. Their origin. 8. Their terminations. 9. Glottis. 10. Cricoid cartilage.

The *sacculus* or *pouch* of the larynx (sinus of Morgagni) is a small conical sac in front of and higher than the ventricle, communicating with it by a narrow opening. It yields a secretion which lubricates the vocal chords.

The *muscles* of the larynx are, the *crico-thyroideus, crico-arytænoideus lateralis, crico-arytænoideus posticus, arytænoideus transversus,* and *thyro-arytænoideus.*

The *crico-thyroid* muscle, on each side, passes upwards and outwards from the front and side of the cricoid cartilage, to the lower and inner border of the thyroid. The *action* of these muscles is, by drawing the thyroid cartilage down, to *elongate* and *make tense* the vocal chords.

The *crico-arytænoideus posticus* passes from the posterior part of the cricoid cartilage, on each side, to the base of the arytenoid. Its action is, by drawing the arytenoid cartilages outwards and backwards, to *open the glottis, making the chords tense* at the same time.

The *crico-arytænoideus lateralis*, on each side, arises near the front of the upper border of the cricoid, and passes upwards and backwards to be inserted into the base of the arytenoid. The action of the *lateralis* is, by rotating the arytenoid cartilages, to approximate their anterior faces, and thus narrow the orifice between the vocal chords.

The *arytænoideus transversus* is a single muscle, crossing upon the back of the two arytenoid cartilages, from one to the other. Some of its fibres are oblique. Its action is to draw the two cartilages together, and thus to narrow the glottis.

The *thyro-arytænoideus* is a broad, flat muscle, within the larynx on each side, lying parallel with the inferior vocal chord, passing from the thyroid back to the front surface of the base of the arytenoid. Its action is to draw the arytenoid forward towards the thyroid, thus *relaxing* the vocal chords.

The *thyro-epiglottidei* muscles depress the epiglottis. The *arytæno-epiglottideus superior* constricts the rima glottidis when the epiglottis closes over it for deglutition. The *arytæno-epyglottideus inferior* compresses the *sacculus laryngis*.

Trachea.

Continuous with the larynx, below it, is this tube, four or five inches in length, to the level of the third dorsal vertebra; there it branches into the right and left *bronchi*. The trachea is composed of eighteen or twenty *imperfect rings* of *cartilage*, completed in the posterior third of each by *muscular* fibres, and united by fibrous and elastic ligaments. The muscular fibres are both longitudinal and transverse; the former most external. They are *unstriped* and *involuntary*. The trachea and bronchi are lined by mucous membrane.

The *black bronchial glands* are lymphatic glands, situated at the bronchial bifurcation.

The *right bronchus* is *shorter* and *larger* in calibre than the left. Both ramify into a multitude of branches or *ramules*, terminating finally in the lungs, in direct communication with the air-vesicles. Fine muscular fibres are discovered by the microscope even in the smaller ramules.

Thyroid Gland.

This is a ductless gland, formed of two lobes, one on each side of the trachea, with an *isthmus* connecting them, across the second and third tracheal rings. Its weight is one or two ounces; color, brownish-red. It is largest in females; increasing a little during menstruation. It is much enlarged in *goitre* or *bronchocele*. The right lobe is somewhat larger than the left.

In structure, this gland is composed of minute closed vesicles invested by a dense capillary network, combined into lobules by connective tissue. The vesicles are almost spherical, and contain a yellowish fluid.

Thymus Gland.

This, too, is a ductless gland. It attains its full size at two years of age; it is then gradually absorbed, and almost ceases to exist at the time of puberty. It is situated in the lower part of the neck and the anterior *mediastinum*, behind the sternum. It consists of two unequal lobes united together, with sometimes an intermediate lobe. Its color is pinkish-gray. At its full development, it is about two inches long, one and a half wide, and a quarter of an inch in thickness.

In *structure*, the lobes of the thymus are composed of numerous lobules, with a common dense capsule. Each lobule (varying in size from that of a pin's head to that of a pea) contains a number of smaller lobules, around a central cavity. These lesser or secondary lobules are also hollow. The latter communicate with the cavities of the primary lobules, and these open into a great central cavity, the *reservoir* of the thymus, which extends through the length of each lateral lobe of the gland. A white milky fluid is found in this reservoir.

In its *development*, the thymus has been shown to begin as a linear tube, with *diverticula* from it.

Lungs.

Each lung is conical; with the apex above, and a concave base, lower behind than in front. The *right* lung is *largest*, being broader and *shorter* than the left; it has *three* lobes, the left but *two*. The fissure is posterior and deep, between the two lobes of the left lung, and between the upper and lower of the right; a shorter fissure separates the middle triangular lobe from the upper one of the right lung.

The *root* of the lungs is the place of their connection with the windpipe and the heart, by the bronchi and the bloodvessels, nerves, &c. *Most anterior* are the pulmonary veins; next the pulmonary artery; behind, the bronchus, for each lung. From above downwards, on the right side, the succession is, bronchus uppermost, then pulmonary artery, lowest pulmonary veins; on the left side, pulmonary artery, bronchus, pulmonary veins.

The lungs together weigh about 42 ounces; less in the female. Their color is pinkish white at birth, mottled with slate-colored patches in the adult, the patches growing black with age. The lung-substance is light, spongy, and elastic, crepitating under pressure, and floating in water. When removed from the chest the lungs collapse.

Each lung has a *serous coat*, a part of the pleura, an *elastic areolar* or *connective tissue*, and the *parenchyma* or proper substance of the lung.

The *parenchyma* is formed of lobules, most easily separable in the fœtus. They vary in size and form. Each consists of the terminal ramifications of a bronchial branch, with their air-cells, and the attendant bloodvessels, lymphatics, and nerves; all united by fibrous connective tissue.

The bronchial ramules are formed of many cartilaginous pieces, with a mucous membrane lined with *ciliated* columnar epithelium. When the size of these ramules becomes reduced almost to $\frac{1}{40}$th of an inch, they become irregular, losing their cylindrical form, and opening in every direction into the air-cells.

Fig. 46.

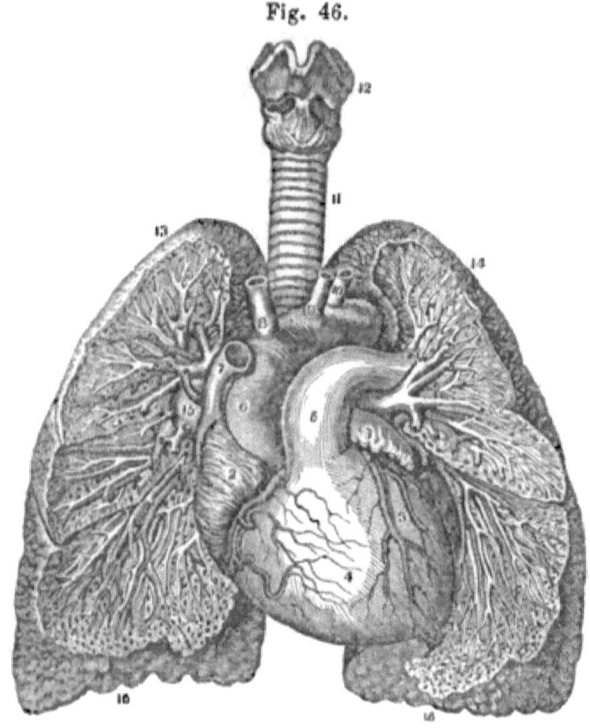

BRONCHI AND BLOODVESSELS.—1. Left auricle. 2. Right auricle. 3. Left ventricle. 4 Right ventricle. 5. Pulmonary artery. 6 Arch of the aorta. 7. Superior vena cava. 8. Arteria innominata. 9. Left primitive carotid artery. 10. Left subclavian artery. 11. Trachea. 12. Larynx. 13. Upper lobe of right lung. 14. Upper lobe of left lung. 15. Trunk of right pulmonary artery. 16. Lower lobes of the lungs.

The *air-cells* are many-sided, divided from each other by their walls or *septa* of a diameter from $\frac{1}{70}$th to $\frac{1}{200}$th of an inch. They, as well as the last bronchial intercellular passages, are lined by a thin mucous membrane, with a *squamous epithelium*. The number of these air-cells in an adult is estimated to be several hundred millions.

By the *pulmonary artery* and its branches, all the venous blood of the body is carried from the heart to the lungs, where it is distributed minutely amongst the air-cells by finely-divided meshes of capillaries with thin walls. From their network arise the pulmonary veins, which return the blood to the heart.

The *bronchial* arteries furnish blood to the lungs for their nutrition. Some of their branches and capillaries terminate in the bronchial, and others in the pulmonary veins.

The *nerves* of the lungs are supplied chiefly from the pneumogastric and sympathetic or ganglionic. The lungs have also superficial and deep *lymphatics*.

Pleuræ.

These are the serous coverings of the lungs, reflected, from their roots, over the inner walls of the thorax. The portion over the lung is the *pleura pulmonalis*, and that over the ribs, *pleura costalis*. Between the two is the *cavity of the pleura*. Like the other serous membranes, the pleura is thus a *closed sac* or double membrane.

The pleura rises, over the apex of each lung, about an inch above the first rib, through the upper orifice of the thorax. Below, it covers the diaphragm.

The *anterior mediastinum* is the interspace between the two pleuræ in front, above and below their point of contact behind the sternum, just above its middle. This cavity is limited behind by the pericardium.

The *middle mediastinum* is a broader inter-pleural cavity, containing the heart, in its pericardial sac, the ascending aorta, descending vena cava, bifurcation of the trachea, pulmonary arteries and veins, and phrenic nerves.

The *posterior mediastinum* is a triangular space, in front of the spinal column, bounded in front by the pericardium and roots of the lungs, and at the sides by the two pleuræ.

CHAPTER V.

URINARY AND GENITAL ORGANS.

Kidneys.

EACH kidney is situated in the lumbar region, from the level of the eleventh rib to near the crest of the ilium; the right one being a little the lowest. It is surmounted by the *supra-renal capsule*, and surrounded by fat. Its position is maintained chiefly by its bloodvessels.

The kidney is somewhat convex in front, and flattened posteriorly; convex on its external border, and concave on the internal margin; the upper end thicker and rounder than the lower.

The right kidney is in contact in front with the liver, duodenum, and ascending colon; the left kidney, with the stomach, spleen, pancreas, and descending colon.

The *hilus* of the kidney is a deep notch in its internal border, where the vessels pass in and out. The *pelvis* of the kidney is the concavity within it.

Each kidney is about four inches long by two in width and one in thickness, in the adult; weight, $4\frac{1}{2}$ to 6 ounces, half an ounce less in the female. The left kidney is longest, thinnest, and lightest by about 2 drachms.

A fibrous capsule surrounds the kidney, and enters at the hilus to invest the bloodvessels and beginnings of the excretory duct.

The general color of the kidney substance is deep red. It is firm on pressure, but easily torn; and is divisible into the *cortical* or external, and the *medullary* substance.

The cortical substance forms an external layer of about $\frac{1}{6}$th of an inch in thickness, with prolongations inwards. Scattered numerously through it are the small red *Malpighian bodies;* around these are small granular cells, convoluted *tubuli uriniferi* (tubes of *Ferrein*), bloodvessels, lymphatics, and nerves.

The Malpighian bodies have a diameter of about $\frac{1}{120}$th of an inch. Each is a *capillary tuft*, inclosed in a membranous capsule, the beginning of one of the uriniferous tubules. The tufts are the networks of small arteries (*vasa afferentia*); and from them and the capillary plexus outside of them, go minute veins (*vasa efferentia*), which also form plexuses around the tubuli uriniferi.

The medullary substance is formed of pale red cones (*pyramids of Malpighi*) about fifteen in number, with their apices (*papillæ* of the kidney) projecting into the central cavity of the gland. Each papilla has about a thousand orifices of tubuli uriniferi. The pyramid of Malpighi is composed of many lesser cones or pyramids (pyramids of *Ferrein*), which are themselves made up of straight tubes (of *Bellini*), the continuations of the convoluted tubuli of the cortical substance.

Over each papilla is an *infundibulum*, or small membranous cup; four or five infundibula together make a *calyx;* and all the calyces, from seven to thirteen in number, open into the *pelvis* of the kidney; from which proceeds the *ureter* or excretory duct.

The ureter is a tube of the diameter of a goose-quill; it has a fibrous, a muscular, and a mucous coat. It passes obliquely inwards and downwards for the length of sixteen or eighteen inches, behind the peritoneum, to enter the base of the bladder. The entrances of the two ureters are about two inches apart.

The *renal artery*, which supplies the kidney, is a large vessel in proportion to the size of the gland. As it enters the hilus it divides into four or five branches. The *renal vein* empties into the vena cava. The kidney has superficial and deep lymphatic vessels.

Supra-Renal Capsules.

These are small, flat, yellowish bodies, behind the peritoneum, one in front of the upper end of each kidney, classed with the "ductless glands." They vary in size in different individuals, but the left is usually the largest—average length, an inch and a half; width, an inch and a quarter; thickness, one-fifth of an inch.

In *structure* the capsule has an *external cortical*, and an *internal medullary* substance. The former is most extensive; and is formed of narrow columns perpendicular to the surface. The medullary substance is soft, pulpy, and brown in color.

Microscopical anatomists differ in their view of the minute character of the columnar masses. The medullary substance is composed of nuclei and granules, amidst which is a plexus of very small veins.

Nerves abound in the supra-renal capsules; derived from the ganglionic or sympathetic plexuses, and the pneumogastric; one observer says also the phrenic. These nerves have a number of small ganglia upon the surface of the capsules.

Bladder.

Situated behind the pubes, in front of the rectum in the male, and of the uterus and vagina in the female, the bladder rises higher in the *infant* than in the adult. In the *adult*, when *empty*, its summit reaches to the upper line of the symphysis pubis; when filled it rises, in a round form, into the hypogastric region, to a greater or less distance above the bony rim of the pelvis according to its distension. Its vertical diameter is the greatest in the male; the transverse in the female; and its capacity is greatest in the latter.

When moderately full, the male adult bladder measures about five inches by three; and contains a pint.

The *summit* of the bladder is connected with the umbilicus by the *urachus*, a fibrous and partly muscular cord, having on each side of it the remnants of the hypogastric arteries, as round fibrous cords. The urachus, in the embryo, is a tube, by which the bladder communicates with the membranous extra-abdominal sac called the *allantois*.

The *body* of the bladder is only covered with peritoneum on its *posterior* surface.

The *vas deferens* runs in a curve, from before backwards, along the side of the bladder towards its base.

The base or *fundus* of the bladder is directed downwards and backwards. Its dimensions vary with the fulness of the organ.

The *neck* or *cervix* of the bladder is the narrow part connected with the *urethra*. It is, in the male, surrounded by the prostate gland.

The *ligaments* of the bladder are called *true* and *false;* five of the former and five of the latter. The true are, the *two anterior*, going to the pubes; *two lateral*, which are broader, and are formed of the fascia between the bladder and rectum; and the *urachus*.

The *false* ligaments are, the *two posterior*, *two lateral*, and *one superior*. The *posterior* pass in the male to the sides of the rectum; in the female, to the side of the uterus. The *lateral* are folds of peritoneum from the iliac fossæ. The *superior* ligament is a fold of peritoneum from the summit of the bladder to the umbilicus.

The bladder has a partial *serous*, a *muscular*, a *cellular*, and a *mucous* coat.

The *muscular* coat consists of two layers of *unstriped* muscular fibre; longitudinal without, and circular within. The former are most abundant on the front and back surfaces; the latter, around the neck. There they form the *sphincter vesicæ*, continuous with the muscular fibres surrounding the prostate gland. Some *oblique* fibres also go from the prostate to the ureters.

The *mucous* coat is smooth and pale-red. It is most closely united to the muscular coat at the neck. It has a number of small cluster-like (racemose) glands, about the neck, and a few scattered mucous follicles. Its epithelium is intermediate between the squamous and the columnar form.

The *vesical triangle* (*trigonum*) is just beneath the opening of the urethra from the neck of the bladder; its apex is forwards. It is bounded in front by the prostate gland, and at the sides by the vasa deferentia and vesiculæ seminales.

The *uvula vesicæ* is a small elevation of the mucous membrane, or thickening of the prostate, projecting into the orifice of the urethra.

Urethra.

In the male, when the penis is relaxed, this tube has the shape of the italic *S*. Its length is eight or nine inches. It is described as having three parts—the *prostatic*, *membranous*, and *spongy* portion.

The *prostatic* is the widest, is spindle-shaped, and has a length of an inch and a quarter; passing through the prostate gland near its upper surface.

The *caput gallinaginis* or *veru montanum* is a narrow ridge on the bottom of the prostatic portion, three-quarters of an inch long and an eighth of an inch in elevation, containing muscular and erectile tissue. On each side of this are numerous orifices of the *prostatic ducts*, from the middle lobe of the gland. The *prostatic vesicle* or *sinus pocularis* is a fossa a quarter of an inch long, just in front of the veru montanum; about its margins are the openings of the seminal ejaculatory ducts. This sinus has fibro-muscular walls, lined by mucous membrane, on which open a number of small glands.

The *membranous* portion is between the prostate and the *bulb* of the urethra. This is the narrowest part of the tube except the orifice. From the projection of the bulb below it, it is three-quarters of an inch long on its concave roof, and half an inch along its convex floor. It passes through the deep perineal fascia; and is invested by a double extension from it, and by the *compressor urethræ* muscle.

The *spongy* portion is the longest, having a length of about six inches, to the *meatus urinarius*. Below the symphysis pubis it ascends a little, and then descends forward. It has an almost uniform diameter of a quarter of an inch; except in the *bulbous* portion, and in the *fossa navicularis* within the *glans penis*, in which localities it is dilated.

The *meatus* or orifice is the narrowest part of the canal; its direction is vertical, with two small elevations or *labia* at its sides.

The lining *mucous membrane* of the urethra has numerous small orifices of mucous glands and follicles, the *glands of Littré*. These openings are sometimes large enough (*lacuna magna*) to detain the end of a catheter. *Cowper's* glands open into the bulbous portion.

The *muscular* coat of the urethra is most abundant in its prostatic part. It consists of an outer longitudinal and an inner circular layer of unstriped fibres.

Prostate Gland.

In shape like a horse-chestnut, and composed of three lobes, this gland lies below and behind the symphysis pubis, around the neck of the bladder and urethra, and upon the rectum. Its transverse diameter at the base is an inch and a half; from before backwards, an inch; depth, three-quarters of an inch.

Its lobes are the two *lateral* and the *middle* lobe. The middle lobe is normally only a transverse band behind the beginning of the urethra; sometimes it is absent. In old men it is frequently much enlarged.

The prostate gland is dense, though friable, and is surrounded by a fibrous capsule. It it perforated by the seminal ducts. Connected with it are circular, unstriped muscular fibres, around the urethra. Its *secretion* is a milky fluid in appearance, with an acid reaction.

Cowper's Glands.

These are two lobulated yellowish bodies, each as large as a pea, in front of and under the membranous portion of the urethra, behind the bulb. Their excretory ducts run for almost an inch obliquely forwards, to open into the bulbous part of the urethra.

Penis.

This organ consists of the *root, body,* and *gland.* The *crura* or fibrous branches connect the root of the penis with the *rami* of the pubes. By the suspensory ligament it is held to the front of the symphysis pubis.

The *glans* penis is of a rounded conical form, with the *meatus urinarius* at its extremity. Behind and below this is the *frænum præputii,* a fold of mucous membrane going backwards to join with the prepuce. The *corona glandis* is a round projecting border at the base of the gland; behind this is the narrowing called the neck or *cervix.* The sebaceous *glands of Tyson* (*glandulæ odoriferæ*) are located around the corona and cervix.

The *body* of the penis is covered by a loose thin tegument, which forms the *prepuce* by doubling upon itself at the neck of the gland. Its internal layer, reflected from the cervix over the glans, is of the character of a mucous membrane.

The body of the penis is composed of three cylinders of fibrous and erectile tissue; the two *corpora cavernosa,* and the *corpus spongiosum.*

Each *corpus cavernosum* is composed of a strong fibrous coat, within which is a network of fibres, containing the vascular erectile tissue. The bands of fibres are called *trabeculæ.* Between the two *corpora* is a *septum,* most complete behind; in front it is comb-like or *pectiniform.*

In the fibrous outer coat and septum are numerous elastic and muscular fibres; some of the latter occur also in the *trabeculæ.*

The *corpus spongiosum* lies below the junction of the corpora cavernosa, inclosing the urethra. It commences between the *crura* of the penis, in the *bulb,* and terminates in the *glans.* The bulb is surrounded by the accelerator urinæ muscle.

The structure of the corpus spongiosum is essentially like that of the cavernosa, but with a thinner and more elastic envelope.

Erectile tissue is principally formed of a close plexus of small veins, freely communicating with each other. They are largest in size in the middle of each corpus cavernosum. Their blood is returned by the **vena dorsalis penis,** and by the prostatic and pudendal veins.

The *arteries* of the penis come from the internal pudic. The *helicine* arteries (whose existence is not universally admitted) are convoluted, tendril-like vessels, most abundant in the back part of the penis.

The *nerves* of the penis are branches of the internal pudic nerve and the hypogastric plexus. Its *lymphatics* are superficial and deep-seated; some going through the inguinal glands, and others joining the deep pelvic lymphatics.

Testes.

These are oval glands, suspended in the *scrotum* by the two *spermatic cords*. Behind each is a long and narrow body, the *epididymis*; of which the upper enlarged end (*globus major*) is connected by its ducts with the upper end of the testicle, and the lower end (*globus minor*) with its lower end and investing tunic. The left testicle is a little the largest.

The *coverings* of the testicle are, the *skin*, *dartos muscle*, *external spermatic fascia*, *cremaster muscle*, *fascia propria* or infundibuliform fascia, and *tunica vaginalis*.

The *dartos* is a thin loose layer of connective tissue mixed with unstriped muscular fibres.

The *cremaster muscle* consists of a few scattered bands of fibres of the internal oblique muscle, carried down in the descent of the testicle.

The *fascia propria* is a similar process of the fascia transversalis.

The *tunica vaginalis* is the serous investment of the testicle, derived from the peritoneum. It is duplicated, over the testis and within the scrotum, the outer layer extending above and below the testis, upon the cord.

The immediate *fibrous* covering the testicle is the *tunica albuginea*. A partial septum formed by the vertical descent of this tunic into the gland is sometimes called the *corpus Highmorianum*, or *mediastinum testis*.

The *tunica vasculosa* is a plexus of bloodvessels lining the tunica albuginea for the supply of blood to the gland.

In *structure*, the testis is composed of from 250 to 400 *lobules*, of unequal size. Each lobule is conical, and formed of from one to three or four convoluted *tubuli seminiferi*. These may be unravelled; and are each several feet in length with a diameter from $\frac{1}{200}$th to $\frac{1}{150}$th of an inch; they number in all from 300 to 800. Each is surrounded by a plexus of minute capillaries. In the posterior apex of each lobule, the tubuli become almost straight, forming the *vasa recta*, twenty or thirty in number; each of which has a diameter of $\frac{1}{50}$th of an inch.

The *vasa recta* anastomose as they ascend, forming the *rete testis;* at the upper end of which are given off from twelve to fifteen

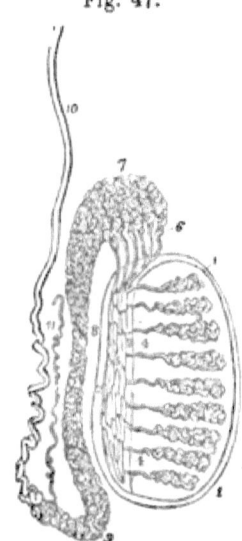

Fig. 47.

MINUTE STRUCTURE OF THE TESTIS.—1, 1. Tunica albuginea. 2, 2. Corpus Highmorianum. 3, 3. Tubuli seminiferi. 4. Vasa recta. 5. Rete testis. 6. Vasa efferentia. 7. Coni vasculosi, the globus major of the epididymis. 8. Body of the epididymis. 9. Its globus minor. 10. Vas deferens. 11. Vasculum aberrans.

ducts, the *vasa efferentia*, which pass through the tunica albuginea to the epididymis. By their convolution and enlargment, in conical masses (*coni vasculosi*), they make the *globus major*. Opposite the bases of the cones (each of which is formed by a single duct, six or eight inches long) their ducts open into a single tube or duct, which is very much convoluted, forming the *globus minor* of the epididymis. The length of this tube when unravelled is more than twenty feet.

The *vas deferens* is the continuation of this duct, upwards from the globus minor, behind the epididymis, and on its inner side. It then goes along, in the *spermatic cord*, through the external and the internal abdominal rings, into the pelvis. Reaching the base of the bladder, it becomes enlarged, and then, narrowing at the prostate, joins the duct of the *vesicula seminalis* to form the *ejaculatory* duct. The vas deferens is a firm tube, about two feet long, with a canal $\frac{1}{24}$th of an inch in diameter. It has a cellular, muscular, and mucous coat.

Vesiculæ Seminales.

These are lobulated membranous pouches, one on each side, between the bladder and rectum. They are reservoirs for semen, besides having a secretion of their own which mixes with it. Each is about two and a half inches long by half an inch wide; but variable in these dimensions.

The *vesicula seminalis* is, in structure, a single tube, convoluted, and giving off irregular branches or *diverticula;* all being held together by fibrous tissue. The tube is of the diameter of a goose-quill, and from four to six inches long; behind, it ends abruptly, as a *cul-de-sac;* in front it is continued as a straight narrow duct, which joins with that of the vas deferens to make the ejaculatory duct.

Each ejaculatory duct (one on each side) is about ¾ths of an inch in length. It commences at the base of the prostate gland, and runs upwards and forwards through its substance, to open into the *sinus pocularis* of the prostatic part of the urethra.

Spermatozoa are minute filaments, found in the liquid of the *semen masculinum* formed in the testis. Each has an oval enlargement at one end, and a caudal prolongation or extremity; which shape, with their undulatory movements, suggested the erroneous idea of their being animalcules. Semen also contains, as shown by the microscope, *seminal granules*, $\frac{1}{4000}$th of an inch in diameter.

Spermatic Cord.

This cord is composed of the vas deferens, three arteries, a plexus of veins, the spermatic plexus of nerves, and some large lymphatic vessels; besides connective tissue, and the cremaster muscle, and fascia. Its course is from the back of the testicle upward to the *external* abdominal ring (see *Hernia*), and through the inguinal or spermatic cord to the *internal* ring, by which it enters the abdominal cavity.

ORGANS OF GENERATION IN THE FEMALE.

The *arteries* of the cord are the *spermatic, cremasteric,* and the *artery of the vas deferens.*

The *nerves* of the cord are derived from the ganglionic or sympathetic plexuses of the abdomen.

Descent of the Testes.

In early fœtal life, the testes are in the abdomen, below the kidneys. At the lower end of each is a cord, the *gubernaculum testis,* connected with the scrotum at its lowest part. In the seventh month of gestation, the testicle descends, in the line of the now shortening gubernaculum, through the *internal ring,* carrying with it a fold of peritoneum; then through the inguinal canal and external ring, to enter the scrotum in the course of the eighth month. Just before birth at full term, the peritoneal pouch is closed above; leaving its fundus to form the *tunica vaginalis testes.* In the female, the *round ligament of the uterus* takes a similar course to that of the *gubernaculum.* At an early period of embryonic life the ovaries and testes correspond in position.

ORGANS OF GENERATION IN THE FEMALE.

The *ovaries* are the *essential* female organs of reproduction, and the *uterus* is that of gestation. Accessory to these are the *Fallopian tubes* and the *external genital organs,* i. e., the *vagina* and its connected parts.

Ovaries.

Each ovary is an oval body, about an inch and a half long, three-quarters of an inch wide, and a third of an inch in thickness. It is situated in the posterior part of the broad ligament of the uterus, behind and below the Fallopian tube, on either side; covered by peritoneum, except on its anterior margin, which is attached to the broad ligament. It has a fibrous coat, *tunica albuginea,* within which is a soft *stroma* or fibro-cellular structure well supplied with bloodvessels.

Graafian vesicles are from five to twenty small, round, transparent sacs, contained in the ovary. Each, before maturation, holds an *ovum,* $\frac{1}{180}$ of an inch in diameter, on the average, surrounded by clear fluid. As it matures, each Graafian vesicle enlarges, approaches the surface of the ovary, and finally bursts; the ovum then escapes through the Fallopian tube into the uterus, and, unless impregnated, is discharged with the menstrual fluid.

Corpus luteum is the name given to the remains of the Graafian vesicle, after its maturation and the discharge of the ovum. Those left *after pregnancy* are peculiar, and are called *true corpora lutea;* those formed at other times, *false.* The former are larger, firmer, yellower, more vascular, and more puckered in appearance.

The *ligament of the ovary* extends from the upper angle of the uterus to the inner end of the ovary.

The *round ligaments* are two cords, each four or five inches long, below and in front of the Fallopian tube, between the layers of the broad ligament. Each passes from the upper angle of the uterus, forwards and outwards, through the internal abdominal ring, and along the inguinal canal, to be lost upon the labia majora.

Fallopian Tubes.

These are two oviducts, one on each side, lying in the free margin of the broad ligament. Each is about four inches long, with a very minute canal, which widens near its outer end, trumpet-like, and then contracts at its termination. The inner end, *ostium internum*, communicates with the uterus; the outer, *ostium abdominale*, opens into the cavity of the abdomen. The latter is *fringed*, and is called the *fimbriated* extremity of the Fallopian tube. It is believed to embrace the ovary during sexual excitement.

The Fallopian tube has a *serous*, *muscular*, and *mucous* coat.

Uterus.

The womb, in the virgin, is pear-shaped, flattened before and behind. It is suspended between the bladder and the rectum, by six ligaments of peritoneum. It is about three inches long, two wide at its upper part, and an inch thick. Its upper end is di-

Fig. 48.

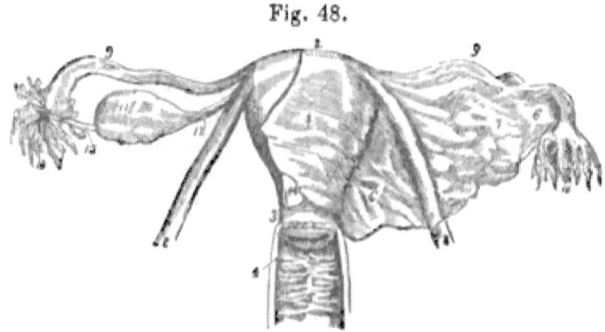

UTERUS AND ITS APPENDAGES.—1. Body of uterus. 2. Fundus. 3. Cervix. 4. Os uteri. 5. Vagina. 6. Broad ligament. 7. Position of ovary. 8. Round ligament. 9. Fallopian tube. 10. Fimbriated extremity of Fallopian tube. 11. Ovary. 12. Ligament of the ovary. 13. Falloplo-ovarian ligament. 14. Peritoneum on anterior surface of uterus.

rected upwards and forwards; its lower end, downwards and backwards, forming an angle with the vagina. Its parts are the *fundus*, *body*, and *neck* or *cervix*.

The *fundus* is the broad base or upper part; the body narrows from the fundus to the neck. The *cervix* communicates with the vagina, which is attached around it, extending upwards farther behind than in front. The *os uteri* or *os tincæ* is the mouth of the uterus, opening into the cavity of the vagina. Obstetricians speak also of the *os internum*, at the upper end of the constriction called the cervix. The cavity of the unimpregnated uterus is quite small. The coats of the uterus are three: *serous*, *muscular*, and *mucous*; the last being its inner lining. It is covered by ciliated columnar epithelium. Numerous follicles and glands exist in the cervix. When distended with fluid, they have been called *ovula Nabothi*.

Ligaments of the Uterus.—Two *anterior* ligaments, or semilunar peritoneal folds, pass between the cervix of the uterus and the posterior surface of the bladder.

Two *posterior* ligaments, also folds of peritoneum, connect the sides of the uterus with the rectum.

The *two broad* ligaments extend from the sides of the uterus to the walls of the pelvis, dividing its cavity into an anterior and a posterior portion. The former contains the bladder, urethra, and vagina; the latter, the rectum. The broad ligaments are connected with the peritoneum, and correspond essentially with it in structure.

Vagina.

This is a musculo-membranous canal, about four inches in length along its anterior wall, longer posteriorly; curved in its direction downwards and forwards from the uterus. Its outlet is called the *vulva*. It consists of an *external muscular*, a *middle erectile*, and an *internal mucous* coat. The last presents two longitudinal (anterior and posterior), and numerous transverse ridges or *rugæ*; also many conical and filiform papillæ, besides mucous glands and follicles, especially at its upper part.

The *pudendum* is a term applied to the vulva and its appendages. The *mons veneris* is the rounded fatty prominence over the pubes in front, covered with hair after puberty. The *labia majora* are longitudinal folds, one on each side of the vulva; reaching from the mons veneris to the perineum. Externally, each is formed of integument; within, of mucous tissue. The two labia are joined in front and behind, by the anterior and posterior *commissures*. Just within the latter is a small transverse fold, the *fourchette*. Between it and the posterior commissure is the *fossa navicularis*.

The *labia minora* or *nymphæ* are smaller mucous folds, within the *majora*; they pass obliquely from the clitoris above to the sides of the vagina below. The *clitoris* is a small erectile organ, analogous to the penis of the male in its cavernous structure. It is situated just above the vagina. The *hymen* is a thin semilunar fold of mucous membrane, across the lower part of the orifice of the vagina. As it is, occasionally, congenitally absent, and has been known to be present in prostitutes, it is not, as was formerly supposed, a test of virginity. Usually, after its rupture, small

Fig 49.

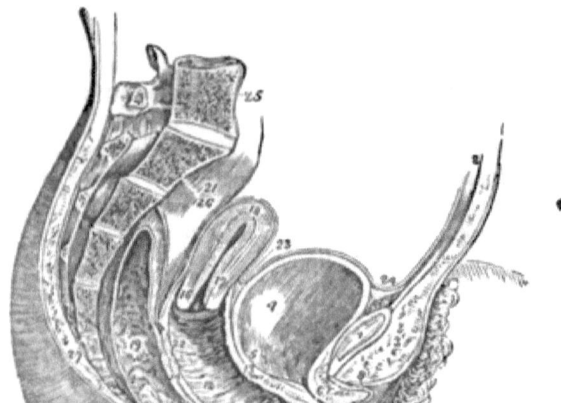

VISCERA OF FEMALE PELVIS.—1. Symphysis pubis. 2. Abdominal parietes. 3. Mons veneris. 4. Bladder. 5. Entrance of ureter. 6. Urethra. 7. Meatus urinarius. 8 Clitoris. 9 Left nympha. 10. Left labium majus. 11. Orifice of vagina. 12, 22. Vaginal canal. 13. Wall between vagina and rectum. 15. Perineum. 16. Os uteri. 17. Cervix. 18. Fundus. 19. Rectum. 20. Anus. 21. Upper part of rectum. 23. Fold of peritoneum. 24. Reflexion of peritoneum. 25. Last lumbar vertebra. 26. Sacrum. 27. Coccyx.

rounded elevations are left around the orifice of the vagina, called *carunculæ myrtiformes*.

Urethra in the Female.

This canal is much shorter than in the male; having an average length of an inch and a half. Its direction, also, is different; being upwards and slightly backwards, with very little curve, behind the pubes. Its external orifice or *meatus* is in the small triangular space (*vestibule*), between the clitoris and the upper end of the entrance of the vagina.

Mammary Glands.

These, as organs of nutrition for offspring, are accessory to the organs of reproduction. In the male they exist, but undeveloped.

In the female, each *mamma* is a true multi-lobular gland. The smallest lobules consist of clusters of rounded vesicles, opening into the smallest branches of the lactiferous (milk-bearing) ducts. These unite into larger ducts, finally making from fifteen to twenty (*tubuli lactiferi*, or *galactophori*), which converge towards the *areola* around the nipple. There they enlarge and form reservoirs for the milk; and then run from the base of the nipple to its summit, where they perforate it with narrow orifices, by which the milk escapes under pressure. The nipple has a somewhat erectile structure.

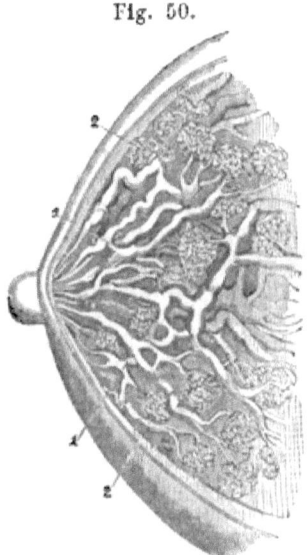

Fig. 50.

SECTION OF MAMMARY GLAND.—1, 1. Galactophorous ducts. 2, 2. Lobuli.

CHAPTER VI.
ORGANS OF CIRCULATION.

HEART.

THE heart is a hollow conoidal muscular organ enveloped by the pericardium, and suspended by its great vessels between the lungs. It is situated behind the lower two-thirds of the sternum, its base above and apex pointing downward and to the left side. Its extent is from the level of the upper border of cartilage of the third rib to the space between the fifth and sixth ribs. It is about five inches long, in the adult, by three and a half inches of width in its broadest part, and two and a half inches in thickness. In the male, it weighs from ten to twelve ounces; in the female, from eight to ten. It increases in size and weight, however, to old age.

The heart is essentially twofold, being divided by a muscular septum into the *right* and *left*, or *respiratory* and *systemic* heart; conjoined, but not communicating after birth. Each half consists of an *auricle* and a *ventricle*. The auricles (named from an ear-like appendage belonging to each) are comparatively thin and weak, the ventricles thick and strong; their internal capacity is about the same, two fluidounces for each cavity. The auricles are above the ventricles. Most of the anterior surface of the heart is formed by the right ventricle, the left making the apex and left border; most of the posterior surface is made by the left ventricle.

The right auricle is a little larger than the left. It consists of a *sinus* or main cavity, and an *auricular appendix*.

The *ascending* and *descending venæ cavæ* both open into the right auricle. Between their terminations, on the right wall of the auricle, is a small projection, the *tuberculum Loweri*.

The *coronary veins*, the largest being sometimes called the *coronary sinus*, open into the right auricle, bringing blood from the substance of the heart. The apertures of the smaller veins are the *foramina Thebesii*. At the mouth of the coronary sinus is the *valve of Thebesius*; sometimes it is double.

Between the front margin of the ascending vena cava and the auriculo-ventricular opening (*ostium venosum*), is the *Eustachian valve*, large in the fœtus, small in the adult; it is semilunar, the free edge being concave.

The *fossa ovalis* is a depression marking the place of the *foramen ovale* in the fœtus, between the right and left auricle. The prominent margin of it is the *annulus ovalis*.

The *musculi pectinati* are comb-tooth-like fleshy columns, which cross the inner part of the auricular appendix and the contiguous portion of the wall of the auricle.

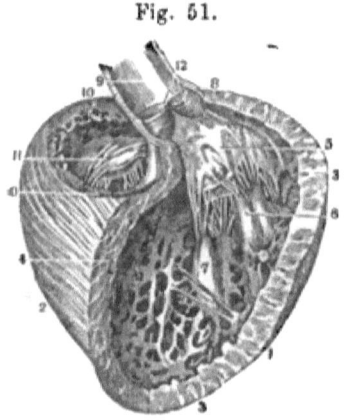

Fig. 51.

THE LEFT VENTRICLE.—1. Outer side of left ventricle. 2. Outer side of right ventricle. 3. Thickness of its outer parietes. 4. Thickness near the right ventricle. 5. Mitral valve. 6, 7. Columnæ carneæ with their chordæ tendineæ. 8. Origin of the aorta. 9. Cavity of the aorta. 10, 10. Superior surface of the right ventricle, showing the ostium venosum and tricuspid valve. 11. Tricuspid valve. 12. Semilunar valves of the aorta.

The *right ventricle* is triangular. It rests below and behind upon the diaphragm. It is separated from the left ventricle by a septum which bulges into the right ventricle. Above, it ends in a small cone (*conus arteriosus*); in which the pulmonary artery begins by a circular opening, guarded by three pocket-like *semilunar valves*. Each of these pockets is a fold of the lining membrane of the heart, with fibres of ligamentous tissue. At its most prominent part is the *corpus Arantii*, a small cartilaginous prominence, which completes the closure of the valves. Behind each valve is a pouch or dilatation; these are the *sinuses of Vulsalva*.

The *columnæ carneæ* are round muscular columns which project from nearly the whole of the inner wall of the ventricle. Some project only as ridges, others are free except at the ends, while three or four of them give attachment to the *chordæ tendineæ*.

These, the tendinous chords, are connected with the free surfaces and margins of the three segments of the *tricuspid* or right auriculo-ventricular valve.

The tricuspid valve is a fold of the lining membrane of the heart, strengthened by fibrous tissue. Its upper margin is attached to the roof of the ostium venosum.

The *left auricle* is cuboidal in form, somewhat smaller but thicker than the right. In structure it resembles the right auricle; having a *sinus, auricular appendix*, and *musculi pectinati*. The *four pulmonary veins* open into it.

The *left ventricle* is more conical and longer than the right; its walls are about twice as thick, except near the apex.

The *aorta* opens from this ventricle by a circular orifice, guarded by three *semilunar valves*, like those of the pulmonary artery, but with larger *corpora Arantii*, and deeper *sinuses of Valsalva*.

The *columnæ carneæ* of the left ventricle resemble those of the right; two of them, only, connect with *chordæ tendineæ* which are attached to the auriculo-ventricular valve.

This valve, the *mitral* valve, consists of but two unequal segments; but is larger and thicker than the tricuspid. Generally two lesser segments also exist at the place of union of the greater. In structure this valve is similar to the tricuspid.

The lining membrane of the heart, the *endocardium*, is a smooth, thin serous membrane, thickest in the auricles. It covers the valves, and is continuous with the inner coat of the bloodvessels.

The *muscular fibres* of the heart are intricately arranged; those of the auricles and those of the ventricles being mutually independent.

The fibres of the *auricles* are in two layers, the *superficial transverse* and the *deep;* the latter being in part *looped*, and in part *annular*. The superficial are common to *both* auricles, the deep are peculiar to *each*.

The *ventricles* also have fibres common to both, and some which are peculiar to each ventricle. The latter are the most common near the base of the heart.

Fig. 52.

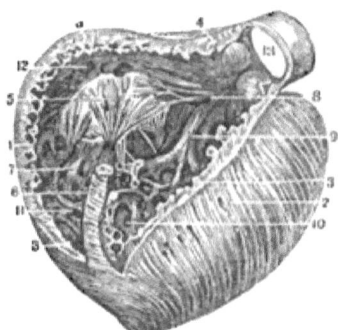

INTERIOR OF THE RIGHT VENTRICLE.—1. Section of the parietes of the right ventricle. 2. Left ventricle. 3. Thickness of the parietes of the right ventricle. 4. Thickness at the commencement of the pulmonary artery. 5. Anterior fold of the tricuspid valve. 6. A portion of the right ventricle. 7, 8. Columnæ carneæ with their chordæ tendineæ. 9. Ventricular septum. 10, 11. Cavities between the bases of the columnæ carneæ. 12. Depression leading to the pulmonary artery. 13. Interior of the pulmonary artery.

The *superficial* fibres are mostly *spiral* in their direction, and placed in layers of unequal thickness. Coiling inwards at the apex of the heart, they there form the *vortex*, and thence ascend again. The *deep* fibres are circular in this direction.

Fibrous rings surround and give fixity to the auriculo-ventricular and arterial openings of the heart. Those of the left side are the strongest.

The heart is supplied with blood for its nutrition by the *two coronary arteries*, anterior and posterior. It has also *lymphatic* vessels.

The *nerves* of the heart are derived from the *cardiac plexus*, which is partly ganglionic or sympathetic, and partly cerebro-spinal in origin. Minute examination displays a great number of small ganglia on the surface and in the substance of the heart.

Pericardium.

This sac, which envelops the heart, is composed of an outer *fibrous* and an inner *serous* membrane. The *fibrous* coat is attached, below, to the diaphragm. At the base of the heart, it is extended for some distance over the *aorta, pulmonary artery, pulmonary veins,* and *descending vena cava; not* upon the *ascending* vena cava. In front, the pericardium lies close to the sternum, covered at the sides by the edges of the lungs, particularly the left. At its sides the pleuræ cover it, with the phrenic nerve and vessels between them on each side. Behind, the pericardial sac rests against the bronchial tubes, the descending aorta, and the œsophagus.

The *serous* coat of the pericardium is double; one layer adhering to the heart, the other lining the fibrous sac. It extends more completely around the aorta and pulmonary artery than around the other great vessels of the heart. Its surface is smooth and moistened with serum.

For the route of the *circulation of the blood*, see PHYSIOLOGY.

ARTERIES.

Aorta.

This, the main trunk of all the arteries of the body, ascends from the upper part of the left ventricle for a short distance, and then forms an arch backwards over the root of the left lung. Thence descending upon the left side of the spinal column, it passes through the diaphragm into the abdomen; and finally divides, opposite to the fourth lumbar vertebra, into the right and left primitive or common iliac arteries. It may be divided in description into the *arch* of the aorta, the *thoracic* and the *abdominal* aorta.

The *arch* extends to the lower part of the third dorsal vertebra. Its ascending portion is included in the pericardium. The descending vena cava is to the right of it, the pulmonary artery to its left.

Below the transverse part of the arch is the left bronchus and the bifurcation of the pulmonary artery. Behind it are the trachea, œsophagus, and thoracic duct.

Five branches go off from the arch of the aorta. From the ascending part, the *right* and *left coronary* arteries. From the transverse portion, the *arteria innominata*, left *carotid*, and *left subclavian*.

In *structure*, the aorta, like all the other arteries, has an *external* coat, of connective and fibrous tissue; a *middle* coat, of muscular, elastic, and connective tissue; and an *internal* coat, of elastic serous membrane. The *muscular* fibres of the middle coat are chiefly transverse; they are of the *unstriped* variety, pale in color and with spindle-shaped cells and persistent nuclei. The proportionate amount of muscular tissue is greatest in the *smallest* arteries, at a distance from the heart. Branches leave arteries generally at an acute angle; but this is variable. *Anastomosis* is the free communication which often occurs between two arteries. All arteries have their own nutritious vessels, or *vasa vasorum*.

Pulmonary Artery.

This vessel conveys *venous* or un-aired blood from the heart to the lungs, for aeration. It is about two inches long, chiefly inclosed in the pericardium; arising in front of the aorta, from the left side of the top of the right ventricle, and passing obliquely upward and backward and to the left, under the arch of the aorta, where it divides into the nearly equal *right* and *left* pulmonary arteries. The *right* pulmonary is rather the larger and longer. This, at the root of the right lung, divides into two branches, one for the upper and middle lobes, and the other and larger, the lower lobe. The *left* pulmonary artery also ends in two branches, one for each lobe of the left lung. Each of these subdivides into a multitude of ramifications, interpenetrating at last the lobules of the lungs with extremely fine networks of capillaries, for aeration of the blood by the air-cells.

Coronary Arteries.

The *right* coronary arises from the aorta just above its valve, and passing in a groove between the right auricle and ventricle, curves around the back of the heart to the posterior groove between the ventricles. There it divides into two branches, which supply the substance of the heart, and anastomose with the branches of the left coronary artery.

The *left* coronary is smaller. It arises above the origin of the right, and descends to the anterior interventricular groove, where it divides, one branch continuing down the groove to the apex, the other winding around to the back of the heart.

Innominata.

This, the largest branch of the arch of the aorta, arises near the left carotid from its transverse portion, and, passing for an inch and a half to two inches obliquely up to the right sterno-clavicular junction, there divides into the right subclavian and right primitive carotid. To the right of the arteria innominata lie the vena innominata, and pneumogastric nerve; behind, it crosses the trachea.

Primitive or Common Carotids.

The *right* carotid arises from the innominata; the left, from the summit of the arch of the aorta. The latter is thus the longer.

Both passing upwards to the neck, their course is thereafter similar. At the level at the top of the thyroid cartilage, each divides into the *internal* and *external* carotid. The common carotids are separated at the lower part of the neck only by the trachea; above, by the pharynx, larynx, and thyroid gland.

In the same fascial sheath with the common carotid, are included the pneumogastric nerve, and, outside of that, the internal jugular vein; each having also its special sheath.

Below, the common carotid lies deeply, having over it the *superficial fascia*, *platysma myoides muscle*, the *deep fascia*, the *sterno-cleido-mastoid*, *sterno-hyoid*, *sterno-thyroid*, and, by the cricoid cartilage, the *omo-hyoid* muscles. Near its bifurcation above, it is bounded within a triangular space, *behind* which is the *sterno-cleido-mastoid*, *above*, the belly of the *digastric* muscle, and *below*, the *omo-hyoid*. A small artery and several veins cross it, and the *descendens noni* nerve lies upon or within its sheath. The *sympathetic* nerve is behind it, between it and the *rectus anticus major* muscle, which rests upon the spine.

At the lower part of the neck, the internal jugular vein of the right side leaves the artery; but that on the left side comes near to and often crosses it.

External Carotid.

From the division opposite the top of the larynx, this vessel curves upwards and forwards and then backwards, to divide between the neck of the condyle of the lower jaw and the external *meatus* or orifice of the ear, into the *temporal* and *internal maxillary* arteries.

At its beginning, the external carotid has *behind* it the *sterno-cleido-mastoid* muscle; *below*, the *omo-hyoid;* and *above*, the *digastric* and *stylo-hyoid* muscles. The *hypoglossal* nerve and *lingual* and *facial veins* cross it, as well as the *digastric* and *stylo-hyoid* muscles. It is *covered*, under the skin, by the *platysma myoid* muscle, the *deep fascia*, and the front edge of the *sterno-cleido-mastoid*.

Above, the external carotid passes into the substance of the *parotid gland;* lying there under the *facial* nerve, and the union of the internal maxillary and temporal veins. Between it and the internal carotid is part of the parotid gland, as well as the *styloglossus* and *stylo-pharyngeus* muscles.

Eight branches leave the external carotid artery; divisible into four sets, as follows : *Anterior,* the *superior thyroid, lingual,* and *facial; posterior,* the *occipital* and *posterior auricular; ascending,* the *ascending pharyngeal;* and *terminal,* the *temporal* and *internal maxillary.* Deviations or variations may occur (*as in all parts of the arterial system*) in the origin and distribution of these vessels ; it is their *normal* or most general course that is described.

Superior thyroid curves upwards and then down to the thyroid gland ; dividing into four branches also, the *hyoid, superficial descending, laryngeal,* and *crico-thyroid.*

Lingual ascends inwards to the greater cornu of the *os hyoides,* and, reaching the tongue, gives off the *hyoid, dorsalis linguæ, sublingual,* and *ranine* branches ; the latter going to the tip of the tongue.

Facial is a tortuous vessel, which, passing up through the submaxillary gland, then crosses the margin of the lower jaw-bone, in front of the insertion of the masseter muscle ; thence it goes across the cheek and up the side of the nose to the inner angle of the eye. Branches of the facial are ten in number ; of which the principal are the *submaxillary, submental, inferior labial, superior* and *inferior coronary* (around the lips), and the *lateral nasal* artery. Muscular branches also go to the masseter and other muscles.

Occipital arises opposite to the facial, and passes behind the mastoid process, and thence upwards tortuously upon the occiput to divide about the vertex into many branches. Not far from its origin the hypoglossal nerve winds around it.

Posterior auricular is small ; it goes beneath the parotid gland between the ear and the mastoid process ; then dividing into the anterior and posterior branches, which anastomose with the temporal and occipital; also giving off the *stylo-mastoid* branch to the foramen of that name, and the *auricular* to the ear and lesser branches.

Ascending pharyngeal is yet smaller but long, and deeply seated. It ascends between the internal carotid and the pharynx. Its branches are *muscular* and *nervous, pharyngeal* and *meningeal.* The latter pass through foramina in the base of the cranium to the dura mater.

Temporal appears as a continuation upwards of the external carotid, from the parotid gland. Two inches above the zygomatic arch it divides into the *anterior* and *posterior* temporal; first giving off the *transverse facial, anterior auricular,* and *middle temporal.*

Internal maxillary is larger than the temporal. It goes inwards at right angles to the latter, within the condyle of the lower jaw-bone; having three sets of branches: 1. from the *maxillary* part, *tympanic, middle meningeal, small meningeal*, and *inferior dental;* 2. from the *pterygoid* part, *deep temporal, pterygoids, masseteric*, and *buccal;* 3. from the *terminal* part, *alveolar, infra-orbital, posterior palatine, vidian, pterygo-palatine,* and *nasal* or *spheno-palatine*.

Internal Carotid.

From the border of the thyroid cartilage this vessel ascends vertically to the carotid *foramen* in the petrous part of the temporal bone. Entering this, it soon winds forwards and inwards through the carotid *canal*, and then, near to the anterior clinoid process of the sphenoid bone, it pierces the dura mater and subdivides into branches.

These are, the *tympanic, anterior meningeal, ophthalmic, anterior cerebral, middle cerebral, posterior communicating,* and *anterior choroid.*

The *ophthalmic* enters the orbit with the optic nerve, through its foramen, and getting on the inner wall of the orbit, passes to the inner angle of the eye, where it divides into the *frontal* and *nasal* branches.

Other branches of the ophthalmic are, the *lachrymal, supra-orbital,* two *ethmoidal, palpebral,* three *ciliary,* and the *central artery of the retina.*

The *supra-orbital* is largest of these; it runs through the supra-orbital foramen of the frontal bone. The *nasal* artery anastomoses with the terminal branch of the facial.

The *anterior cerebral* artery, leaving the internal carotid at the base of the brain (near the fissure of Sylvius), runs forwards in the fissure between the cerebral hemispheres; the *two* anterior cerebral, right and left, having a short connecting trunk, the *anterior communicating* artery. Then they curve over the front edge of the corpus callosum, and upon its upper surface, to connect with the posterior cerebral.

The *middle cerebral* is larger. It goes obliquely outwards along the fissure of Sylvius, in which it divides into three branches.

The *posterior communicating* artery runs back from the internal carotid to anastomose with the posterior cerebral.

The *anterior choroid* goes to the choroid plexus, and the parts of the brain near it.

Subclavian Artery.

On the *right* side, the subclavian comes from the *innominata;* on the *left*, from the *aorta.* For a short distance, they differ; then, their description becomes the same.

The *right* subclavian passes from its origin opposite the sterno-clavicular articulation upwards and outwards to the inner edge of the *scalenus anticus* muscle. It is here covered, in *front*, by the *skin, superficial fascia, platysma myoides, deep fascia, sterno-cleido-mastoid* muscle, *sterno-hyoid* and *sterno-thyroid* muscles. The internal jugular and verterbral veins *cross* it; as also do the pneumogastric, phrenic, and some branches of the sympathetic nerves. *Beneath* it is the pleura; *behind* it, the longus colli muscle, the sympathetic, and the transverse process of the third cervical vertebra. The recurrent laryngeal winds around its lower part.

The *left* subclavian arises opposite the second dorsal vertebra, from the transverse portion of the arch of the aorta; and ascends to the first rib, behind the insertion of the scalenus anticus muscle. In *front* of it are the pleura, left lung, pneumogastric and phrenic nerves and cardiac branches of the sympathetic, left carotid artery, left internal jugular and innominata veins, sterno-hyoid, sterno-thyroid and sterno-cleido-mastoid muscles. *Behind* it, the œsophagus, thoracic duct, inferior cervical ganglion of the sympathetic, longus colli muscle, and spinal column. *Outside* of it is the pleura; on its *inner* side, the œsophagus, trachea, and thoracic duct.

Reaching the scalenus anticus muscle, and passing over the first rib between that muscle and the scalenus medius, the right and left subclavian arteries thenceforth have the same course; both being, like other vessels, subject to occasional variation or anomaly.

The most superficial part of the subclavian lies in a triangle, whose base in front is the scalenus anticus; one side above, being the omo-hyoid muscle, and the other the clavicle, below. The external jugular vein crosses it on the inner side, receiving there two venous branches. The subclavian vein lies below the artery, behind the clavicle.

The *four branches* of the subclavian artery are, the **vertebral,** *internal mammary, thyroid axis*, and *superior intercostal*.

Vertebral Artery.

This, the largest branch of the subclavian, enters the foramen in the transverse process of the *sixth* cervical vertebra, and passes through the corresponding foramina of the upper five vertebræ, to enter the head through the foramen magnum occipitis. In front of the medulla oblongata it forms, by union with the opposite one, the *basilar* artery. Before this, it gives off *lateral spinal* and *muscular* branches. Within the cranium, the vertebral sends off the *posterior meningeal, anterior* and *posterior spinal*, and *inferior cerebellar*. The last is the largest.

The *basilar* artery extends from the posterior to the anterior border of the pons Varolii; there it divides into the two *posterior cerebral* arteries. On each side it gives off the *transverse, anterior*, and *superior cerebellar*.

The *circle of Willis* is the anastomosis, at the base of the brain, between the basilar artery and the internal carotid and its branches.

Thyroid Axis.

This is a large but short vessel, dividing into the *inferior thyroid, supra-scapular*, and *transversalis colli*.

The *inferior thyroid* goes to the thyroid gland, giving off as branches the *laryngeal, tracheal, œsophageal*, and *ascending cervical*. The *supra-scapular* and *transversalis colli* have principally a muscular distribution.

Internal Mammary.

This artery descends from the subclavian behind the clavicle to the inner surface of the costal cartilages, near the sternum; between the sixth and seventh cartilages, it divides into the *musculophrenic* and *superior epigastric*. Besides these, its branches are the *superior phrenic, mediastinal, pericardiac, sternal, anterior intercostal*, and *perforating* arteries.

Superior Intercostal.

Last branch of the subclavian, this goes backwards, sending off the *deep cervical* to descend in front of the first and second ribs, at their necks, and anastomoses with the first aortic intercostal. It sends a branch also along the first intercostal space, and one for the second, which joins on the aortic branch. Each of these sends branches to the spinal cord.

Axillary Artery.

The continuation of the subclavian, this vessel passes from the lower edge of the first rib, to become the *brachial* as it passes the border of the tendons of the latissimus dorsi and teres major muscles. Very deep at first, it is afterwards almost superficial.

In the former position, it has in *front* of it the pectoralis major muscle, costo-coracoid fascia, and cephalic vein. *Behind* it, the first intercostal muscle, serratus magnus muscle, and posterior thoracic nerve. *Outside* of it is the brachial nervous plexus. *Inside* of it, the thoracic vein.

Next, it passes under the pectoralis minor muscle, having, in *front*, also, the pectoralis major; *behind*, the subscapularis; *inside* of it, the axillary vein.

Lastly, the axillary lies still below the pectoralis minor, with the lower edge of the pectoralis major covering it, in *front*, only above; the skin and fascia, only below. There, it has *behind* it the subscapularis muscle, and tendons of the latissimus dorsi and teres major; *outside* of it the median nerve, and part way, the musculo-cutaneous nerve; on the *inner* side, the ulnar, internal cutaneous and lesser internal cutaneous nerves; *behind*, the musculo-spiral and circumflex nerves.

Branches of the axillary are, *first*, the *superior thoracic* and *acromial thoracic; secondly,* the *thoracica longa* and *thoracica alaris;* lastly, the *subscapular* and *anterior* and *posterior circumflex* arteries.

The *thoracica longa* goes downwards and inwards to the muscles at the side of the chest. The *subscapular* is the largest branch of the axillary. The *circumflex* arteries wind around the neck of the humerus.

Brachial Artery.

From the margin of the teres major tendon this passes down on the inside of the humerus, coming forward gradually near the elbow, to divide there into the *radial* and *ulnar* arteries. It is superficial throughout, covered in *front* by the skin and superficial and deep fasciæ, it has the basilic vein to lie near its line, the median nerve to cross its middle, and the bicipital fascia to separate it, near the elbow, from the median basilic vein.

Behind the brachial, the long and inner heads of the triceps extensor muscle, and the superior profunda artery and musculo-spiral

Fig. 53.

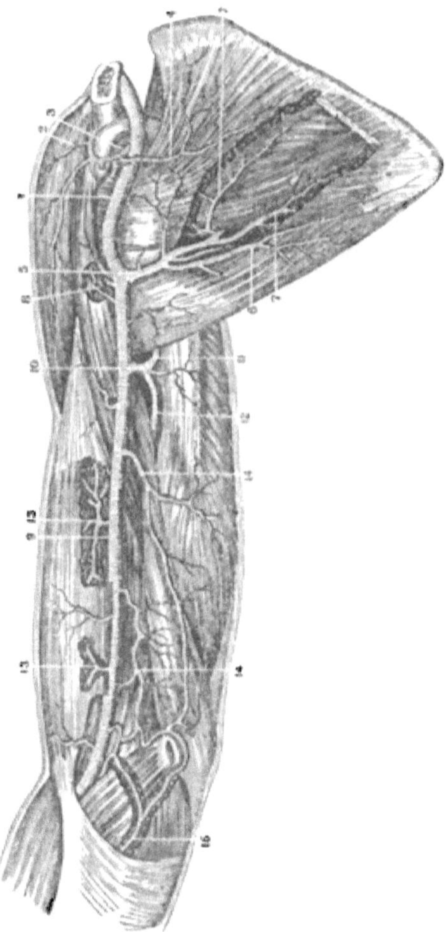

ARTERIES OF THE ARM AND SHOULDER.—1. Axillary artery. 2. Thoracica acromialis. 3. Thoracica superior. 4. Sub-scapularis branch. 5. Inferior scapulæ. 6, 7. Branches to the teres and sub-scapularis muscles. 8. Anterior circumflex. 9. Brachial artery. 10. Profunda major humeri. 11. Posterior circumflex. 12. Main trunk of the profunda major. 13. Muscular branches. 14. Branches to the brachialis internus. 15. Recurrens ulnaris anastomosing with the anastomotica of the brachial.

nerve, come between it and the humerus. *Outside* of it, besides the median nerve, lie the biceps flexor and coraco-brachialis muscles; over the insertion of the latter of which, and the brachialis anticus, the bifurcation of the artery occurs. The place of this (bifurcation), however, is especially subject to anomaly.

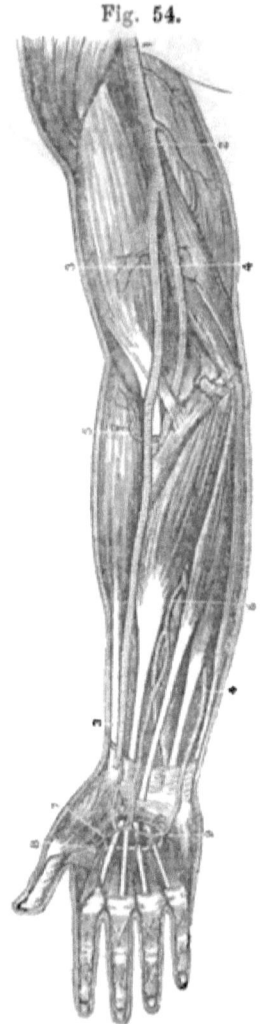

Fig. 54.

ONE OF THE ANOMALIES IN THE BRACHIAL ARTERY.—1. Termination of the axillary artery. 2. Brachial artery 3, 3. Radial artery. 4, 4. Ulnar artery. 5. A recurrent branch. 6. Anterior interosseous artery. 7. Superficial palmar arch. 8. Deep-seated palmar arch. 9. Anastomosis of the two arteries.

Inside of the brachial are, above, the internal cutaneous and ulnar nerves; below, the median nerve. *Accompanying* it in its course with crossing branches, are the two *venæ comites*.

It is, usually, opposite to the coronoid process of the ulna that the brachial, having sunk there into a triangular space, divides into the radial and ulnar. If deviation exist, it is, most frequently, *above* the normal point.

The branches of the brachial are, the *superior* or *major profunda, nutritious* artery, *profunda minor* or *inferior, anastomotic* and *muscular* branches.

Radial Artery.

Smaller, and more superficial than the ulnar, the radial passes along the outer side of the forearm, guarded by muscles and tendons; especially by the supinator radii longus and flexor carpi radialis. At the wrist, it winds around the carpus, under the extensor tendons of the thumb, to the interosseous space behind. Two *venæ comites*, and a filament of the musculo-cutaneous nerve, accompany the radial artery.

In the *forearm*, the branches of the radial are, the *radial recurrent, muscular, superficialis volæ,* and *anterior carpal.*

In the *wrist*, the *posterior carpal, metacarpal, dorsales pollicis,* and *dorsalis indicis*.

In the *hand*, the *princeps pollicis, radialis indicis, perforating*, and *interosseous* arteries. The *deep palmar arch* is the termination of the radial; it joins with a communicating branch of the ulnar.

Ulnar Artery.

This runs along the ulnar border of the forearm, becoming near the wrist more deep-seated than the radial; it ends in the palm of the hand by forming the *superficial palmar arch;* this anastomoses with the *superficialis volæ*.

The branches of the ulnar artery are, in the *forearm*, the *anterior* and *posterior ulnar recurrent*, the *anterior* and *posterior interosseous*, and muscular branches.

In the *wrist*, the *anterior* and *posterior carpal*. In the *hand*, the *communicating* or *deep* branch, and the *digital* arteries.

The *recurrent* arteries, both radial and ulnar, anastomose with vessels proceeding from above towards the hand.

The *digital* arteries are four; going off from the *superficial* palmar arch, to the *sides* of the index, middle, ring, and little fingers; lying beneath the digital nerves. At the middle of the last phalanx of each finger, an arch is formed by the meeting of these with the *interosseous* arteries, from the *deep* palmar arch. Branches from this anastomosis pass to the matrix of the finger nail.

Thoracic Aorta.

From the lower edge of the third dorsal vertebra, on the left side of the spinal column, the aorta descends to the diaphragm; which it penetrates near the last dorsal vertebra. This course is in the posterior mediastinum; having near it, the left lung, pleura, pulmonary artery, and bronchus, the pericardium and œsophagus, the thoracic duct, the vena azygos major and vena azygos minor.

Branches of the aorta in the thorax are, the *pericardiac, bronchial, œsophageal, posterior mediastinal,* and *intercostal*.

The *bronchial* arteries, various in number, usually one for the right side and two for the left, are the nutritious vessels for the lungs. They go to join and follow the bronchial tubes in their distribution.

The *intercostal* arteries are normally ten on each side, longest on the right. Going to the intercostal spaces, below the first, each divides into an anterior and a posterior branch; small fibrous arches protect these vessels from pressure by the intercostal muscles.

Abdominal Aorta.

From the aortic opening in the diaphragm, the aorta goes down on the left side of the spinal column to the fourth lumbar vertebra. There it divides into the *right* and *left* common or *primitive iliac* arteries. The ascending vena cava lies to the right of it, with the vena azygos, thoracic duct, and right semilunar ganglion; on its left are the sympathetic nerve and left semilunar ganglion.

The branches here are separable into 1, *visceral;* and 2, *parietal,* i.e., for the abdominal walls. The first are the *cœliac axis, superior* and *inferior mesenteric, supra-renal, renal,* and *spermatic.* The second group, the *phrenic, lumbar,* and *median sacral* arteries.

Cœliac Axis.

This arises near the margin of the diaphragm, and, after a course of half an inch, divides into the *gastric, hepatic,* and *splenic arteries.*

The *gastric* is smallest. It goes up, and to the left, to the cardiac end of the stomach, and then along its lesser curvature, some of its branches joining those of the hepatic and splenic; others go to the œsophagus.

The *hepatic* is next in size in the adult. It passes to the right side, into the transverse fissure of the liver, where it divides into a right and a left branch, whose ramifications accompany those of the portal vein into the substance of the liver. Its branches are the *pyloric, gastro-duodenal,* and *cystic.*

The *splenic* is a large and tortuous artery. Besides supplying the spleen, it sends blood to the cardiac end of the stomach, by the *vasa brevia* and the *gastro-epiploica sinistra,* and to the pancreas, by the *pancreatica magna* and *pancreaticæ parvæ.*

Superior Mesenteric.

By this vessel nearly the whole of the small intestine is supplied, as well as the cæcum and colon. Going between the pancreas and duodenum, it crosses and descends in an arched form to end in the right iliac fossa. The superior mesenteric vein and plexus of nerves accompany it. Besides a branch to the pancreas and duodenum, and twelve or more branches to the small intestine, it gives off also the *ileo-colic, colica dextra,* and *colica media.*

Inferior Mesenteric.

This branch of the aorta supplies the descending colon, its sigmoid flexture, and most of the rectum. It terminates in the *superior hemorrhoidal* artery, sending off, also, the *colica sinistra* and *sigmoid.*

The *superior hemorrhoidal* divides into small branches only when it has reached, at the posterior part of the rectum, a distance of about four inches from the anus.

Supra-renal Arteries.

These are, though large in the fœtus, small in the adult. There is one for the right and one for the left supra-renal capsule.

Renal or Emulgent Arteries.

Each kidney receives a large trunk from the aorta, leaving it almost at a right angle. The left one arises above the right, but the right is somewhat the longest. Before entering the kidney, each artery divides into several branches.

Spermatic.

In the male, this artery on each side goes to the testicle; in the female, to the ovary. It is longer in the male, and quite tortuous, attending the spermatic cord in its course through the abdominal rings.

Phrenic.

Varying a good deal in its origin, the *phrenic* artery chiefly supplies the diaphragm; some branches, however, going from it to the supra-renal capsule, the spleen on the left side, and the liver on the right.

Lumbar Arteries.

Four of these go on each side, nearly at right angles from the aorta, and outwards and backwards, around the lumbar vertebræ. Between the transverse processes, each divides into an *abdominal* and a *dorsal* branch.

Middle Sacral Artery.

A small branch, from the bifurcation of the aorta, passing down in front of the last lumbar vertebra and the middle of the sacrum, to anastomose, about the upper end of the coccyx, with the lateral sacral arteries. Some of its ramifications go to the rectum, and others enter the anterior sacral foramina.

Common Iliac.

The primitive or *common iliac* artery, on each side, is about two inches in length; the right being somewhat the largest. They diverge from the aorta at the fourth lumbar vertebra, outwards and downwards; dividing, opposite the junction of the last lumbar vertebra and the sacrum, into the *external* and *internal* iliacs. Two common iliac veins accompany each. The lateral branches of the common iliac arteries are small and local.

Internal Iliac.

This is a short but thick artery, an inch and a half long. At the top of the great sacro-sciatic foramen, it divides into a large *anterior* and a similar *posterior* branch. From this bifurcation, the cord-like remainder of the *hypogastric* artery of the fœtus extends to the bladder. From the *anterior* trunk pass off, as branches, the *superior, middle,* and *inferior vesical, middle hemorrhoidal, obturator, internal pudic,* and *sciatic;* and, in the female, the *uterine* and *vaginal* arteries. From the *posterior* trunk go the *gluteal, ilio-lumbar,* and *lateral sacral.*

The *middle hemorrhoidal* goes to the rectum; anastomosing with the other hemorrhoidal vessels.

The *obturator* artery sometimes arises from the *posterior* branch of the internal iliac; generally from the anterior. Passing out of the pelvis through the obturator foramen, it divides into an *internal* and an *external* branch. Its other branches are small.

The *internal pudic* supplies the external genital organs. Besides *muscular, nervous*, and *visceral* branches given off within the pelvis, it sends off after its emergence through the great sacro-sciatic notch or foramen, the *inferior hemorrhoidal, superficial,* and *transverse perineal*, the *artery of the bulb*, the *artery of the corpus cavernosum*, and the *dorsalis penis*. Remark, however, that the internal pudic *re-enters* the pelvic cavity by the lesser sacro-sciatic foramen; afterwards, crossing to the ramus of the ischium, to ascend it and run along the inner margin of the ramus of the pubes. Its *terminal* branches are the artery of the corpus cavernosum and the dorsalis penis.

The *transverse perineal* artery accompanies the transversus perinei muscle.

Sciatic Artery.

This is the larger of the two terminal branches of the anterior trunk of the internal iliac; the internal pudic being the other.

The *sciatic* supplies the posterior muscles of the pelvis.

Gluteal Artery.

The gluteal is the largest branch of the internal iliac. Short and thick, it leaves the pelvis above the pyriformis muscle; then dividing into a *superficial* and a *deep* branch. The *superficial* passes under the gluteus maximus muscle, and subdivides. The *deep* branch goes between the gluteus medius and gluteus minimus muscles, and divides into two.

Ilio-Lumbar.

This branch of the posterior trunk of the internal iliac ascends beneath the psoas muscle and internal iliac artery and vein, to divide, in the iliac fossa, into a *lumbar* and an *iliac* branch.

Lateral Sacral.

These are commonly two on each side; the *superior* and the inferior lateral sacral, branches of the posterior trunk of the internal iliac.

External Iliac Artery.

Larger in the adult than the internal iliac, the *external iliac* artery passes downwards and outwards along the psoas muscle to the margin of the pelvis, half way between the anterior superior spine of the ilium and the symphysis pubes. There emerging from the pelvis to enter the thigh, it becomes the *femoral artery*. At the femoral arch, the femoral vein lies at its inner side; the anterior crural nerve is outside and in front of it.

The most important branches of the *external iliac* are the *epigastric* and the *circumflex iliac*.

Epigastric Artery.

This vessel arises a few lines above Poupart's ligament (border of the tendon of external oblique muscle) and after a short descent it passes obliquely upwards and inwards between the transversalis fascia and the peritoneum, to the margin of the rectus abdominis muscle. Penetrating its sheath it ascends behind that muscle, subdividing, and anastomosing finally with branches of the internal mammary and intercostal arteries.

The branches of the *epigastric* artery are the *cremasteric*, *pubic*, and *muscular* branches.

The occasional *variations* in the origin of the epigastric are important in the surgical anatomy of hernia.

Circumflex Iliac.

Arising from the outside of the external iliac, nearly opposite to the epigastric, this artery runs to the crest of the ilium, to anastomose afterwards with the gluteal and ilio-lumbar arteries.

Femoral Artery.

This continuation of the external iliac, passing under Poupart's ligament down the front and inside of the thigh, runs, at the junction of the middle and lower third of the thigh, through an opening in the adductor magnus muscle, to become the *popliteal* artery. Its course may be marked by a line drawn from a point half way between the anterior superior spine of the ilium and the symphysis pubis to the internal condyle of the femur; to which line the femoral artery lies parallel.

The *femoral* is superficial in the upper third of the thigh. Its location is there sometimes called *Scarpa's triangle*. The apex of this space is below; the inner side is the line of the adductor longus muscle, the outer, the sartorius muscle, and the base, Poupart's ligament. The iliacus, psoas, pectineus, and adductor longus muscles principally form the floor of the triangular space, which is bisected by the femoral vessels, from the base to the apex. The femoral artery and vein are inclosed together in a strong fibrous sheath, made partly by a process of the *fascia lata* of the thigh; but each vessel has also its thinner sheath.

In the middle third of the thigh the femoral is more deeply

118   ANATOMY.

Fig. 55.

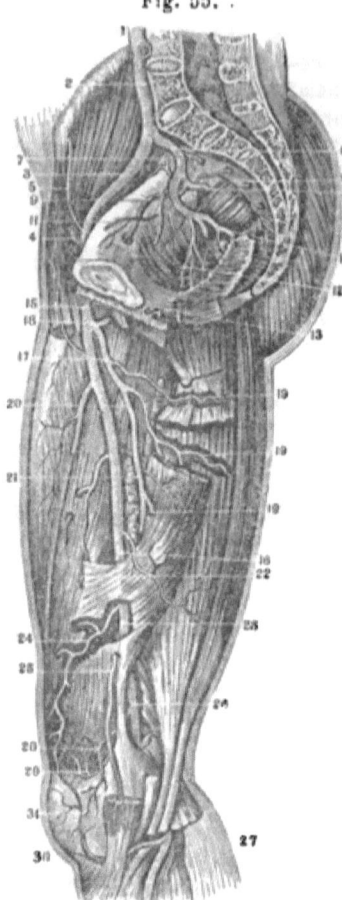

ARTERIES OF THE PELVIS AND THIGH.—
1. Inferior extremity of abdominal aorta.
2. Right primitive iliac. 3. Right external iliac. 4. Epigastric artery. 5. Circumflexa Ilii. 6. Internal iliac. 7. Ileo lumbar. 8. Gluteal. 9. Obturator. 10. Lateral sacral. 11. Vesical arteries cut off. 12. Middle hemorrhoidal. 13. Internal pudic. 14. Ischiatic. 15. Origin of femoral artery. 16. Point where it passes through the adductor muscles. 17. Profunda major. 18. Internal circumflex.

seated. Over it is the sartorius muscle, inside of it the adductor longus and adductor magnus, and outside of it the vastus internus muscle.

The femoral vein here lies to the outside of it; and just beyond that is the long or internal saphenous nerve.

The branches of the *femoral* artery are the *superficial epigastric, superficial circumflex iliac, superficial external pudic, deep external pudic, profunda femoris, muscular,* and *anastomotica magna.*

The *superficial epigastric* ascends through the saphenous opening of the fascia lata, to pass over the external oblique muscle of the abdomen, in the superficial fascia, to the umbilicus. Its subdivisions anastomose with those of the deep epigastric and internal mammary arteries.

The *profunda femoris,* nearly as large as the femoral, arises from its outer and back part, an inch or two below Poupart's ligament. Passing to the inner side of the femur, it goes through the adductor magnus muscle to the back of the thigh. Its main branches are the *external* and *internal circumflex,* and the *three perforating* arteries.

The *external circumflex* goes to the muscles on the front of the thigh. The *internal circumflex* winds round the femur to supply a number of muscles. The three *perforating* arteries go through the tendons of the adductor brevis and adductor magnus muscles.

The *muscular* branches of the femoral, from two to seven, go principally to the sartorius and vastus internus.

The *anastomotica magna* is the last to leave the femoral before it becomes the popliteal. It divides into two branches; the deeper of which has some ramifications reaching to the knee-joint.

Popliteal Artery.

From the opening in the adductor magnus muscle the popliteal goes obliquely outwards and downwards, behind the knee, to divide, at the lower edge of the popliteus muscle, into the *anterior* and *posterior tibial* arteries. Over the popliteal, at first, lies the semi-membranosus muscle; lower, it is covered by the gastrocnemius, soleus, and plantaris muscles, the popliteal vein, and internal popliteal nerve.

Branches of the popliteal are the *superior muscular, inferior muscular or sural, cutaneous, superior external and internal articular, azygos articular, inferior external and internal articular*.

Anterior Tibial Artery.

This vessel passes from the bifurcation of the popliteal, between the heads of the tibialis posticus muscle, and between the tibia and fibula, to the deep part of the front of the leg. It then goes down on the interosseous ligament and tibia, becoming more superficial in front of the ankle, as the *dorsalis pedis*. It has two *venæ comites*, one on each side.

Its branches are the *recurrent tibial, muscular, internal malleolar*, and *external malleolar*.

Its terminal continuation, the *dorsalis pedis* artery, runs along the tibial side of the foot to the first inter-metacarpal space; there it divides into the *dorsalis hallucis* and the *communicating artery*. Other branches of it are the *tarsal, metatarsal*, and *interosseal* vessels.

The *dorsalis hallucis* divides into two branches, one for the inner side of the great toe, and the other for the adjoining sides of the great and second toes.

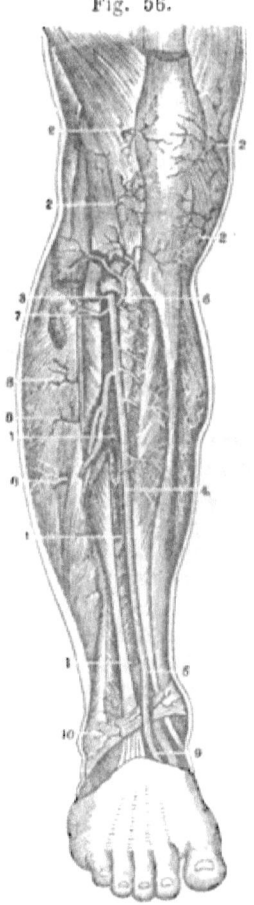

Fig. 56.

ANTERIOR TIBIAL ARTERY.— 1, 1. Extensor proprius pollicis pedis muscle and tendon. 2, 2. Articular arteries. 3. Anterior tibial artery. 4, 5. The same artery. 6. Recurrent branch. 7. Branch to muscles. 8, 8. Other muscular branches. 9. Pedal artery, or continuation of the anterior tibial on the foot. 10. External malleolar artery.

The *communicating artery* dips down into the sole of the foot, to anastomose with the external plantar artery, forming the *plantar arch*, and giving off two digital branches, for the great and second toes.

Posterior Tibial.

From the popliteus muscle this artery goes along the tibial side of the leg, behind the tibia, to the fossa between the inner malleolus and the heel; there to divide into the *internal* and *external plantar*. Covered above by the gastrocnemius and soleus muscles, in its lower third it is covered only by the skin and fascia, on the inner side of the tendo Achillis. It has two companion veins. The posterior tibial nerve lies above, on the inner side of the artery; soon it crosses it, and lies chiefly on its outer side.

The branches of the *posterior tibial* artery are the *peroneal, muscular, nutritious, communicating*, and *internal calcanean*.

Peroneal Artery.

This vessel leaves the posterior tibial about an inch below the popliteus muscle. It goes obliquely outwards to the fibula, and descends along its inner border to the lower third of the leg; there it sends off the *anterior peroneal*, which pierces the interosseous ligament to pass down the front of the leg to the tarsus. The other branches of the peroneal are muscular.

Plantar Arteries.

These are the terminal branches of the posterior tibial. The *external* plantar is the larger. It goes, from the space between the inner ankle and the heel, outwards and forwards to the base of the last metatarsal; then it turns inwards to the space between the first and second metatarsal bones, to complete the plantar arch by joining the communicating branch of the dorsalis pedis.

The *internal plantar* passes forwards along the inner side of the foot, and inner border of the great toe; anastomosing with its digital branches.

From the *plantar arch* the largest branches of many, are the three *posterior perforating*, and the four *digital* vessels. The latter, at the bifurcation of the toes, send off the *anterior perforating* branches, to join the *interosseous* branches of the *metatarsal* artery.

Both sides of the three outer toes, and the outer side of the second toe, are supplied by branches from the plantar arch; both sides of the great toe, and the inner side of the second, by the *dorsalis pedis* artery.

VEINS.

The capacity of the venous system is nearly three times as great as that of the arterial; the veins being both larger and more numerous than the arteries. The veins communicate very freely with each other. Each vein has *three* coats; *internal, middle*, and *external*. The *first*, like that of the arteries, is composed of connective tissue and epithelium, resembling serous tissue. The *middle* coat contains less muscular and elastic tissue than that of the arteries. The *muscular* tissue is greatest in amount in the *larger* veins, near the heart. The *external* coat is much like that of the arteries, but thinner, and contains longitudinal muscular fibres.

Most veins have *valves*, at intervals along their course, formed of projections or folds of the middle and inner coats, semilunar in shape. They open only towards the heart. Valves are most numerous in the veins of the lower limbs. There are *no* valves in the *smallest* veins, nor in the *venæ cavæ*, *hepatic* vein, *portal* vein, *pulmonary*, *cerebral*, *spinal*, *renal*, *uterine*, and *ovarian* veins; nor in the *umbilical* vein of the fœtus. Veins have nutritious vessels, or *vasa vasorum*, like the arteries.

Exterior Veins of the Head.

These are the *facial, temporal, internal maxillary, temporo-maxillary, posterior auricular*, and *occipital* veins.

The *facial* vein runs from the inner angle of the eye obliquely across the face to the front edge of the masseter muscle. Its origin is in the *frontal* vein, which descends near the middle of the forehead; those of the two sides, at first parallel, uniting at the root of the nose by the transverse trunk called the *nasal arch*.

The *facial* vein, lying outside of the facial artery, crosses the jaw, and in the neck joins with a branch from the temporo-maxillary vein to empty into the internal jugular. The facial receives as branches, some from the pterygoid plexus, and the inferior palpebral, labial, buccal, masseteric, submental, inferior palatine, submaxillary, and ranine veins.

The *temporal* vein begins on the side of the head in a minute plexus, the trunk from which, above the zygoma, is joined by the *middle temporal* vein. The temporal vein goes down between the ear and the condyle of the jaw, enters the parotid gland, and joins the *internal maxillary* vein to form the *temporo-maxillary*. The branches of the temporal are small, except the *transverse facial*, from the side of the face.

The *internal maxillary* is a considerable vein, whose branches correspond with those of the internal maxillary artery; being the *middle meningeal, deep temporal, pterygoid, masseteric, buccal, palatine*, and *inferior dental*.

The *temporo-maxillary* vein, formed by the union of the last two, divides in the parotid gland into two branches, one of which joins the *facial*, and the other the *external jugular*.

The *posterior auricular*, beginning on the side of the head, by a plexus, descends behind the ear to empty into the *temporo-maxillary* vein.

The *occipital* vein commences in a plexus at the back of the head, and follows the course of the occipital artery, beneath the muscles of the neck, to, and generally in the *internal jugular;* sometimes in the *external jugular*. By the *mastoid* vein, it communicates with the *lateral sinus* of the dura mater.

Veins of the Neck.

These are the *external jugular, posterior external jugular, anterior jugular, internal jugular,* and *vertebral*.

The *external jugular* vein is a continuation of the temporo-maxillary and posterior auricular, from the parotid gland down the neck to the middle of the clavicle. There it penetrates the deep fascia to join the subclavian vein. Sometimes it is double.

The *posterior external jugular* descends at the back of the neck to empty into the external jugular.

The *anterior jugular* goes down on the inner side of the anterior edge of the sterno-cleido-mastoid muscle, to join the subclavian. It varies in size.

The *internal jugular* vein is formed in the jugular foramen at the base of the skull, by the junction of the lateral and inferior petrosal sinuses. It passes down the neck, outside of the carotid artery, to join the subclavian at the root of the neck, thus forming the *vena innominata*. Its branches are the *facial, lingual, pharyngeal, superior* and *middle thyroid,* and *occipital* veins.

The *vertebral* vein, beginning in the occipital region, goes through the foramina of the transverse processes of the upper six cervical vertebræ; then emptying into the vena innominata.

Veins of the Interior of the Skull.

The *diploë* of the cranium has a number of tortuous canals, containing large veins with thin walls, enlarged at intervals into pouches. They communicate with the sinuses of the dura mater, and with the exterior veins of the head.

The *cerebral* vein has thin coats, without muscular tissue, and without valves. They are *superficial* and *deep;* eight or ten in number. They and the *cerebellar* veins empty into the sinuses of the dura mater.

The *sinuses of the dura mater* are channels for venous blood, having for their outer coat the dura mater itself. They are twelve; five at the upper and back part of the skull, and seven at the base of the skull.

The first named are the *superior* and *inferior longitudinal,* the *straight, lateral,* and *occipital* sinuses.

The *superior longitudinal* sinus runs backwards from the crista galli of the ethmoid bone, making a groove in the frontal, the junction of the parietal, and the occipital bones. At the crucial edge of the last, it divides into the *two lateral* sinuses. The *glands of Pacchioni* are small white bodies projecting from the inner surface of this sinus. The junction of the three sinuses just named is the *torcular Herophili.*

The *inferior longitudinal* sinus or vein occupies the back and free part of that fold of the dura mater called *falx cerebri.* It ends in the *straight* sinus.

The *straight* sinus is at the junction of the *falx* with the *tentorium.* It joins the other sinuses at the *torcular Herophili.*

The *lateral* are two large sinuses, passing from the *torcular* first outwards, and then coming downwards and inwards to empty into the jugular vein at the jugular foramen. The right one is larger than the left.

The *occipital* sinuses are the smallest. They are in the attached margin of the *falx cerebelli*, and end at the *torcular Herophili.*

The sinuses at the base of the cranium are the *cavernous, circular, superior* and *inferior petrosal,* and *transverse.* The *circular* and *transverse* communicate, near the middle of the base of the skull, between the sinuses of the two sides. The *cavernous* are short and longitudinal; with a reticulated structure.

Fig. 57.

Veins of the Arm and Hand.

These are *superficial* and *deep.* The first begin chiefly on the back of the head. The last are the *venæ comites* of the arteries.

Superficial are the *anterior ulnar, posterior ulnar, basilic, radial, cephalic, median, median basilic,* and *median cephalic* veins.

The *anterior ulnar* and *posterior ulnar,* named from their situation, join at the

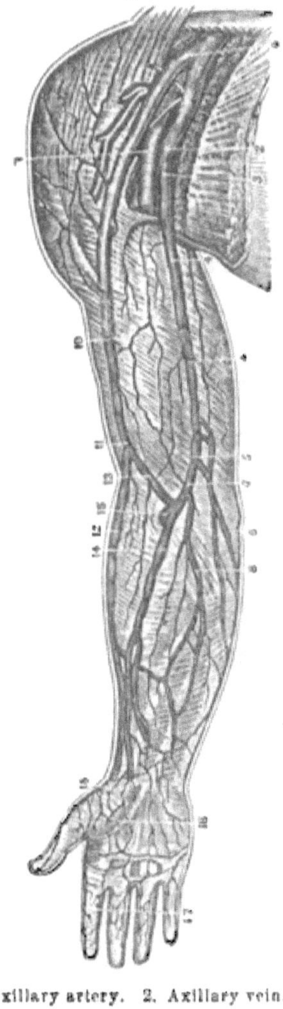

SUPERFICIAL VEINS OF THE UPPER EXTREMITY.—1. Axillary artery. 2. Axillary vein. 3. Basilic vein. 4, 4. Basilic vein. 5. Point where the median basilic joins the basilic vein. 6. Posterior basilic vein. 8. Anterior basilic vein. 9. Point where the cephalic enters the axillary vein. 10. A portion of the same vein. 11. Point where the median cephalic enters the cephalic vein. 12. Lower portion of the cephalic vein. 13. Median cephalic vein. 14. Median vein. 15. Anastomosing branch. 16. Cephalica-pollicis vein. 17. Subcutaneous veins of the fingers. 18. Subcutaneous palmar veins.

bend of the elbow to form the *basilic* vein. This runs up, receiving the *median basilic* in an oblique direction, and pierces the deep fascia to join one of the *venæ comites* or the axillary vein.

The *radial* runs up on the outside of the forearm from the back of the thumb and index finger. Receiving obliquely the *median cephalic* at the bend of the elbow, it becomes the *cephalic* vein. This runs outside of the biceps muscle to the upper third of the arm; then passes between the pectoralis major and deltoid muscles, to end in the axillary vein below the clavicle.

The *median* vein runs up from the palm of the hand, near the middle of the forearm, to divide at the bend of the elbow into the *median cephalic* and the *median basilic;* of which, as said above, the one joins the cephalic and the other the basilic vein. Branches of the external cutaneous nerve pass behind the median cephalic, and filaments of the same nerve pass both behind and in front of the median basilic vein.

Veins of the Thorax.

The *axillary* vein, the continuation of the basilic, increasing in size in the axilla by the supply of the *venæ comites* of the arteries of the arm, becomes the *subclavian* vein under the clavicle at the margin of the first rib.

The *subclavian* vein, at the inner end of the sterno-clavicular junction, unites with the internal jugular to form the *vena innominata*. The subclavian receives as branches the external, anterior, and internal jugular veins. It is separated from the subclavian artery by the scalenus anticus muscle.

The *right vena innominata* is about an inch and a half long. It goes almost directly downwards, to join the *left* vena innominata to form the *descending vena cava*, just below the cartilage of the first rib. It is external to, and nearer the surface than the arteria innominata. At the angle of junction of this vein and the subclavian, enters the vertebral vein. Lower down it receives the internal mammary, inferior thyroid, and superior intercostal veins.

The *left vena innominata* is longer and larger than the right. It passes obliquely across the upper anterior part of the cavity of the chest, to joint the right in forming the *vena cava*.

The *vena cava descendens* is a large trunk, two and a half to three inches long, vertical in its direction, entering the pericardium an inch and a half above the heart, and then emptying into the upper part of the right auricle.

The *right* and *left azygos* veins, of which the *right* is the larger, connect the descending and ascending *venæ cavæ*. The *right* azygos receives nine or ten lower *intercostal* veins, the right *bronchial*, and other veins. The *left lower azygos* receives four or five lower intercostal and a few other small veins. The *left upper* azygos receives the upper intercostal veins except the superior intercostal.

The *left bronchial* vein empties into the *left superior intercostal*. The bronchial veins carry back the blood from the tissue of the lungs.

Spinal Veins.

These form venous *plexuses*, as follows: *dorsi-spinal* veins, outside of the spinal column; *meningo-rachidian* veins, within the canal, outside of the membranes of the cord; *venæ basis vertebrarum*, of the bodies of the vertebræ; *medulli-spinal*, or the veins of the spinal cord.

Those of the *first* set, in the intervals between the vertebræ terminate, in the neck, in the vertebral veins; in the chest, in the intercostals; in the lumbar region in the lumbar and sacral veins.

Of the second set, there are two plexuses, the *anterior* and the *posterior longitudinal* spinal veins. The terminating connections are like those of the first set.

The veins of the *third* set join the transverse trunks which convert the anterior longitudinal spinal veins of the two sides. Those of the *fourth* set unite with the other veins of the spinal canal.

Veins of the Lower Extremity.

These are *superficial* and *deep*; the latter being the *venæ comites* of the arteries.

The *superficial* veins of the lower extremity are the *long internal saphenous*, and the *short external saphenous* vein.

The *internal saphenous* vein ascends, from a plexus of small veins on the back and inner side of the foot, in front of the inner malleolus, up the inside of the leg, behind the inner margin of the tibia. Going backwards gradually, at the knee it is behind the internal condyle of the femur; thence it passes up the inside of the thigh to the *saphenous opening*, where it penetrates the facia lata and ends in the *femoral* vein, an inch and a half below Poupart's ligament. At the saphenous opening it receives the *superficial epigastric, superficial circumflex iliac*, and *external pudic* veins. Its other branches are cutaneous. The internal saphenous nerve accompanies it.

The *external saphenous* vein ascends, from a plexus on the back and *outer* side of the foot, behind the outer malleolus, outside of the tendo Achillis, and then across it, to the middle of the back of the leg. Passing upwards, it penetrates the fascia in the lower part of the popliteal region, and ends in the *popliteal* vein, between the heads of the gastrocnemius muscle. The external saphenous nerve accompanies it.

The arteries of the leg are attended by *venæ comites* or deep

Fig. 58.

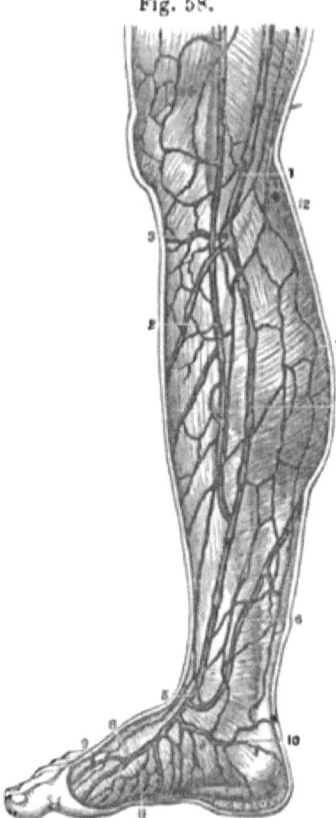

SUPERFICIAL VEINS OF THE LEGS.—1. Saphena major. 2. Collateral branch. 3. Anastomosis of veins. 4. Internal saphena. 5. Origin of the saphena vein. 6. Anastomosing branch. 7. Branches on the back of the leg. 8. The great internal vein of the foot. 9. Arch of veins on the metatarsal bones. 10. Branch from the heel. 11. Branches on the sole of the foot.

veins. These are supplied, particularly, by the *external* and *internal plantar*, and the *peroneal* veins.

The *popliteal* vein is formed by the union of the anterior and posterior tibial *venæ comites*. Passing up through the popliteal space to the opening in the tendon of the adductor magnus muscle, it becomes the *femoral* vein. *Below*, it is on the *inner* side of the artery; then it lies between it and the integument; above, it is *outside* of it.

The *femoral* vein runs with the femoral artery. *Below*, it is *outside* of it; then, *behind* it; near Poupart's ligament, it is on the *inner* side of it. Its principal branch is the *profunda femoris*.

The *external iliac* vein is the continuation of the femoral, above the crural arch. Going along the margin of the pelvis it unites, near the sacro-iliac symphysis, with the *internal iliac* to form the *common iliac* vein.

The *internal iliac* vein is supplied by the *venæ comites* of the branches of the internal iliac artery. Besides other branches, important ones are those of the *hemorrhoidal* and *vesico-prostatic*, and, in the female, *uterine* and *vaginal* plexuses. The *hemorrhoidal* plexus surrounds the lower end of the rectum.

The *vena dorsalis penis* is a large vein which returns the blood of the penis, along its back, to the prostatic plexus.

The two *common iliac* veins unite, opposite the space between the fourth and fifth lumbar vertebræ, to form the *ascending vena cava*. The *right* common iliac is shorter and more nearly vertical in its course than the *left*.

The *inferior or ascending vena cava* lies upon the spinal column to the *right* of the aorta. Passing through a groove in the posterior border of the liver, it penetrates the tendon of the diaphragm, and enters the pericardium, covered by its serous lining, to empty into the lower and posterior part of the right auricle.

Branches of the ascending vena cava are the *lumbar, right spermatic, renal, supra-renal, phrenic,* and *hepatic* veins.

The *lumbar* veins are three or four in number. The *spermatic* veins come, along the spermatic cords, from the testes. The *left* spermatic empties into the *renal.* The *ovarian* veins in the female correspond with them.

The *renal* veins are large; the *left* is longer than the *right* renal.

The *hepatic* veins, three in number, one from the right lobe, one from the left, and one from the middle of the liver, bring the blood from its substance, supplied by the portal vein and hepatic artery.

Portal System.

This is composed of *four* large veins, collecting blood from the digestive organs, and joining to make the *vena portæ,* which enters and ramifies within the liver. The four veins are the *superior* and *inferior mesenteric,* the *splenic,* and the *gastric* veins.

The *superior mesenteric* is supplied from the small intestines, cæcum, and transverse colon.

The *inferior mesenteric* receives its branches from the descending colon, sigmoid flexure, and rectum. It empties into the *splenic* vein. By anastomoses of its hemorrhoidal branches with the branches of the internal iliac, the portal system and the general venous circulation are connected.

The *splenic* vein receives the blood of five or six venous trunks from the spleen, and, also, the *vasa brevia* from the stomach, the *left gastro-epiploic* vein, *pancreatic, pancreatico-duodenal,* and *inferior mesenteric* veins.

The *gastric* vein is small, accompanying the gastric artery along the upper curvature of the stomach.

The *portal* vein, from the junction of the *splenic* and *superior mesenteric,* in front of the *vena cava,* behind the pancreas, goes up to the transverse fissure of the liver. Enlarged there into a *sinus,* it divides into two trunks, which ramify through the substance of the liver, communicating with the branches of the hepatic artery. The portal vein is about four inches in length.

Pulmonary Veins.

These are four, two for each lung, returning blood from the capillaries around the air cells. They have no valves, are but little larger than the pulmonary arterial trunks, and carry *arterial* blood. There is a venous trunk for each pulmonary *lobe;* making three for the right lung and two for the left. The pulmonary veins empty into the left auricle of the heart.

Cardiac Veins.

The blood from the heart's substance is returned by the *great, anterior* and *posterior cardiac* veins, and the *venæ Thebesii.* The *great cardiac* and *posterior cardiac* make the *coronary sinus,* which lies in the posterior part of the left auriculo-ventricular groove, and empties into the right auricle. The *anterior cardiac* and the *venæ Thebesii* open into the right auricle directly. The coronary sinus receives also a small *oblique* vein from the left auricle, the remainder of a trunk of the fœtal circulation.

LYMPHATICS.

These are delicate vessels with three coats, almost transparent, constricted at intervals so as to have a *beaded* appearance, passing

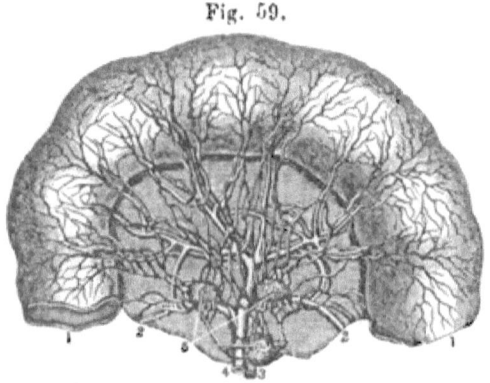

Fig. 59.

LYMPHATICS OF THE JEJUNUM AND MESENTERY, INJECTED: THE ARTERIES ARE ALSO INJECTED—1. Section of the jejunum. 2. Section of the mesentery. 3. Branch of the superior mesenteric artery. 4. Branch of the superior mesenteric vein. 5. Mesenteric glands receiving the lymphatics of this intestine.

through numerous *lymphatic glands,* to combine and empty at last into the *left* or *greater,* or into the *right* or *lesser thoracic duct.* All organs have lymphatics *except* the brain, spinal cord, eyeball, cartilages, tendons, nails, hair, cuticle, umbilical cord, placenta, and membranes of the ovum.

Lymphatic *glands* are most numerous in the neck, axilla, groin, and abdomen.

The lymphatic vessels of the *small intestine* are called *lacteals,* from the milky character of the chyle which they absorb. The glands which they pass through are the *mesenteric* glands.

Left Thoracic Duct.

This principal trunk receives most of the lymphatics and all the lacteals. It extends from the second lumbar vertebra to near the neck. Its beginning is the *receptaculum chyli;* which lies behind and to the right of the aorta. The thoracic duct passes up through the aortic foramen in the diaphragm, goes up to the left behind the arch of the aorta, near the œsophagus; then near the seventh cervical vertebra it curves downwards, to end at the angle of union of the left internal jugular and the subclavian veins. It has numerous valves. Its diameter is not uniform, but is about that of a small quill.

Right Thoracic Duct.

This is but an inch or so in length. It receives lymphatics from the right side of the head, neck, chest, right upper extremity, right

Fig. 60.

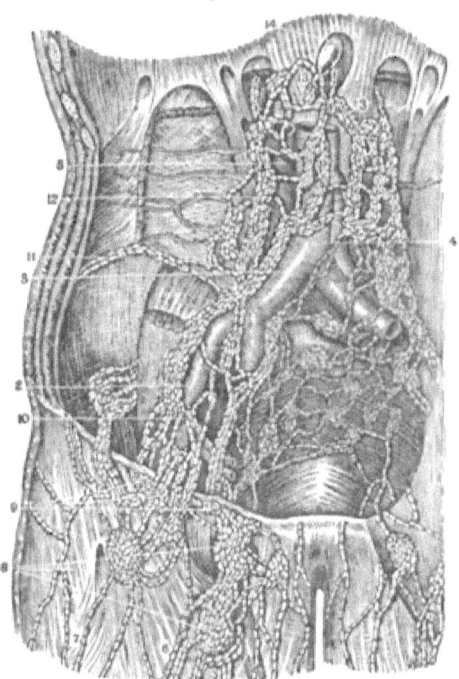

FEMORAL ILIAC AND AORTIC LYMPHATIC VESSELS AND GLANDS.—1. Saphena magna vein. 2. External iliac artery and vein. 3. Primitive iliac artery and vein. 4. Aorta. 5. Ascending vena cava. 6, 7. Lymphatics. 8. Lower set of inguinal lymphatic glands. 9. Superior set of inguinal lymphatic glands. 10. Chain of lymphatics. 11. Lymphatics which accompany the circumflex iliac vessels. 12. Lumbar and aortic lymphatics. 13. Origin of the thoracic duct. 14. Thoracic duct at its commencement.

lung, right side of the heart, and upper surface of the liver. It ends in the angle of junction between the right internal jugular and right subclavian veins.

CHAPTER VII.
THE SKIN.

The skin consists of two layers; the *cutis vera, derma, chorion*, or true skin, and the *cuticle, epidermis* or scarf-skin.

The *cutis vera* may be described as composed of the deep layer or *corium*, formed of fibro-connective (white fibrous and yellow elastic) tissue, and the *papillary* layer.

The latter consists of numerous small conical eminences or *papillæ*, $\frac{1}{100}$th of an inch in length and $\frac{1}{250}$th of an inch in diameter at the base. On the palm of the hand and sole of the foot they are larger and are in close parallel lines or ridges. Each *papilla* contains one or more capillary loops and one or more nerve-fibres, whose mode of termination is not yet ascertained.

On the lips and fingers, the nervous filaments of the papillæ have attached to them oval shaped bodies of minute size, the *Pacinian corpuscles*.

The *epidermis* is an insensitive epithelial tissue, of various thickness; composed of layers of *flattened cells*, most rounded and columnar in the deepest layers.

The color of the negro and other dark races, and in less degree that of all persons except *albinos*, depends upon the presence of pigment in the deeper layer (*rete mucosum*) of the epidermis.

The *lymphatics* of the skin are most abundant around the nipple and in the scrotum.

The Nails.

These are horny structures growing from the skin. The nail is planted in a groove or doubling of skin (the *matrix*), by its root; it has also a body and a free edge. The *lunula* is the white part of the nail just in advance of the root. The nail grows constantly forwards by the formation of new cells from its root and the under surface of its body.

The Hair.

Hairs are also growths from and modifications of epidermic tissue. Each hair is composed of a *root* and a pointed *shaft*. The root is lodged in a depression of the skin, the *hair-follicle*. The *shaft* of the hair consists of a central, deep-colored and opaque *medulla*, a fibrous surrounding to this, and an outer part or *cortex*. Around the roots of the hairs a few unstriped muscular fibres are found in the skin.

Glands of the Skin.

These are the *sebaceous* and the *sudoriferous* glands.

The *sebaceous* glands abound most in connection with the hair-follicles, into which their ducts open. They are largest on the face,

Fig. 61.

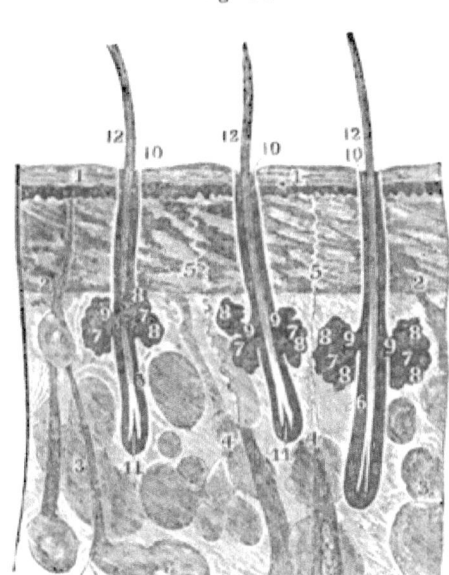

SEBACEOUS GLANDS AND FOLLICLES OF HAIRS IN THE SKIN OF THE AXILLA.—1. Epidermis. 2. Cutis vera. 3. Adipose tissue. 4, 4. Two perspiratory follicles. 5, 5. Their spiral canals. 6, 6. Follicles of hairs. 7, 7. Sebaceous glands. 10, 10. The orifices of the follicles of the hairs. 11, 11. Their roots. 12, 12. The hairs as seen under the microscope.

especially about the nose. Those of the eyelids are called *Meibomian* glands.

The *sudoriferous* or sweat glands are distributed over the skin in great numbers. On the palm of the hand there are estimated to be 2800 of them on a square inch.

Each gland consists of a convoluted tube, passing spirally through the skin from the deepest part of the corium or subcutaneous areolar tissue. It opens on the surface by a slightly enlarged valve-like orifice.

Fig. 62.

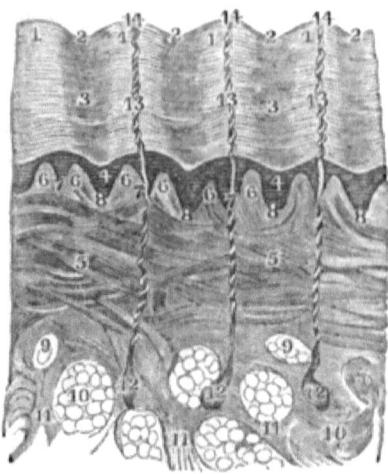

SUDORIFEROUS ORGANS OF THE SOLE OF THE FOOT.—12. The Sudoriferous follicles. 13. The spiral or sudoriferous canals. 14. The infundibular-shaped pores or orifices of these canals.

Connective Tissue.

Cellular, areolar, or *connective* tissue is the general *packing* or interstitial material of all parts of the body, except certain organs of peculiar structure. It abounds under the skin, and between the fibres and layers of muscles. It is white and silky in appearance, and is in nature allied to fibrous tissue, but more delicate in its subdivisions. Virchow asserts it to be a sort of generative tissue for other fabrics of the body, and to contain or consist of a number of *corpuscles* with pointed or linear processes by which they are joined together. Though adopted by many, this statement is not yet considered to be proved by all anatomists.

The areolar or connective tissue is moistened by a serosity or transudation from the bloodvessels; the excessive accumulation of this constitutes œdema or anasarca.

Fat.

The adipose tissue of the body is found by the retention of an almost liquid oleaginous material (olein, *margarin*, and a little stearin) in cells or spaces like those of the areolar tissue. It abounds under the skin, especially over the abdomen. More of it exists in general distribution under the skin of the female than under that of the male. Fat is also placed in some of the cavities of the body, as in the orbit of the eye, around the heart and around the kidneys. The marrow of the bones is also in part a similar substance.

CHAPTER VIII.
MUSCLES.

Structure.—There are two sorts of muscular tissue; that of the quickly acting and mostly voluntary muscles, and that of the involuntary muscles of the organs of nutritive functions, whose action is slower and more continuous, or *peristaltic.*

The *muscles of voluntary or animal life* are made up of bundles of red fibres, and these of *primitive fasciculi*, each of which is inclosed in a *sarcolemma* or sheath. The primitive fasciculus is formed by the binding together of filaments or *primitive fibrils*. Each of these is a flattened cylinder $\frac{1}{18000}$ of an inch in thickness, with wavy parallel lines around it (seen by aid of the microscope) transversely. It is hence called *striated* or *striped* muscular tissue. Most anatomists consider each primitive fibril to consist of a row of connected, almost rectangular, contractile *cells* or *corpuscles;* which widen and shorten in the contraction of the muscle.

The *unstriped muscles of organic life* are composed of pale flattened bands, or bundles of spindle-shaped fibres, $\frac{1}{4700}$ to $\frac{1}{3100}$ of an inch in diameter. Every fibre has an elongated *nucleus*, visible by aid of the microscope.

The *striped* muscle exists in all the voluntary muscles; in those also of the larynx, pharynx, upper half of the œsophagus, heart, and largest of the veins.

Unstriped muscular tissue is found throughout the alimentary canal, in the trachea and bronchial tubes, gall-bladder, bile-duct and other gland-ducts, calyces and pelvis of the kidney, uterus, bladder and urethra, iris, ciliary muscle, skin, all arteries, lymphatics and veins. In the male, it is also present in the scrotum, epididymis, vas deferens, vesiculæ seminales, and prostate gland. In the female, in the uterus, Fallopian tubes, broad ligaments, and vagina. Also, in the *corpora cavernosa* of the penis in the male, and of the clitoris in the female.

The voluntary muscles vary much in form. Each is enveloped in its *sheath*, and terminates at one or both ends, or, as in the diaphragm, at its centre, in a white fibrous *tendon*. The tendon may be of any shape, but is most generally a rounded cord. All muscles are well supplied with nerves and bloodvessels.

Aponeuroses are fibrous membranes, pearly white, connected with muscles and also with bones, cartilages, ligaments, &c.

Fasciæ are aponeurotic layers of various thickness, inclosing and protecting as well as retaining in their places the organs in all parts of the body. They are divisible chiefly into the *superficial* and the *deep* fasciæ.

The *superficial fascia* extends under the integument over the whole body.

The deep fascia is more immediately connected with the muscles; giving a sheath to each of them, and also sheaths to the nerves and bloodvessels.

MUSCLES OF THE HEAD AND FACE.

Occipito-frontalis.—*Origin*, from the occiput, by two flat bellies; *course*, over the cranium; on the frontal bone are formed two other flat bellies, one on each side; *insertion*, into the nasal bones, and os frontis near them, and into the upper edge of the *corrugator supercilii* and *orbicularis oculi* muscles. *Action*, to raise or depress the eyebrows and skin of the forehead; sometimes that of the occiput.

Attollens aurem, Attrahens aurem, Retrahens aurem.—Three small *rudimentary* or imperfectly developed muscles, seldom capable of use in man. They are immediately under the skin; the *attollens* above the ear, the *attrahens* above and in front of it, the *retrahens* behind and below it. The first two originate in the occipito-frontalis muscle; the third, in the mastoid process of the temporal bone. The first, in action, would raise the ear; the second would draw it forward; the third would draw it backward.

The muscles of the *tympanum* will be described with the organ of hearing.

Corrugator supercilii.—*Origin*, internal angular process of the os frontis. *Course*, obliquely upwards and outwards for a short distance. *Insertion*, the junction of the *occipito-frontalis* and *orbicularis oculi* muscles. *Action, frowning.*

Orbicularis oculi.—*Origin*, upper end of nasal process of superior maxillary bone, internal angular process of frontal bone, and upper margin of palpebral ligament. *Course*, entirely around the eye or edge of the orbit. *Action*, to close the eyelids.

Levator palpebræ superioris.—*Origin*, upper and back part of the orbit of the eye. *Course*, forward over the eyeball, widening as it goes. *Insertion*, the upper eyelid. *Action*, to raise the lid and open the eye.

Tensor tarsi, or *Horner's muscle.*—*Origin*, os unguis. *Course*, forwards a quarter of an inch, when it *bifurcates*. *Insertions*, one branch into each lachrymal duct. *Action*, to dilate the lachrymal ducts, and keep the lids close upon the eyeball.

Muscles of the Eyeball—Rectus superior.—*Origin*, upper part of optic foramen in the rear of the orbit, and inner margin of sphe-

MUSCLES OF THE HEAD AND FACE.

Fig. 63.

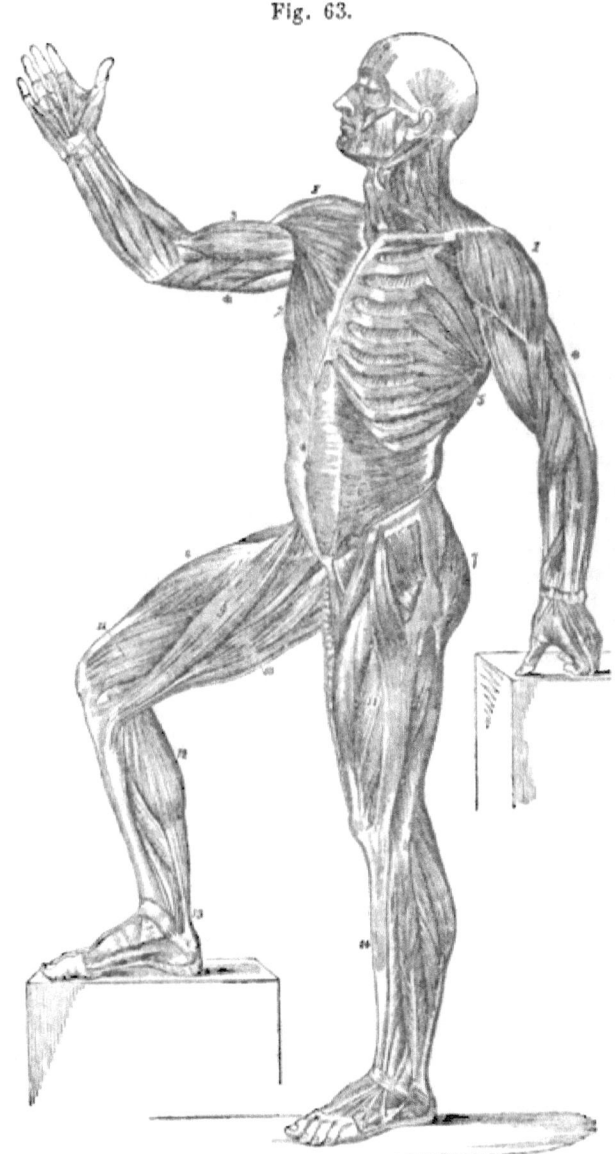

MUSCLES—FRONT VIEW.—On the right half, the superficial muscles. Left half, deep-seated muscles.

noidal fissure. *Course*, forward. *Insertion*, upper part of the eyeball, a little behind the cornea. *Action*, to raise the eye.

Rectus inferior.—*Origin*, from the lower margin of optic foramen; by the *ligament of Zinn*, a tendinous beginning for it and the internal and external recti muscles. *Course*, forwards. *Insertion*, lower part of eyeball behind the cornea. *Action*, to depress the eye.

Rectus internus.—*Origin*, ligament of Zinn, and sheath of the optic nerve (with which the other recti muscles also have some connection). *Course*, forwards. *Insertion*, eyeball behind the cornea. *Action*, to draw the eye inwards towards the nose.

Rectus externus.—*Origin*, by two heads, one from the common ligament or tendon of Zinn, the other from the margin of the optic foramen. Between these two heads pass the third and sixth nerves and the nasal nerve. *Course*, forwards. *Action*, to draw the eye outwards.

Obliquus superior or *Trochlearis.*—*Origin*, margin of optic foramen and sheath of nerve. *Course*, forwards to the *trochlea* or pulley-like process near the internal angle of the os frontis through which its round tendon passes, thence turning downwards and backwards. *Insertion*, sclerotic coat near the entrance of the optic nerve. *Action*, to roll the eye downwards and outwards. It is sometimes called *musculous patheticus*.

Obliquus inferior.—*Origin*, inner margin of upper maxillary bone. *Course*, beneath the *rectus inferior*, outwards and somewhat backwards. *Insertion*, outer and posterior part of eyeball. *Action*, to roll the eye obliquely inwards and downwards.

Pyramidalis nasi.—*Origin*, lower margin of occipito-frontalis. *Course*, downwards. *Insertion*, upper edge of *compressor naris* muscle. *Action*, to draw down the skin of the forehead, or draw up that of the nose.

Compressor naris.—*Origin*, root of ala nasi on one side. *Course*, across the dorsum of the nose. *Insertion*, into the corresponding muscle of the other side. *Action*, to compress the nostril.

Levator labii superioris alæque nasi.—*Origin*, by two slips, from nasal and orbitar processes of upper maxillary bone. *Course*, downwards. *Insertions*, one slip into the upper lip, the other into the ala or wing of the nose. *Action*, to raise the upper lip and dilate the nostril.

Depressor labii superioris alæque nasi.—*Origin*, alveoli of front teeth. *Course*, upwards. *Insertion*, upper lip and ala nasi. *Action*, to draw down the upper lip and the nose.

Levator anguli oris.—*Origin*, upper maxillary bone under the orbit of the eye. *Course*, downwards and inwards. *Insertion*, angle of the mouth. *Action*, to raise the corner of the mouth.

Zygomaticus major, and *Zygomaticus minor.*—*Origin*, malar part of zygoma. *Course*, downwards and forwards. *Insertion*, angle of mouth and upper lip. *Action*, to raise and draw out the corner of the mouth.

MUSCLES OF THE HEAD AND FACE.

Orbicularis oris.—A *circular* muscle, surrounding the mouth, in the lips. *Action*, to close the mouth.

Depressor anguli oris.—*Origin*, base of lower jaw at the side of the chin. *Course*, obliquely, some fibres vertically, upward to an apex. *Insertion*, corner of the mouth. *Action*, to draw down the corner of the mouth.

Depressor labii inferioris.—*Origin*, base of lower jaw at the side of the chin. *Course*, upwards and inwards. *Insertion*, the lower lip. *Action*, to draw the lower lip downwards.

Levator menti.—*Origin*, alveoli of front teeth of the lower jaw. *Course*, downwards and forwards. *Insertion*, lower lip and contiguous tegument. *Action*, to raise the chin and lower lip.

Buccinator.—*Origin*, ridge between the last molar tooth and the coronoid process of the lower jaw and upper jaw between the last molar and the pterygoid process of the sphenoid bone. *Course*, horizontally forwards. *Insertion*, the corner of the mouth. *Action*, to draw the corner of the mouth backwards or outwards.

Masseter.—*Origin*, by an *outer* layer, from the tuberosity of the upper jaw-bone, the lower edge of the malar bone and the zygoma; by an *inner* layer, from the posterior surface of the zygoma. *Course*, outer layer, downwards and backwards; inner layer, downwards and forwards. *Insertion;* outer plane, ramus and angle of the lower jaw; inner plane, coronoid process of the same bone. *Action*, to draw up the lower jaw. The outer plane or layer acting alone, protrudes the jaw; the inner alone, draws it backwards.

Temporalis.—*Origin*, the side of the head from the lower part of the parietal bone to the zygoma. *Course*, downwards and somewhat forwards. *Insertion*, tendinous, passing under the zygoma, into the coronoid process of the lower jaw.

Pterygoideus internus.—*Origin*, fossa formed by the internal face of the pterygoid process of the sphenoid and palate bones. *Course*, downwards and outwards. *Insertion*, inner face of the angle of the lower jaw. *Action*, to close the jaw, or, if one acts at a time, to give it a lateral grinding motion.

Pterygoideus externus.—*Origin*, outside of the pterygoid process of the sphenoid bone, and its spinous and temporal processes. *Course*, outwards and a little backwards. *Insertion*, neck of the lower jaw. *Action*, to draw the lower jaw forwards; or, when the two act alternately, to produce an oblique or grinding motion.

Hyo-glossus.—*Origin*, hyoid bone. *Course*, upwards and outwards. *Insertion*, side of the tongue. *Action*, to depress the side of the tongue, making its dorsum convex.

Genio-hyo-glossus.—*Origin*, back of the chin. *Course*, fan-like, backwards, to the middle of the tongue and hyoid bone. *Insertion*, the whole length of the tongue, and the base of the hyoid bone. *Action*, according to the fibres of it engaged, to draw the tongue forwards, backwards or downwards; or, to draw the os hyoides forwards and upwards.

Stylo-glossus.—*Origin*, styloid process of the temporal bone. *Course*, in slender rounded mass, downwards and forwards. *Insertion*, root and side of the tongue. *Action*, to draw the tongue backwards and sideways.

Lingualis.—*Origin*, the side of the root of the tongue. *Course*, forwards, between the hyo-glossus and the genio-hyo-glossus. *Insertion*, apex of the tongue. *Action*, to raise the apex of the tongue, shorten it and curve it backwards.

Other muscles of the substance of the tongue will be described hereafter with it.

Circumflexus palati.—*Origin*, spinous process of sphenoid bone, Eustachian tube, and internal pterygoid process of sphenoid. *Course*, over the hook of the internal plate of the pterygoid process as a pulley, after which it widens. *Insertion*, velum palati (soft palate), and edge of palate bones. *Action*, to extend the velum palati.

Levator palati.—*Origin*, apex of petrous portion of temporal bone, and Eustachian tube. *Course*, downwards, forwards and inwards. *Insertion*, into the soft palate as far as the uvula. *Action*, to raise the soft palate and draw it backwards.

Constrictor isthmi faucium.—*Origin*, side of the tongue near its roots. *Course*, upwards between the folds of the anterior half arch of the palate. *Insertion*, soft palate at the base of the uvula. *Action*, to bring the tongue and palate together, and close the opening from the mouth into the pharynx.

Palato-pharyngeus.—*Origin*, posterior half arch of the palate. *Course*, downwards, in a curve, the convexity outwards. *Insertion*, upper and back edge of the thyroid cartilage, and the pharynx between its middle and lower constrictor muscles. *Action*, to depress the palate and force it down over the pharynx.

Azygos uvulæ.—*Origin*, spinous process of the palatal suture. *Course*, downwards. *Insertion*, the whole length of the uvula. *Action*, to contract the uvula and draw it upwards.

MUSCLES OF THE NECK.

Platysma myoides.—*Origin*, connective tissue below the clavicle. *Course*, flat and thin (with pale and imperfectly developed fibres), upwards under the skin of the front and side of the neck. *Insertion*, the side of the lower jaw and skin of the face. *Action*, rudimentary or almost *null* in man; in the ox, horse, &c., as the *cutaneous muscle*, to shake forcibly the skin of the neck. If it act in man, it must draw down the jaw or lower lip, or raise the skin of the neck.

Sterno-cleido-mastoideus.—*Origin*, upper end of sternum, and sternal end of clavicle. *Course*, by two separate heads, at first, which, not far above the clavicle, join into one muscle, to pass obliquely upwards and outwards. *Insertion*, mastoid process of the temporal bone. *Action*, if only one contracts, to draw the head down to one side; if both act together, to draw the head down toward the chest.

Digastricus.—*Origin*, the mastoid process. *Course*, first fleshy, downwards and forwards, then tendinous through a perforation in the stylo-hyoideus muscle, then fleshy again, upwards and forwards. *Insertion*, base of lower jaw at its median line or symphysis. *Action*, to draw the jaw down and open the mouth; or, to raise the hyoid bone and throat.

Stylo-hyoideus.—*Origin*, styloid process of temporal bone. *Course*, downwards and forwards, perforated by the *digastricus*. *Insertion*, os hyoides. *Action*, to raise the hyoid bone and draw it backwards.

Mylo-hyoideus.—*Origin*, broad and thin from inside of lower jaw, from the middle of the chin to the last molar tooth. *Course*, downwards and forwards. *Insertion*, os hyoides. *Action*, to draw the os hyoides upwards and forwards, and protrude the tongue. It forms the floor of the mouth.

Genio-hyoideus.—*Origin*, tubercle inside of lower jaw near its middle line. *Course*, downwards and backwards. *Insertion*, os hyoides. *Action*, to raise the hyoid bone and draw it forwards.

Omo-hyoideus.—*Origin*, upper margin of scapula. *Course*, obliquely upwards and forwards, as a long, slender muscle, tendinous where it passes under the sterno-cleido-mastoid. *Insertion*, os hyoides. *Action*, to draw down the hyoid bone, or draw it to one side.

Sterno-hyoideus.—*Origin*, first bone of the sternum, part of clavicle, and cartilage of first rib. *Course*, upwards. *Insertion*, os hyoides. *Action*, to draw down the hyoid bone.

Sterno-thyroideus.—*Origin*, margin of first bone of sternum, and cartilage of first rib. *Course*, upwards on the front and side of the trachea and thyroid gland. *Insertion*, lower edge of the thyroid cartilage. *Action*, to draw the larynx downwards.

Thyro-hyoideus.—*Origin*, side of thyroid cartilage. *Course*, upwards. *Insertion*, base and corner of os hyoides. *Action*, to approximate the thyroid cartilage and hyoid bone.

Constrictor pharyngis inferior.—*Origin*, sides of thyroid and cricoid cartilages of the larynx. *Course*, superior fibres, obliquely upwards, covering part of the constrictor medius; inferior fibres, horizontally, over the upper part of the œsophagus. *Insertion*, meeting its fellow of the opposite side at the median line of the back of the pharynx. *Action*, to constrict the lower portion of the pharynx and draw it upwards and backwards.

Constrictor pharyngis medius.—*Origin*, corner of os hyoides and ligament between it and the thyroid cartilage. *Course*, spreading, and dividing into two terminal portions. *Insertion*, into its fellow at the back of the pharynx, and into the cuneiform process of the occiput, before the foramen magnum. *Action*, to constrict the middle of the pharynx, and to draw the hyoid bone upwards and backwards.

Constrictor pharyngis superior.—*Origin*, cuneiform process of occiput, pterygoid process of sphenoid bone, and upper and lower jaws near the last molar tooth; also, the buccinator muscle, palate, and root of the tongue. *Course*, almost horizontal, its lower edge covered by the constrictor medius. *Insertion*, meeting its fellow at the median line of the pharynx behind. *Action*, to constrict the upper portion of the pharynx.

Stylo-pharyngeus.—*Origin*, styloid process. *Course*, downwards and forwards. *Insertion*, the side of the pharynx between the upper and middle constrictors, and into the posterior edge of thyroid cartilage. *Action*, to raise the pharynx and larynx, and open the pharynx above.

The muscles belonging to the *larynx* itself have been described in the account of it as a part of the apparatus of respiration.

Longus colli.—*Origin*, sides of the bodies of the three upper dorsal vertebræ, and transverse processes of four lower cervical vertebræ. *Course*, upwards and slightly forwards. *Insertion*, front of bodies of all the vertebræ of the neck. *Action*, to bend the neck forwards or to one side.

Rectus capitis anticus major.—*Origin*, transverse processes of third, fourth, fifth and sixth vertebræ of the neck. *Course*, upwards and a little inwards. *Insertion*, cuneiform process of occiput, in front of condyle. *Action*, to depress the head.

Rectus capitis anticus minor.—*Origin*, front of the atlas. *Course*, upwards. *Insertion*, occiput in front of condyle. *Action*, to depress or bow the head.

Rectus capitis posticus major.—*Origin*, spinous process to the second cervical (axis or dentata) vetrebra. *Course*, obliquely upwards, widening as it ascends. *Insertion*, inferior semicircular ridge of the occiput. *Action*, to draw back or rotate the head.

Rectus capitis posticus minor.—*Origin*, posterior tubercle of the atlas vertebra. *Course*, obliquely upwards, widening as it ascends. This muscle is *within* the last described, the two *minores* being thus between the two *majores*. *Insertion*, occiput, along part of the semicircular ridge and the space between it and the foramen magnum. *Action*, to draw the head backwards.

Obliquus capitis inferior.—*Origin*, spinous process of vertebra dentata. *Course*, upwards and outwards. *Insertion*, transverse process of atlas. *Action*, to rotate the atlas, and head upon it, on the axis.

Obliquus capitis superior.—*Origin*, transverse process of the atlas. *Course*, upwards and slightly inwards. *Insertion*, outer end of the lower semicircular ridge of the occiput. *Action*, to draw the head backwards.

Rectus capitis lateralis.— *Origin*, transverse process of the atlas. *Course*, obliquely upwards. *Insertion*, occiput outside of the condyle. *Action*, to draw the head to one side.

Fig. 64.

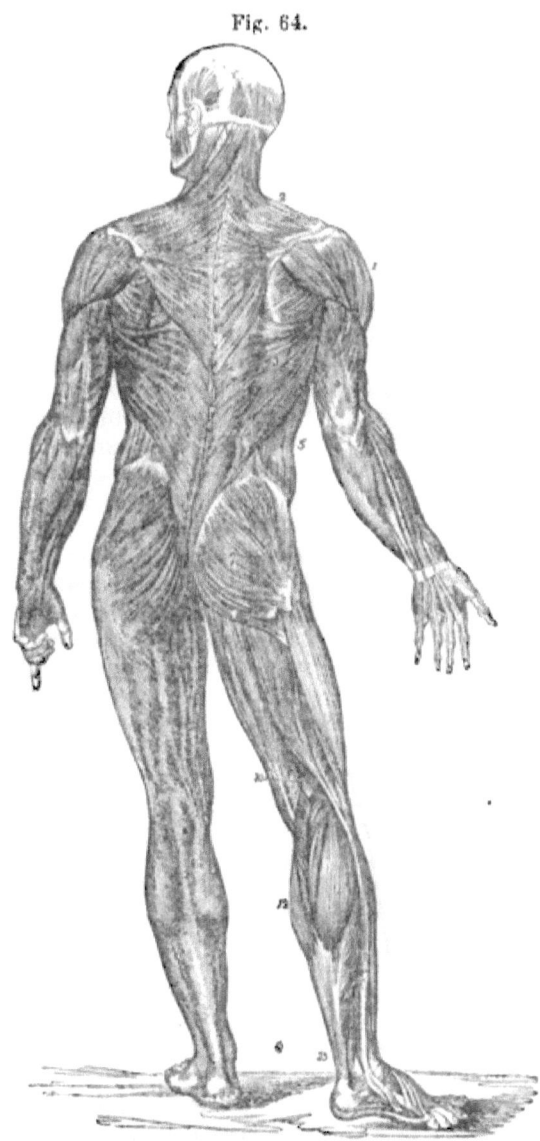

Muscles—Back View.—The fascia is left upon the left limb, removed from the right.

Scalenus anticus.—*Origin*, first rib near its cartilage. *Course*, upwards, inwards, and backwards. *Insertion*, transverse processes of the fourth, fifth, and sixth vertebræ of the neck, by three separate tendons. *Action*, to draw the neck to one side, or lift the first rib.

Scalenus medius.—*Origin*, first rib, upper and outer part. *Course*, upwards, inwards, and backwards; separated from the scalenus anticus below by the subclavian artery, above by the cervical nerves. *Insertion*, by separate tendons, into the transverse processes of all the cervical vertebræ. *Action*, to draw the neck to one side, or lift the rib.

Scalenus posticus.—*Origin*, second rib near its tubercle. *Course*, upwards, inwards, and backwards. *Insertion*, transverse processes of fifth and sixth cervical vertebræ. *Action*, like that of the two preceding muscles.

Trapezius.—*Origin*, occipital protuberance, spinous processes of five first vertebræ of the neck, by the ligamentum nuchæ, and spinous processes of two last cervical and all the dorsal vertebræ. *Course*, by *converging* fibres, some downwards, some horizontally, and others upwards, the whole outwards. *Insertion*, outer half of the clavicle, the acromion process, and whole length of spine of the scapula. *Action*, according to its partial or total contraction, to draw the shoulder upwards, backwards, or downwards; or, if the upper portion act when the shoulder is fixed, to draw the head to one side. This muscle is just beneath the skin, covering the other dorso-cervical muscles.

Rhomboideus major.—*Origin*, spinous processes of the last cervical and first four dorsal vertebræ. *Course*, downwards and outwards. *Insertion*, the whole base of the scapula, below its spine. *Action*, to draw the scapula upwards and backwards.

Rhomboideus minor.—*Origin*, spinous processes of three lower cervical vertebræ. *Course*, obliquely downwards and outwards. *Insertion*, base of the scapula opposite its spine. *Action*, same as that of rhomboideus major.

Levator scapulæ.—*Origin*, from transverse processes of four or five upper cervical vertebræ, by distinct tendons. *Course*, downwards and outwards. *Insertion*, base of the scapula above the spine. *Action*, to raise the scapula.

Cervicalis descendens.—*Origin*, upper margin of first four ribs, by as many tendons. *Course*, upwards and inwards. *Insertion*, transverse processes of fourth, fifth, and sixth cervical vertebræ. *Action*, to draw the neck backwards.

Splenius.—*Origin*, spinous processes of the five lower cervical and four upper dorsal vertebræ. *Course*, upwards and outwards. *Insertion*, into transverse processes of three or four cervical vertebræ; and also, into the mastoid process and occiput.

Complexus.—*Origin*, four lower cervical and seven upper dorsal vertebræ, and spinous process of first dorsal. *Course*, upwards.

MUSCLES OF THE TRUNK. 143

Insertion, on the inner side of splenius, into the occiput, between the upper and lower semicircular ridges. *Action*, to draw the head backwards or to one side.

Trachelo-mastoideus.—*Origin*, transverse processes of five lower cervical and three upper dorsal vertebræ.. *Course*, upwards and outward. *Insertion*, mastoid process. *Action*, same as that of the complexus.

MUSCLES OF THE TRUNK.

Pectoralis major.—*Origin*, sternal half of clavicle, whole length of upper and middle bones of the sternum, and cartilages of fifth

Fig. 65.

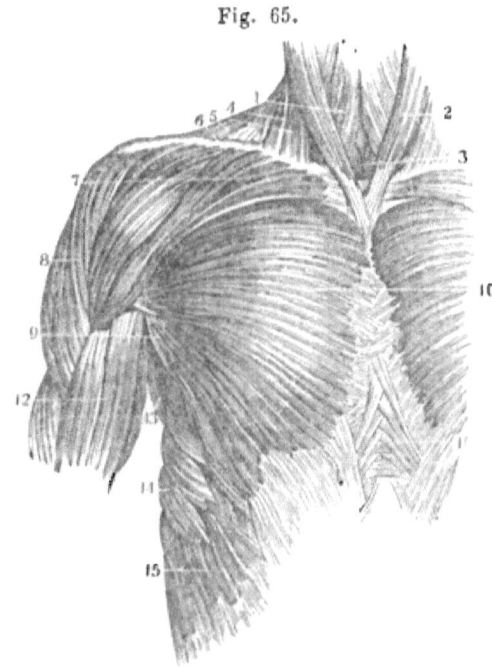

SUPERIOR MUSCLES OF THE UPPER FRONT OF THE TRUNK.—1. Sterno-hyoid. 2. Sterno-cleido-mastoid. 3. Sterno-thyroid. 4. Sterno-cleido-mastoid. 5. Edge of the trapezius. 6. Clavicle. 7. Clavicular origin of the pectoralis major. 8. Deltoid. 9. Fold of pectoralis major on the anterior edge of the axilla. 10. Middle of the pectoralis major. 11. Crossing and interlocking of fibres of the external oblique of one side with those of the other. 12. Biceps flexor cubiti. 13. Teres major. 14. Serratus major anticus. 15. Superior heads of external oblique interlocking with serratus major.

and sixth ribs. *Course*, outwards, *converging*, and downwards to the axilla. *Insertion*, by a flat, broad, twisted tendon, into the humerus, in front of the bicipital groove. *Action*, to draw the arm inwards and forwards, or downwards if raised. This muscle lies under the skin.

Pectoralis minor.—*Origin,* third, fourth, and fifth ribs near their cartilages. *Course,* upwards and outwards, converging; it lies beneath the pectoralis major. *Insertion,* coracoid process of the scapula. *Action,* to depress the shoulder; or, to raise the ribs.

Subclavius.—*Origin,* cartilage of the first rib. *Course,* outwards, under the clavicle. *Insertion,* under margin of the clavicle almost its whole length. *Action,* to draw the clavicle downwards.

Serratus magnus, vel *anticus.*—*Origin,* by tooth-like digitations, from the first nine ribs. *Course,* upwards and backwards, in front of and beneath the subscapularis. *Insertion,* whole length of the base of the scapula. *Action,* to draw the scapula forwards and downwards, or to raise the ribs.

Intercostales, externi et *interni.*—*Origin,* lower margin of each rib except the last. *Course,* oblique; the fibres of the external intercostals going downwards and forwards, those of the internal downwards and backwards, so as to cross each other. *Insertion,* upper edge of each rib except the first. *Action,* to elevate the ribs. Between the two layers or sets of muscles pass the intercostal vessels and nerves. In contact with their inner surface is the pleura.

Sterno-costalis.—*Origin,* middle and last piece of the sternum. *Course,* upwards and outwards. *Insertion,* cartilages of third, fourth and fifth, and sometimes sixth ribs. *Action,* to draw down the ribs and diminish the cavity of the chest.

Levatores costarum.—*Origin,* transverse processes of the last cervical and eleven upper dorsal vertebræ. *Course,* downwards and outwards, twelve in number on each side. *Insertions,* into the rough surfaces of the ribs between their tubercles and angles. *Action,* to elevate the ribs.

Diaphragma major.—*Origin,* by fleshy slips from the ensiform cartilage of the sternum, and from the inner face of the cartilages of the last six ribs. *Course,* converging, by all its fibres, to a broad central tendon. *Insertion,* into the cordiform tendon, which is notched in shape at the vertebral column, and pointed near the sternum. Through the *foramen quadratum,* near the spine, the vena cava ascends.

Diaphragma minor.—*Origin,* by four pairs of fleshy and tendinous slips, of which the longest arise from the third and fourth lumbar vertebræ. The second pair come from the ligament between the second and third lumbar vertebræ. The third pair from the sides of the second, and the fourth pair from the base of the transverse process of the same vertebra. *Course,* by muscular bands, upwards, in two columns, one on each side. *Insertion,* into the back and notch of the cordiform tendon. Through this muscle, the lesser diaphragm, pass through one foramen (œsopha-

geum), the œsophagus and pneumogastric nerve; through another (*hiatus aorticus*) the aorta, thoracic duct, and great splanchnic nerve. *Action*, that of the greater and lesser diaphragm, often described and generally named as one muscle, is, by *descending* in its contraction, to increase the cavity of the chest in inspiration. The diaphragm is also a septum or partition between the chest and the abdomen.

Obliquus externus abdominis.—Origin, by digitations from the last eight ribs, near their cartilages; in apposition to portions of the pectoralis major, serratus magnus, and latissimus dorsi. *Course*, downwards and inwards over the abdomen; a few fibres crossing the median line just above the pubes. This muscle is next beneath the skin and superficial fascia. *Insertion*, into the median line or *linea alba*, where it meets its fellow; into the anterior half of the crest of the ilium; anterior superior spine of the ilium, and pubes at and near the symphysis. The tendinous cord which reaches from the spine of the ilium to the pubes is called *Poupart's ligament*. This divides into two bands anteriorly, one passing to the symphysis and the other to the spine of the pubes. The latter is reflected outwards and backwards along the *linea ileo-pectinea* for about an inch; the reflection being called in surgical anatomy *Gimbernat's ligament*. *Action*, to sustain and compress all the contents of the abdomen and force them upwards towards the diaphragm, or downwards towards the perineum.

Obliquus internus, vel *Ascendens abdominis.—Origin*, posterior face of sacrum, spinous processes of three lower lumbar vertebræ, crista of the ilium, and Poupart's ligament. *Course*, upwards and inwards to the *linea semilunaris*, where the tendon separates into an anterior and posterior layer. The former joins the tendon of the obliquus externus and passes in front of the *rectus* muscle to the linea alba. The posterior layer joins the tendon of the *transversalis* muscle to go behind the rectus; except that, from about half way between the umbilicus and the pubes downwards, it passes with the anterior portion in front of the rectus muscle. *Insertion*, into the six lower ribs at their cartilages, the side of the ensiform cartilage, and the *linea alba* or median vertical tendinous line of the abdominal superficies. *Action*, the same as that of the external oblique.

Transversalis abdominis.—Origin, cartilages of six or seven lower ribs, transverse processes of the last dorsal and upper four lumbar vertebræ, almost the whole length of the crest of the ilium, and anterior half of Poupart's ligament. *Course*, across the front of the abdomen. *Insertion*, into the whole length of the linea alba. *Action*, like that of the last two muscles.

Rectus abdominis.—Origin, anterior face of ensiform cartilage, and cartilages of the fifth, sixth, and seventh ribs. *Course*, downwards on each side of the linea alba, in a flat band about three inches in width, narrowing below and becoming tendinous. *Insertion*, the pubes, at and near the symphysis. *Action*, to compress the contents of the abdomen, or to bend the body or raise the anterior part of the pelvis.

The *linea semilunaris* is a white curved line, with the convexity outwards, extending, on each side, downwards from the cartilage of the eighth rib to the pubes. It is formed by the tendon of the internal oblique muscle at its division.

The *linea transversæ* are three or four tendinous lines crossing the rectus muscle at right angles to the direction of its fibres at the distance of a few inches from each other.

Fig. 66.

The anatomy of *hernia* will be given by itself hereafter.

Pyramidalis.—This muscle is often absent. *Origin*, upper margin of the pubes. *Course*, upwards, in a pyramidal form, within the sheath of the rectus. *Insertion*, the linea alba and inner edge of the rectus. *Action*, to make tense the rectus muscle, and support the lower part of the abdomen.

Quadratus lumborum.—*Origin*, posterior and upper margin of ilium, for two inches from the spine. *Course*, upwards and inwards. *Insertion*, into the transverse processes of all the lumbar and the side of the last dorsal vertebræ, and into the last rib near the spine. *Action*, to move the loins to either side, or if both act, to move the pelvis forward. It may also depress the last rib.

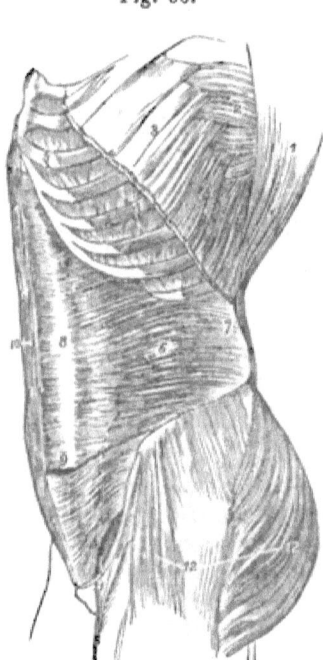

LATERAL VIEW OF THE MUSCLES OF THE TRUNK.—1. Latissimus dorsi. 2. Serratus major anticus. 3. External oblique. 4. Two external intercostal muscles. 5. Two internal intercostal muscles. 6. Transversalis abdominis. 7. Fascia lumborum. 8. Sheath of the rectus. 9. Rectus abdominis cut off. 10. Rectus abdominis of right side. 11. Crural arch. 12. Gluteus magnus—medius and tensor vaginæ femoris covered by the fascia lata.

Psoas magnus.—*Origin*, bodies and transverse processes of the last dorsal and all the lumbar vertebræ. *Course*, in an oblong form downwards and forwards, under or behind Poupart's ligament and over the pubes. The peritoneum covers it in front. *Insertion*, into the trochanter minor, and an inch or so of the shaft of the femur. *Action*, to

MUSCLES OF THE TRUNK.

flex the thigh on the pelvis and rotate it a little outwards; or when the thigh is fixed, to bend the body forwards.

Psoas parvus.— *Origin*, sides of last dorsal and first lumbar vertebræ, and intervertebral ligament. *Course*, along the internal side of the psoas magnus. *Insertion*, linea ilio-pectinea at the junction of the pubes and ilium. *Action*, to bend the spine upon the pelvis; and to draw up the femoral vessels in their sheath.

Iliacus internus.— *Origin*, transverse process of the last lumbar vertebra, whole inner edge of crest of the ilium, and the same bone between the anterior superior spine and the acetabulum; also, from the whole *venter* or concavity of the ilium. *Course*, downwards, and somewhat forwards, over the edge of the pubes behind Poupart's ligament. *Insertion*, with the psoas magnus into the trochanter minor of the femur. *Action*, to flex the thigh upon the pelvis, or the body towards the thigh.

Fig. 67.

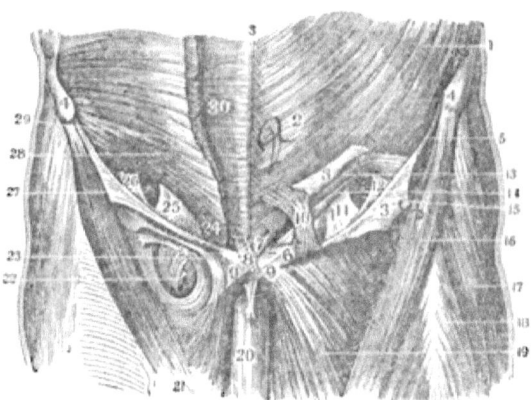

ABDOMINAL MUSCLES AND INGUINAL CANAL.—1. External oblique muscle. 2. Its aponeurosis. 3. Its tendon slit up and turned back to show the canal. 4. Anterior superior spinous processes. 5. Poupart's ligament. 6. External column of external ring. 7. Internal column of external ring. 8. Intercrossing of the tendons of each side. 9. Body of the pubes. 10. Upper boundary of the external abdominal ring—the line points to the ring. 11, 12. Fascia transversalis. 13. Fibres of internal oblique turned up. 14. Fibres of transversalis muscle. 15. Internal ring enlarged for demonstration. 16. Sartorius. 17. Fascia lata femoris. 18. Rectus femoris. 19. Adductor longus. 20. Penis. 21. Fascia lata of the opposite thigh. 22. Point where the saphena vein enters the femoral. 23. Fascia lata as applied to the vessels. 24. Insertion of transversalis muscle. 25, 26. Fascia transversalis. 27. Poupart's ligament turned off from the internal muscles. 28. Transversalis abdominis. 29. Internal oblique. 30. Rectus abdominis.

Latissimus dorsi.— *Origin*, spinous processes of the last seven dorsal and all of the lumbar vertebræ and the sacrum; also the outer margin of the sacrum and posterior part of the ilium, and the last four ribs. *Course*, by converging fibres, upwards and horizontally outwards, towards the axilla, passing over the inferior angle of the scapula. Except above, where the trapezius covers it, this muscle lies next under the skin. *Insertion*, by a flat, strong tendon, into the humerus just behind the bicipital groove. *Action*, to draw the arm downwards and backwards, and to roll it inwards.

Serratus posticus superior.—*Origin*, from the ligamentum nuchæ, over the last three cervical and two upper dorsal vertebræ. *Course*, obliquely downwards and outwards. *Insertion*, by fleshy slips, into the second, third, fourth, and fifth ribs, beyond their angles. *Action*, to raise the ribs.

Serratus posticus inferior.—*Origin*, lumbar fascia, and spinous processes of the last two dorsal and first three lumbar vertebræ. *Course*, upwards and outwards. *Insertion*, by fleshy slips, into the last four ribs, near their cartilages.

Interspinales.—*Origin*, in the cervical and lumbar regions, from the upper part of each spinous process. *Course*, upwards, in quadrilateral form, but of small size. *Insertion*, into the spinous process of the vertebra next above its origin. *Action*, to draw the spinous processes together, and sustain the spine in the erect position.

Transversalis cervicis.—*Origin*, transverse processes of five upper dorsal vertebræ. *Course*, upwards, between the splenius and the trachelo-mastoideus. *Insertion*, into the transverse processes of all the cervical vertebræ except the first and the last. *Action*, to draw the neck backwards or to one side.

Intertransversalis.—*Origin*, each transverse process. *Course*, vertical. *Insertion*, the transverse process next above or below. *Action*, to bend the spine to one side.

Semi-spinalis colli.—*Origin*, transverse processes of first six dorsal vertebræ. *Course*, upwards. *Insertion*, spinous processes of all the cervical vertebræ except the first and last. *Action*, to draw the neck obliquely backwards.

Semi-spinalis dorsi.—*Origin*, transverse processes of the seventh, eighth, ninth, and tenth dorsal vertebræ. *Course*, upwards. *Insertion*, spinous processes of last two cervical and five or six upper dorsal vertebræ.

Multifidus spinæ.—*Origin*, back of the sacrum and adjacent part of crest of the ilium, from the oblique and transverse processes of all the lumbar vertebræ, and from the transverse processes of all the dorsal and the last four cervical vertebræ. *Course*, upwards and inwards. *Insertion*, last five or six cervical spinous processes, and all those of the dorsal and lumbar vertebræ. *Action*, to draw the spine backwards or to one side; or, both acting, to sustain the spine in the erect position.

Spinalis dorsi.—*Origin*, spinous processes of three lower dorsal and two upper lumbar vertebræ. *Course*, upwards. *Insertion*, into the spinous processes of eight or nine of the upper dorsal spinous processes, excluding the first. *Action*, to sustain the spine in the erect position.

Longissimus dorsi.— *Origin*, from the back of the whole sacrum, the posterior part of the crest of the ilium, and the spinous and

Fig. 68.

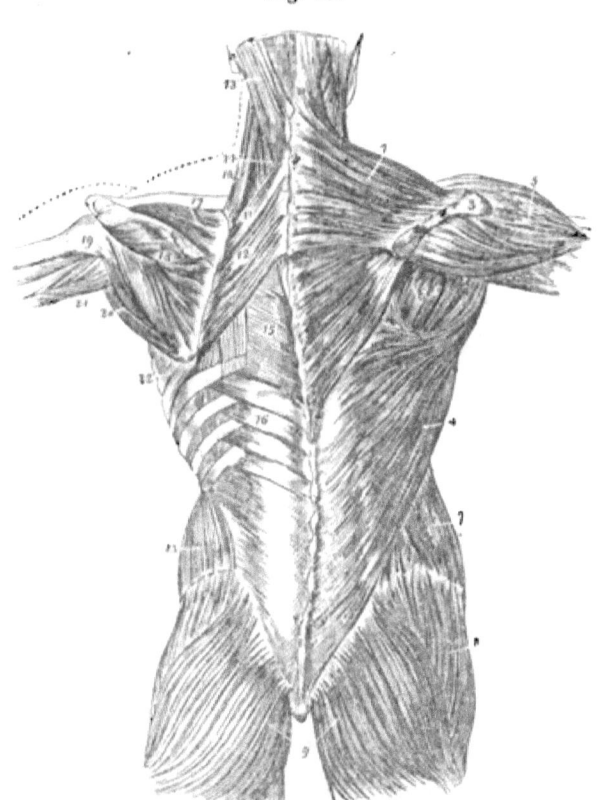

SECOND LAYER OF MUSCLES OF THE BACK.—1. Trapezius. 2. A portion of the tendinous ellipse formed by the trapezius on both sides. 3. Spine of scapula. 4. Latissimus dorsi. 5. Deltoid. 6. Infra-spinatus and teres minor. 7. External oblique. 8. Gluteus medius. 9. Gluteus magnus. 10. Levator scapulæ. 11. Rhomboideus minor. 12. Rhomboideus major. 13. Splenius capitis. 14. Splenius colli. 15. Portion of origin of latissimus dorsi. 16. Serratus inferior posticus. 17. Supra-spinatus. 18. Infra-spinatus. 19. Teres minor. 20. Teres major. 21. Long head of triceps extensor cubiti. 22. Serratus major anticus. 23. Internal oblique.

transverse processes of all the lumbar vertebræ. *Course*, upwards; filling the space between the spine and the angles of the ribs.

Insertion, into the transverse processes of all the dorsal vertebræ, and the lower edge of each rib, except the last two, near their tubercles. *Action*, to erect, or keep erect, the spinal column.

Sacro-lumbalis.—*Origin*, in common with the longissimus dorsi; and, also, by the *musculi accessorii*, from the last six or eight ribs, *Course*, obliquely upwards. *Insertion*, into the angles of all the ribs. *Action*, with the last two, to raise the trunk and to keep it erect. The three muscles are sometimes together called the erector spinæ.

Levator ani.—*Origin*, pubes near the symphysis and arch, and upper margin of thyroid foramen; spine of the ischium, and connected aponeurosis. *Course*, by converging fibres, downwards and inwards. *Insertion*, last two bones of the coccyx, semi-circumference of the rectum just above the sphincter ani, and the side of the prostate gland and membranous part of the urethra. *Action*, to form the floor of the pelvis, to dilate the anal orifice, and to retract the bowel after defecation.

Coccygeus.—*Origin*, spine of the ischium. *Course*, gradually expanding over the inside of the posterior sacro-ischiatic ligament, inwards and backwards. *Insertion*, into the whole length of the side of the os coccygis. *Action*, to draw forwards the coccyx, and to aid the levator ani in forming a floor to the pelvis.

Sphincter ani.—*Origin*, point of the os coccygis. *Course*, forwards and around the anus just beneath the skin. *Action*, to close the anus.

Cremaster.—*Origin*, by an *outer* and *inner* fasciculus: the former, from Poupart's ligament; the latter, from the spine of the pubes. *Course*, downwards, over the testis. *Insertion*, into the tunica vaginalis testis and scrotum. *Action*, to draw up the testis. The lower portion of this muscle is generally pale and indistinct.

Erector penis.—*Origin*, tuberosity of the ischium. *Course*, upwards, to surround the crus penis. *Insertion*, into the membrane of the corpus cavernosum. *Action*, to compress the corpus cavernosum and detain blood in it during erection.

Accelerator urinæ.—*Origin*, ramus of the pubes and crus penis. *Course*, downwards; broad and thin in form. *Insertion*, into the anterior median line of the bulb of the urethra, and also into the anterior margin of the sphincter ani. *Action*, to propel urine or semen into and along the urethra.

Transversus perinei.—*Origin*, tuberosity of the *ischium*. *Course*, across the perineum. *Insertion*, into the anterior margin of the sphincter ani. *Action*, to dilate the bulbous portion of the urethra; or, to hold it in its position.

Erector clitoridis, of the female.—*Origin* and *course* as in the *erector penis* of the male; *insertion*, into the clitoris; *action*, to assist in its erection.

Sphincter vaginæ, in the female.—*Origin*, anterior margin of the sphincter ani, and neighboring connective tissue. *Course*, around the orifice of the vagina. *Insertion*, into the clitoris, meeting its fellow of the other side. *Action*, to contract the external orifice of the vagina.

A further account of the surgical anatomy of the *perineum* will be given hereafter.

MUSCLES OF THE SHOULDER.

Supra-spinatus.—*Origin*, the whole fossa above the spine of the scapula, and the spine itself. *Course*, with a strong tendon, under the acromion process. *Insertion*, into the great tubercle of the humerus. *Action*, to raise the arm and turn it outwards.

Infra-spinatus.—*Origin*, spine of scapula and fossa infraspinata. *Course*, with a strong tendon, under the acromion process. *Insertion*, into the great tubercle of the humerus, middle face. *Action*, to roll the humerus outwards and backwards, and to sustain it when raised.

Teres major.—*Origin*, inferior angles of scapula. *Course*, upwards and outwards, with the latissimus dorsi. *Insertion*, by a broad tendon, into the ridge at the inner margin of the bicipital groove of the humerus. *Action*, to draw the arm downwards and backwards, and to roll it inwards.

Subscapularis.—*Origin*, base and under surface of the scapula. *Course*, upwards and outwards, its fibres converging. *Insertion*, into the lesser tubercle of the head of the os humeri. *Action*, to draw down the arm and roll it inwards.

Deltoides.—*Origin*, outer third of the clavicle, acromion process, and inferior edge of spine of scapula opposite to the trapezius muscle. *Course*, converging to make a covering for the shoulder, and triangularly down upon the outside of the arm to near its middle. *Insertion*, into a rough surface near the centre of the humerus. *Action*, to raise the arm, and move it either forwards or backwards according to the fibres used.

MUSCLES OF THE ARM.

Coraco-brachialis.—*Origin*, coracoid process of the scapula. *Course*, downwards. *Insertion*, inner side of humerus near its middle. *Action*, to draw the arm upwards and forwards.

Biceps flexor cubiti.—*Origin*, by two heads: the longer, by a round slender tendon, from the upper margin of the glenoid cavity of the shoulder-joint; the shorter, from the coracoid process of the scapula. *Course*, the two heads uniting into a thick and long muscle, downwards upon the front of the humerus. *Insertion*, into the tubercle at the upper and anterior part of the radius. *Action*, to bend the forearm upon the arm.

Brachialis anticus.—*Origin*, from the middle of the front part of the humerus, on each side of the insertion of the deltoid. *Course*, downwards. *Insertion*, into a depression at the base of the coronoid process of the ulna. *Action*, to flex the forearm upon the arm.

Triceps extensor cubiti.—*Origin*, by three heads. The longest comes from the scapula near the glenoid cavity. The second head, from the back part of the upper end of the humerus. The third, from the inner side of the humerus near the insertion of the teres major. *Course*, the three heads uniting above the middle of the humerus, downwards upon the back of the arm. *Insertion*, into the olecranon process, ridge of the ulna, and condyles of the humerus. *Action*, to extend the forearm.

Fig. 69.

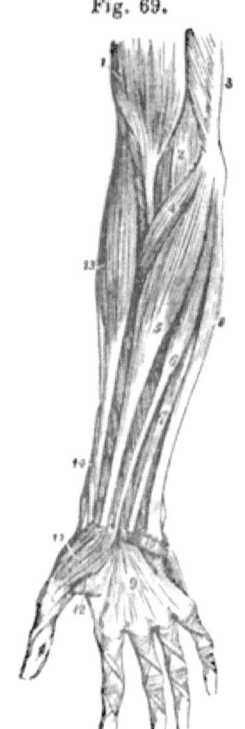

OUTER LAYER OF MUSCLES ON THE FRONT OF THE FOREARM.—1. Biceps flexor cubiti. 2. Brachialis internus. 3 Triceps. 4. Pronator radii teres. 5. Flexor carpi radialis. 6. Palmaris longus. 7. Flexor sublimis digitorum. 8. Flexor carpi ulnaris. 9. Palmar fascia. 10 Palmaris brevis muscle. 11. Abductor pollicis manus. 12. Flexor brevis pollicis manus. 13. Supinator longus. 14. Extensor ossis metacarpi pollicis.

MUSCLES OF THE FOREARM.

Anterior, Superficial Layer.

Pronator radii teres—*Origin*, internal condyle of os humeri, and coronoid process of ulna. *Course*, obliquely across the forearm. *Insertion*, posterior part of the middle of the radius. *Action*, to roll the radius inwards, and *pronate* the hand or turn the palm backwards or downwards.

Flexor carpi radialis.—*Origin*, inner condyle of humerus, and upper front part of the ulna, between the pronator radii teres and the flexor digitorum sublimis. *Course*, downwards along the radius, ending in a long tendon which goes over the trapezium under the annular ligament of the wrist. *Insertion*, metacarpal bone of the forefinger, in front of its upper end. *Action*, to bend the hand at the wrist.

Palmaris longus.—*Origin*, inner condyle of humerus. *Course*, soon becoming tendinous, downwards. *Insertion*, annular ligaments of the wrist, and palmar aponeurosis. *Action*, to aid in bending the hand, or to make tense the tegument of the palm.

Flexor carpi ulnaris.—*Origin*, inner condyle of humerus, olecranon process, and inner edge of ulna to within three or four inches of the wrist. *Course*, downwards. *Insertion*, pisiform bone, and base of metacarpal of little finger. *Action*, to bend the hand towards the ulna.

Flexor digitorum sublimis.—*Origin*, inner condyle of humerus, coronoid process of ulna, and upper front of radius. *Course*, downwards; dividing above the wrist into four bellies, each of which sends off a tendon; all the tendons pass under the annular ligament. *Insertions;* each tendon is attached to the second phalanx of a finger; being first *perforated* by the tendon of the flexor profundus. *Action*, to bend the finger at the second phalanx, and the hand on the forearm.

Deep-Seated Anterior Layer.

Flexor digitorum profundus.—*Origin*, upper and outer part of ulna, coronoid process, interosseous ligament, and half way down the ulna. *Course*, beneath the sublimis downwards; also, dividing above the wrist and giving off four tendons. *Insertions*, into the third phalanges of the fingers, perforating first the tendons of the sublimis.

Flexor longus pollicis.—*Origin*, front of radius below its tubercle, middle two-thirds of the same bone, and also from the inner condyle of the humerus. *Course*, downwards. *Insertion*, base of second phalanx of the thumb. *Action*, to bend the last joint of the thumb, and aid in bending the hand.

Pronator quadratus.—*Origin*, from a ridge on the inner and under part of the ulna. *Course*, in quadrangular form, two inches wide, across the forearm. *Insertion*, into the front of the radius. *Action*, to pronate the forearm and hand.

Posterior, Superficial Layer.

Extensor digitorum communis.—*Origin*, outer condyle of humerus, and contiguous fascia. *Course*, downwards, upon the back of the forearm, dividing above the wrist into four tendons, which pass under the annular ligament in a groove of the radius. *Insertions*, into the whole length of the posterior faces of the fingers. *Action*, to extend the fingers.

Extensor carpi ulnaris.—*Origin*, external condyle of humerus, middle of ulna, and fascia. *Course*, downwards, ending in a round tendon which passes through a groove on the back of the ulna. *Insertion*, into the base of the metacarpal of the little finger. *Action*, to extend the hand at the wrist.

Extensor minimi digiti.—*Origin*, in common with the extensor communis digitorum. *Course*, downwards, its tendon going through a separate ring of the annular ligament. *Insertion*, with that of the tendon of the extensor communis, into the back of the little finger. *Action*, to extend the little finger.

154 ANATOMY.

Fig. 70.

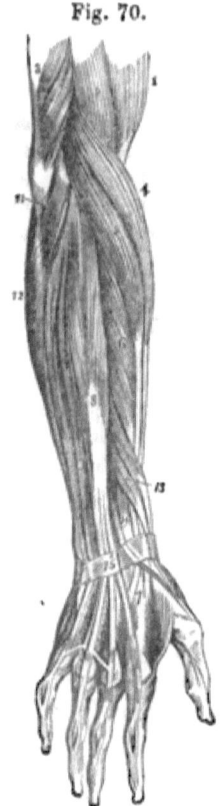

OUTER LAYER OF MUSCLES ON THE BACK OF THE FOREARM.—1. Biceps Flexor. 2. Brachialis Internus. 3. Triceps extensor. 4. Supinator radii longus. 5. Extensor carpi radialis longior. 6. Extensor carpi radialis brevior. 7. Tendinous insertions of these muscles. 8. Extensor communis digitorum. 9. Extensor communis digitorum. 10. Extensor carpi ulnaris. 11. Anconeus. 12. Flexor carpi ulnaris. 13. Extensor minor pollicis. 14. Extensor major pollicis. 15. Posterior annular ligament.

Anconeus.—*Origin*, back of the outer condyle of the humerus. *Course*, upper fibres, horizontally, lower ones obliquely, across the back of the forearm near the elbow. *Insertion*, into the olecranon and adjacent part of the ulna. *Action*, to aid in extending the forearm or to rotate it inwards.

Extensor carpi radialis longior.—*Origin*, humerus, just above the outer condyle. *Course*, downwards, along the back of the radius, and, by a long flat tendon, under the annular ligament. *Insertion*, metacarpal of the forefinger. *Action*, to extend or draw backward the wrist and hand.

Supinator radii longus.—*Origin*, the ridge of the humerus, above the outer condyle almost to the middle of the bone. *Course*, downwards, upon the outside of the radius. *Insertion*, into the radius, above its styloid process. *Action*, to roll out the radius, and turn the palm of the hand forwards.

Extensor carpi radialis brevior.—*Origin*, outer condyle of humerus. *Course*, downwards along the back of the radius. *Insertion*, metacarpal of the middle or second finger. *Action*, to extend the hand and wrist.

Posterior Deep-Seated Layer.

Supinator radii brevis.—*Origin*, external condyle of humerus, upper and outer part of ulnar and interosseous ligament. *Course*, obliquely downwards, over the outer edge of the radius. *Insertion*, upper and outer part of radius, and its tubercle. *Action*, to supinate the foramen and hand.

Extensor major pollicis.—*Origin*, back of ulna above its middle, interosseous ligament, and adjacent part of radius. *Course*, downwards on the back of the radius. *Insertion*, second phalanx of the thumb. *Action*, to extend the last bone of the thumb.

Extensor minor pollicis.—*Origin*, back of ulna below its middle, and interosseous ligament. *Course*, downwards on the back of the radius. *Insertion*, first phalanx of the thumb. *Action*, to extend the first bone of the thumb.

Extensor ossis metacarpi pollicis.—*Origin*, posterior surface of the middle of the ulna, interosseous ligament and radius. *Course*, downwards, its tendon passing through a groove of the radius. *Insertion*, into the trapezium, and the metacarpal bone of the thumb. *Action*, to extend the metacarpal, and with it the whole thumb.

Indicator.—*Origin*, back of ulna near its middle, and interosseous ligament. *Course*, downwards. *Insertion*, with tendon of extensor communis, into the whole back of the forefinger. *Action*, to extend the forefinger, as in pointing.

MUSCLES OF THE HAND.

Abductor pollicis.—*Origin*, trapezium, trapezoid, and annular ligament. *Insertion*, base of first phalanx of thumb. *Action*, to draw the thumb away from the forefinger.

Opponens pollicis.—*Origin*, point of the trapezium, and the annular ligament. *Insertion*, radial side of metacarpal of the thumb. *Action*, to draw the metacarpal and thumb towards the palm of the hand.

Flexor brevis pollicis.—*Origin*, by two heads; one, from the trapezium and trapezoid; the other, from the magnum, unciform, and metacarpal of the middle finger. Between the two passes the tendon of the flexor longus pollicis. *Insertions*, into the two sesamoid bones of the first joint of the thumb. *Action*, to flex the first phalanx of the thumb.

Adductor pollicis.—*Origin*, ulnar side of second metacarpal bone; its fibres converging thence towards the insertion. *Insertion*, base of first phalanx of thumb. *Action*, to draw the thumb towards the forefinger.

Palmaris brevis—*Origin*, annular ligament of wrist in front, and palmar aponeurosis. *Insertion*, into the tegument on the inner side of the hand. *Action*, to draw up the skin on the palm.

Lumbricales.—*Origin* (four in number), each from the outside of a tendon of the flexor profundus digitorum. *Insertion*, with the tendons of the extensor communis, into the middle of the backs of the first phalanges of the fingers. *Action*, to hold in place the flexor and extensor tendons, and to flex the fingers.

Abductor minimi digiti.—*Origin*, pisiform bone and annular ligament. *Insertion*, inner side of first phalanx of little finger. *Action*, to remove the little finger from the next.

Adductor metacarpi minimi digiti.—*Origin*, hook of the unciform bone. *Insertion*, metacarpal of little finger. *Action*, to flex the little finger, and approximate it to the rest.

Flexor parvus minimi digiti.—*Origin*, hook of the unciform.

Insertion, first phalanx of little finger. *Action*, to bend that finger.

Interossei.—These are seven in number; *three palmar* and *four dorsal*. They arise from the sides of the metacarpal bones, and are inserted, with the lumbricales, into the sides of the first phalanges. They all act either as adductors or abductors of the fingers according to their position.

MUSCLES OF THE PELVIS AND THIGH.

Gluteus maximus.—*Origin*, from the posterior part of the crest of the ilium, and the dorsum near it; the side of the sacrum and coccyx; and the posterior sacro-sciatic ligament. *Course*, forwards and somewhat downwards, going over the great trochanter of the femur. *Insertion*, into the upper third of the linea aspera of the femur. *Action*, to draw the thigh backwards; or to maintain the balance of the body upon the lower extremity.

Gluteus medius.—*Origin*, anterior two-thirds of crest of ilium, and dorsum ilii between the crest and the semi-circular ridge. *Course*, under the gluteus maximus, converging, to a broad strong tendon. *Insertion*, into the great trochanter and adjacent part of the shaft of the femur. *Action*, to draw the thigh backwards and outwards.

Gluteus minimus.—*Origin*, dorsum ilii between the semi-circular ridge and the hip-joint. *Course*, downwards, converging under the gluteus medius. *Insertion*, into the great trochanter. *Action*, to draw the thigh outwards and rotate it.

Pyriformis.—*Origin*, anterior face of the sacrum. *Course*, in a conical form, leaving the pelvis through the sacro-sciatic foramen. *Insertion*, into the great trochanter. *Action*, to rotate the thigh outwards.

Gemelli.—*Origin*, by two heads; one from the spine and the other from the tuberosity of the ischium. The tendon of the obturator goes between the two. *Course*, forwards and outwards. *Insertion*, into the great trochanter. *Action*, to rotate the thigh outwards.

Obturator internus.—*Origin*, from the margin of the thyroid foramen, except where the vessels pass through it. *Course*, converging to a round tendon which goes over the spine and tuberosity of the ischium as a pulley. *Insertion*, into the great trochanter. *Action*, to rotate the thigh outwards.

Obturator externus.—*Origin*, outer margin of the thyroid foramen. *Course*, converging, outwards. *Insertion*, great trochanter. *Action*, to rotate the thigh outwards.

Quadratus femoris.—*Origin*, tuberosity of the ischium. *Course*, transversely outwards. *Insertion*, the ridge between the two trochanters. *Action*, to rotate the thigh outwards.

MUSCLES OF THE PELVIS AND THIGH. 157

Fig. 71.

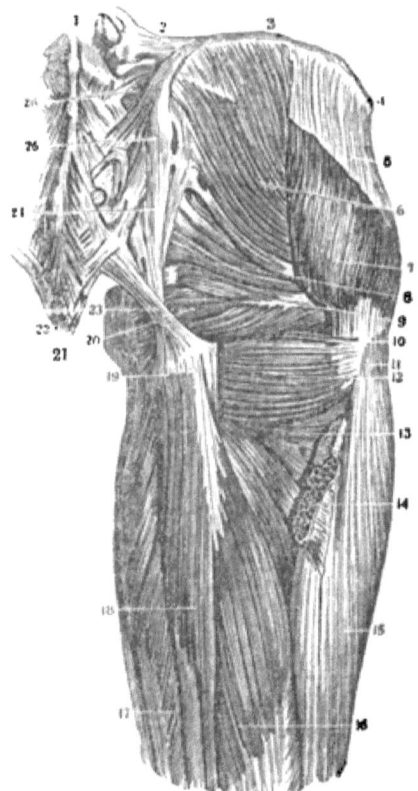

DEEP-SEATED MUSCLES ON THE POSTERIOR PART OF THE HIP-JOINT.—1. Fifth lumbar vertebra. 2. Ilio-lumbar ligament. 3. Crest of the ilium. 4. Anterior superior spinous process. 5. Origin of the fascia femoris. 6. Gluteus medius. 7. Its lower and anterior portion. 8. Pyriformis. 9. Gemini. 10. Trochanter major. 11. Insertion of the gluteus medius. 12. Quadratus femoris. 13. Part of the adductor magnus. 14. Insertion of the gluteus magnus. 15. Vastus externus. 16 Long head of the biceps. 17. Semi-membranosus. 18 Semi-tendinosus. 19. Tuber ischii. 20. Obturator internus. 21. Point of the coccyx. 22. Posterior coccygeal ligament. 23, 24. Greater sacro-sciatic ligament. 25. Posterior superior spinous process of ilium. 26. Posterior sacro-iliac ligaments.

Sartorius.—*Origin*, anterior superior spinous process of the ilium. *Course*, downwards and inwards, crossing the length of the thigh; it being the longest muscle in the body. *Insertion*, inner side of the tubercle at the head of the tibia. *Action*, to bend the leg and draw it across the other.

Tensor vaginæ femoris.—*Origin*, anterior superior spinous process of the ilium. *Course*, downwards and a little backwards.

14

Insertion, into the fascia femoris on the outside of the thigh. *Action*, to make the fascia tense, and rotate the thigh inwards.

Rectus femoris.—*Origin*, anterior inferior spine of ilium, and from the same bone above the acetabulum. *Course*, downwards over the front of the thigh. *Insertion*, into the upper edge of the patella. *Action*, by the ligamentum patellæ, to straighten or extend the leg.

Crureus.—*Origin*, front of femur and its sides to the linea aspera. *Course*, downwards upon the femur. *Insertion*, into the patella. *Action*, to extend the leg.

Vastus externus.—*Origin*, outer part of femur below the trochanter major, and the whole length of the linea aspera. *Course*, downwards on the outside of the thigh. *Insertion*, the patella. *Action*, to extend the leg.

Vastus internus.—*Origin*, front of femur and whole length of linea aspera. *Course*, downwards, on the inside of the thigh. *Insertion*, the patella. *Action*, to extend the leg.

The last four muscles are sometimes called, together, the *quadriceps femoris*.

Pectineus.—*Origin*, the upper part of the pubes, between the linea ilio-pectinea and the ridge above the thyroid foramen. *Course*, obliquely, in flattened form, downwards and outwards. *Insertion*, into the linea aspera, below the lesser trochanter. *Action*, to draw the thigh inwards and upwards, and to rotate it outwards.

Adductor longus.—*Origin*, pubes near the symphysis. *Course*, downwards and outwards. *Insertion*, middle third of linea aspera. *Action*, to draw the thigh inwards and upwards.

Adductor brevis.—*Origin*, pubes, below the last named. *Course*, downwards and outwards. *Insertion*, upper third of linea aspera. *Action*, as the adductor longus.

Adductor magnus.—*Origin*, body and ramus of pubes, and ramus of ischium. *Course*, downwards and outwards. *Insertion*, whole length of linea aspera, and internal condyle. *Action*, the same as the last two muscles.

Gracilis.—*Origin*, pubes near the symphysis. *Course*, downwards and outwards.—*Insertion*, the tubercle of the tibia. *Action*, to flex the leg, and adduct the thigh.

Semi-tendinosus.—*Origin*, tuberosity of ischium. *Course*, downwards; becoming tendinous about four inches above the knee. *Insertion*, by a round tendon, into the inner side of the tibia, below its tubercle. *Action*, to bend the leg on the thigh.

Semi-membranosus.—*Origin*, tuberosity of the ischium. *Course*, downwards. *Insertion*, the inner and back part of the head of the tibia. *Action*, to bend the leg.

The three last-named muscles constitute the inner ham-string.

Biceps flexor cruris.—*Origin*, by two heads; one, from the tuberosity of the ischium; the other, from the linea aspera, high up. *Course*, downwards. *Insertion*, into the head of the fibula. *Action*, to bend the leg.

This muscle forms the outer hamstring.

MUSCLES OF THE LEG.

Tibialis anticus.—*Origin*, head of the tibia, and upper half of the interosseous ligament. *Course*, downwards upon the outer face of the tibia, sending its tendon over the astragalus in front of the internal malleolus. *Insertion*, the front of the internal cuneiform bone on the sole of the foot. *Action*, to raise the foot towards the leg, and turn the sole inwards.

Extensor longus digitorum.—*Origin*, head of the tibia, interosseous ligament, and head, and nearly the whole length of the fibula. *Course*, downwards, giving off four tendons which pass under the annular ligament. *Insertion*, each tendon into nearly the whole length of one of the toes, leaving out the great toe. *Action*, to extend the toes.

Extensor proprius pollicis pedis.—*Origin*, from the fibula, beginning three or four inches below its head. *Course*, downwards, its tendon going under the annular ligament. *Insertion*, into the whole length of the great toe. *Action*, to extend the great toe.

The last two muscles will also aid in raising the foot towards the front of the leg.

Peroneus longus.—*Origin*, head and shaft of the fibula to within three or four inches of the ankle. *Course*, downwards, its tendon passing through a groove in the external malleolus, the sinuosity of the os calcis, and a groove of the cuboid bone, to the middle of the sole. *Insertion*, outside of the base of the first metatarsal, and internal cuneiform bone. *Action*, to depress the foot and incline the sole outwards.

Peroneus brevis.—*Origin*, outside of fibula from just above its middle to the external malleolus. *Course*, downwards, through the same groove of the malleolus with the peroneus longus, and

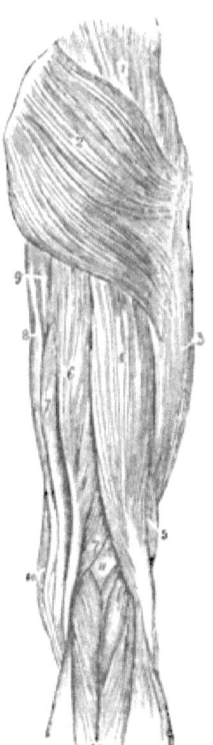

Fig. 72.

MUSCLES OF THE BACK OF THE THIGH.—1. Gluteus medius. 2. Gluteus magnus. 3. Fascia lata. 4, Long head of biceps. 5. Short head of biceps. 6. Semi-tendinosus. 7, 7. Semi-membranosus. 8. Gracilis. 9. Adductor magnus. 10. Sartorius. 11. Popliteal space. 12. Gastrocnemius.

Fig. 73.

MUSCLES OF THE FRONT OF THE LEG.—1. Tendon of quadriceps. 2. Spine of tibia. 3. Tibialis anticus. 4. Extensor communis digitorum. 5. Extensor proprius pollicis. 6. Peroneus tertius. 7. Peroneus longus. 8. Peroneus brevis. 9, 9. Soleus. 10. Gastrocnemius. 11. Extensor brevis digitorum.

through a fossa on the outer surface of the os calcis. *Insertion*, base of metatarsal of the little toe. *Action*, same as peroneus longus.

Peroneus tertius.—*Origin*, middle of the fibula. *Course*, downwards to the outer malleolus, sending a tendon under the annular ligament. *Insertion*, base of metatarsal of the little toe. *Action*, to raise the foot towards the front of the leg.

Gastrocnemius.—*Origin*, by two heads; one from the inner condyle of the femur and ridge leading to the linea aspera; the other from the outer condyle and adjacent ridge. *Course*, as a double-bellied muscle, downwards, forming the outer part of the calf of the leg. *Insertion*, with the next muscle, by the tendo Achillis, into the os calcis, behind and below. *Action*, to raise the heel, and thus depress the foot; this action is called, in anatomy, the *extension* of the foot.

Soleus.—*Origin*, beneath the last named, by two heads; one from the head and upper part of the fibula, the other from the back of the tibia, for some inches, below the popliteus muscle. *Course*, downwards to the Achillis tendon. *Insertion*, with the gastrocnemius, into the os calcis.

Plantaris.—*Origin*, external condyle of femur, and capsular ligament of the knee. *Course*, soon becoming tendinous and slender, downwards. *Insertion*, os calcis, below the tendo Achillis. *Action*, to depress or extend the foot. It is a feeble muscle, and sometimes absent.

Popliteus.—*Origin*, external condyle of femur, and capsular ligament of the knee. *Course*, inwards and downwards behind the knee. *Insertion*, a ridge at the inner and upper part of the tibia just below its head. *Action*, to bend the leg slightly, and rotate it inwards, and to draw tense the capsular ligament.

Flexor longus pollicis pedis.—*Origin*, back of the tibia, from about three inches below the head almost to the ankle. *Course*, downwards, through a groove in the back of the tibia and of the astragalus. *Insertion*, into the last phalanx of the great toe. *Action*, to flex the great toe.

Flexor longus digitorum pedis.—*Origin*, back and inside of tibia from below the popliteus almost to the ankle; also from the outer edge of the tibia above the ankle. *Course*, downwards, in contact with the tibialis posticus; its tendon passing behind the inner malleolus, and through the sinuosity of the os calcis to the middle of the sole of the foot. There it receives a slip from the flexor longus pollicis, and divides into four tendons, which perforate those of the flexor brevis digitorum. *Insertion*, each tendon into the last phalanx of one of the four lesser toes. *Action*, to flex the toes and depress or extend the foot.

Tibialis posticus.—*Origin*, upper front of tibia, and, going through the interosseous ligament, also from the back of the tibia and fibula most of their length. *Course*, downwards, the tendon going through a groove of the inner malleolus. *Insertion*, inner face of os naviculare, and under surface of the tarsus, one slip reaching to the middle metatarsal bone. *Action*, to extend or depress the foot and turn the toes inwards.

Extensor brevis digitorum pedis.—*Origin*, front and outer part of the os calcis. *Insertion*, by four tendons, into all the toes but the last, under the insertions of the extensor longus digitorum. *Action*, to extend the toes.

Flexor brevis digitorum pedis.—*Origin*, great tuberosity of os calcis, and plantar aponeurosis. *Insertion*, by four tendons, perforated by those of the flexor longus, into the second phalanges of the four lesser toes. *Action*, to flex the toes.

The remaining muscles of the foot are small, and similar to the corresponding ones in the hand; so that their description may be here omitted. Their names are as follows: four *dorsal* and three *plantar interossei*, abductors and adductors of the toes; *abductor pollicis pedis*, *abductor minimi digiti*, *flexor accessorius*, four *lumbricales pedis*, *flexor brevis pollicis*, *adductor pollicis pedis*, *flexor brevis minimi digiti*, *transversalis pedis*.

CHAPTER IX.
NERVOUS SYSTEM.

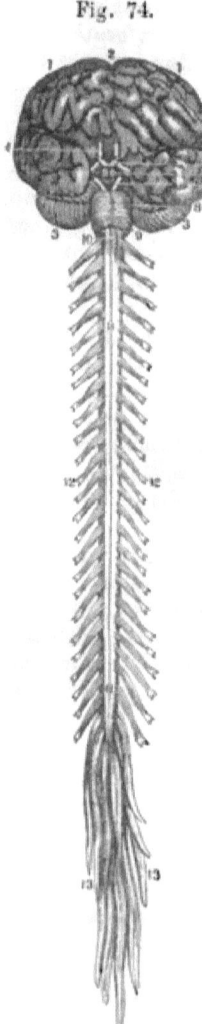

Fig. 74.

Portions —Cerebro-spinal axis, ganglia, and nerves.

Minute Structure.—Two sorts of nervous tissue exist: the *white* (fibrous or) *tubular*, and the (cineritious or) *gray, vesicular*. The former is seen in the proper tubular fibres, prevailing in the cerebro-spinal system, and the gelatinous fibres, most common in the ganglionic system. In the *tubular*, under the microscope, the nerve is seen to consist of the central transparent *axis cylinder*, and the peripheral *white substance of Schwann*. In the *gelatinous* fibres, the white substance is almost absent. The diameter of the tubular fibres is about $\frac{1}{25000}$ of an inch. The gelatinous are less than half as large.

Vesicular nerve-substance is found in the brain and ganglia. It is formed of cells, each with a nucleus or central vesicle, and within that a clear nucleolus. Some are small and round or oval; others larger, and caudated or stellated, the processes sometimes dividing into minute branches.

Chemical Composition.—Nervous tissue contains albumen, or an albuminoid material, with fatty matter (cerebric and oleophosphoric acids, cholesterin, olein, and margarin) and salts (phosphates and lactates).

ANTERIOR VIEW OF THE BRAIN AND SPINAL MARROW.—1, 1. Hemispheres of the cerebrum. 2. Great middle fissure. 3. Cerebellum. 4. Olfactory Nerves. 5. Optic nerves. 6. Corpora albicantia. 7. Motor oculi nerves. 8. Pons Varolii. 9. Fourth pair of nerves. 10. Lower portion of the medulla oblongata. 11, 11. Medulla spinalis in its whole length. 12, 12. Spinal nerves. 13. Cauda equina.

Connections and Terminations.—Doubt yet exists as to these. Sometimes, at the centres or ganglia, a nerve-tube seems to dilate and receive a nerve-cell or corpuscle within it. The processes of caudate vesicles are described as extending into nerves. Whether nerves ever terminate, peripherally, by free ends, or always by loops or meshes, is undecided. Beale insists that every nerve filament makes part of a completed *circuit*.

Nerves are round or flattened cords, each containing a number of filaments or tubules inclosed in a sheath (neurilemma), and connecting a nerve-centre with some other part. In their course, they branch frequently, and sometimes form *plexuses;* but no two filaments ever truly unite or inosculate.

THE BRAIN.

Membranes.

These are, the *dura mater, arachnoid,* and *pia mater.*

The dura mater is a thick fibrous membrane, with a smooth epithelial lining. It adheres to the skull, especially at its base and along the sutures; and is continuous with the dura mater of the spinal cord, and with the sheath of the optic and other cephalic nerves. Three *processes* pass inwards from it; the falx cerebri, falx cerebelli, and the tentorium. The *falx cerebri* descends vertically between the hemispheres of the brain. In front, it connects with the crista galli of the ethmoid bone; behind, it widens, and joins the tentorium. Above, it is broad, containing the longitudinal sinus. In its lower curved edge is the inferior longitudinal sinus.

The *falx cerebelli* is a smaller triangular process, between the two lobes of the cerebellum. It passes from the under and posterior part of the tentorium to the occiput.

The *tentorium* is an arched layer of dura mater covering the cerebellum, beneath the posterior lobes of the cerebrum. It is connected behind with the occiput, at the sides with the temporal bones, and on its middle line above, with the edge of the falx cerebri. The anterior border is free and concave, with a large oval passage for the crura cerebri.

The *arteries* of the dura mater are, principally, the anterior meningeal arteries, from the ethmoidal and internal carotid; middle and small meningeal, from the internal maxillary; the posterior meningeal from the occipital, and the posterior meningeal from the vertebral.

Its *veins*, which, like its arteries, are also those of the contiguous bones, anastomose with the diploic veins, terminating in the sinuses, with two minor exceptions which attend the middle meningeal artery.

The *nerves* of the dura mater are the recurrent of the fourth, and filaments from the ophthalmic, ganglion of Casser, and sympathetic.

Glandulæ Pacchioni are small whitish granulations found on both the outer and inner surfaces of the dura mater, near the superior longitudinal sinus, and on the pia mater of the same region. They are fibro-cellular in structure; absent in infancy, they increase gradually in number after the seventh year; but are sometimes wanting.

Arachnoid Membrane.

This is the middle serous membrane of the brain, described by most anatomists as double, one layer lining the dura mater, and the other investing the brain. It is very thin; thickest at the base of the hemispheres. It does not descend between the convolutions, but passes over them. The *sub-arachnoid space* is between the arachnoid and the pia mater. It contains the serous *cerebrospinal fluid*.

Pia Mater.

A fine but extended plexus of bloodvessels, held together by delicate connective tissue, investing the whole brain, and dipping between the convolutions, receives this name. The pia mater is extended into the interior of the cerebrum, making the velum interpositum and the choroid plexuses of the fourth ventricle. Some long straight vessels pass from it through the white substance.

Brain or Encephalon.

We divide this into the *cerebrum, cerebellum, medulla oblongata,* and *pons Varolii*. The average weight of the whole mass in the adult male is nearly fifty ounces; in the female, less than forty-five ounces. The maximum is about sixty-five ounces. Up to nearly forty years of age, in both sexes, it increases; after that, it loses about an ounce of weight with each ten years of age.

Cerebrum.

As no description can enable the student to understand the anatomy of the brain without repeated *dissections*, we shall attempt but a very brief statement—especially of the terms applied to its parts.

The cerebrum is an ovoidal mass, divided into the right and left *hemispheres;* which are partly separated by the great longitudinal fissure. The surface of each hemisphere, under the pia mater, is marked by *convolutions;* these being different in different brains, and even upon the two hemispheres in the same subject. Gray vesicular nerve-substance predominates in the convolutions, although alternating thin layers of white substance exist.

CEREBRUM.

Fig. 75.

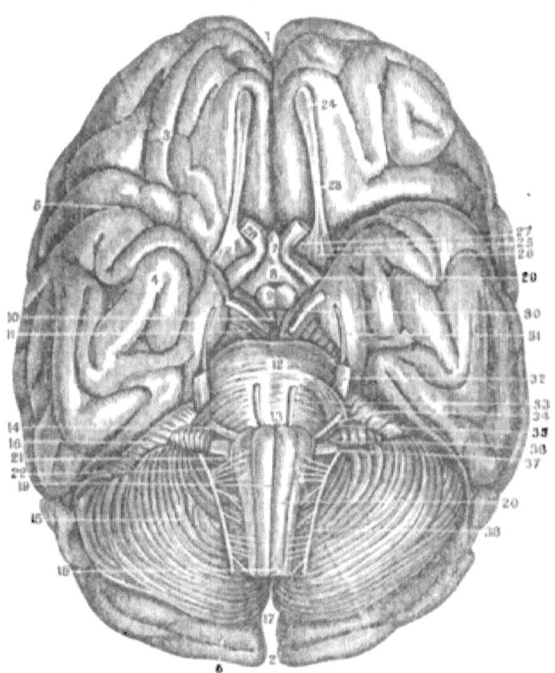

BASE OF THE CEREBRUM AND CEREBELLUM.—1. Fissure of the hemispheres 2. Posterior extremity of the same fissure. 3. Anterior lobes of the cerebrum. 4. Its middle lobe. 5. Fissure of Sylvius. 6. Posterior lobe of the cerebrum. 7. Infundibulum. 8. Its body. 9. Corpora albicantia. 10. Cineritious matter. 11. Crura cerebri. 12. Pons Varolii. 13. Medulla oblongata. 14. Posterior prolongation of the pons Varolii. 15. Middle of the cerebellum. 16. Anterior part of the cerebellum. 17. Its posterior part and fissure. 18. Medulla spinalis. 19. Middle fissure of the medulla oblongata. 20. Corpus pyramidale. 21. Corpus restiforme. 22. Corpus olivare. 23. Olfactory nerve. 24. Its bulb. 25. Its external root. 26. Its middle root. 27. Its internal root. 28. Optic nerve beyond the chiasm. 29. Optic nerve before the chiasm. 30. Third pair of nerves. 31. Fourth pair. 32. Fifth pair. 33. Sixth pair. 34. Facial nerve. 35. Auditory. 36, 37, 38. Eighth pair of nerves.

The base of each hemisphere presents a division into the *anterior, middle,* and *posterior lobes.* The *fissure of Sylvius,* on each side, separates the middle from the anterior lobe. The posterior lobe rests upon the tentorium.

The fissure of Sylvius lodges the middle cerebral artery. The *island of Reil* is the name given to some convolutions inclosed within the sides of the fissure.

Laying the brain over to examine its basal surface, the order of location of parts is as follows: from before backwards, longitudinal fissure; on each side of this, bulb and trunk of *olfactory nerve; pituitary body*, resting upon the sella Turcica of the sphenoid bone; attached to the body, the *infundibulum;* connected with this, also, back of the *chiasm* or union of the optic nerves, the *tuber cinereum; corpora albicantia*, or eminentia mamillares; *crura cerebri; pons Varolii; medulla oblongata;* with the lobes of the *cerebellum* at its sides.

The *anterior perforated space*, or locus quadratus, is at the inner end of the fissure of Sylvius, at the entrance of the branches of the olfactory nerve.

The *tuber cinereum* is a small prominence of gray nerve substance, between the optic commissure and the corpora albicantia; it forms part of the floor of the third ventricle.

The *pituitary body* is a small reddish, vascular, oval mass; having two lobes; it is proportionally larger in the fœtus than in the adult. It has a cavity, leading through the infundibulum to the third ventricle. Its structure resembles that of the ductless glands.

The *corpora albicantia* are two small round bodies, of the size of peas, just back of the tuber cinereum.

The *pons Tarini* or *posterior perforated space*, lies back of the corpora albicantia. Minute bloodvessels pass through it.

The *crura cerebri* are bundles of white nerve-substance diverging from the pons Varolii into the hemispheres, and widening as they

Fig. 76.

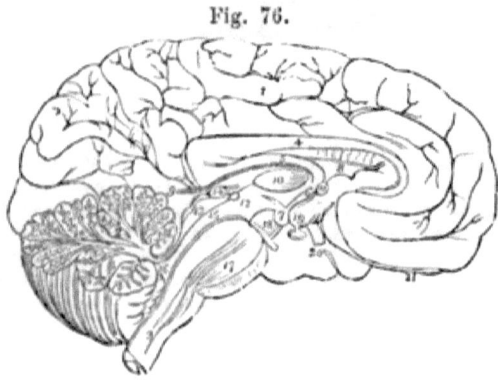

LONGITUDINAL SECTION OF THE BRAIN.—1. Left hemisphere. 2. Cerebellum. 3. Medulla oblongata. 4. Corpus callosum. 5. Fornix. 6. Crus of fornix. 7. Corpus albicans. 8. Septum lucidum. 9. Velum Interpositum. 10. Middle commissure. 12. Posterior commissure. 13. Corpora quadrigemina. 15. Aqueduct of Sylvius. 17. Pons Varolii. 18. Crus cerebri. 19. Tuber cinereum. 20. Optic nerve. 21. Olfactory nerve.

pass forwards. In the interior of each crus is the dark gray *locus niger*. The third nerve (motor oculi) comes out from the crus; the fourth winds around it.

To examine the interior of the brain, it may be placed upon its base, and sliced away above the level of the *corpus callosum* or great transverse commissure joining the hemispheres. The mass of white substance, *centrum ovale*, is thus displayed. The corpus callosum is continuous behind the *fornix*. Removing or cutting through the corpus callosum, the *lateral ventricles* are exposed; with the *septum lucidum* for their thin dividing partition. The floor of each lateral ventricle is formed of the *corpus striatum, tænia semi-circularis, thalamus opticus, choroid plexus, corpus fimbriatum,* and *fornix*. Their roof is the corpus callosum.

The *fornix* is a triangular plane of white nerve-substance, of which the point is forwards. It is about the twelfth of an inch in thickness. It is supported by its anterior and posterior *crura* or curved pillars, which pass into the other parts of the brain. Its base, between the posterior crura, is continuous with the corpus callosum.

The *foramen of Monro* is a bifurcating opening or passage, from

Fig. 77.

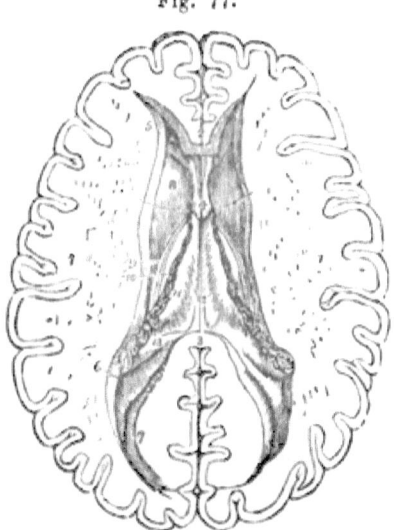

LATERAL VENTRICLES OF THE CEREBRUM.—1, 1. The two hemispheres cut down. 2. A small portion of the corpus callosum. 3. Its posterior boundary. 4. Septum lucidum. 5. Anterior cornu. 6. Middle cornu. 7. Posterior cornu. 8. Corpus striatum. 9. Tænia striata. 10. Thalamus opticus. 11. Plexus choroides. 12. Fornix. 13 Hippocampus major.

the third ventricle below, upwards into the two lateral ventricles; just behind the anterior pillars of the fornix. The choroid plexus is prolonged through it.

The *septum lucidum*, between the two lateral ventricles, is formed of two laminæ or layers, between which is the cavity called the *fifth ventricle*.

The *cornua* of the lateral ventricles are the *anterior, middle,* and *posterior*. The *anterior* cornu is a curved triangular cavity passing outwards and forwards in the substance of the anterior lobe, in front of the corpus striatum.

The *middle* cornu descends tortuously to terminate in the middle lobe. It contains within it the hippocampus major, pes hippocampi, pes accessorius, corpus fimbriatum, choroid plexus, fascia dentata, and transverse fissure.

The *hippocampus major*, or cornu ammonis, is a true convolution of the lateral edge of the hemisphere. The *corpus fimbriatum*, or tænia hippocampi, is a tape-like band of white nerve-substance attached to the inner border of the hippocampus. The *pes hippocampi* is a series of knotted elevations at the termination of the hippocampus major in front. *Pes accessorius* is behind the hippocampus major, between it and hippocampus minor.

The *fascia dentata*, or corpus denticulatum, is a narrow serrated layer of gray nerve-substance, displayed by raising the edge of the corpus fimbriatum. The *posterior cornu* curves into the posterior lobe of the hemisphere. Its floor has a prominent cord-like elevation, the *hippocampus minor*.

The *corpora striata* are two elongated pear-shaped masses, making parts of the floor of the lateral ventricles. Externally, they are of gray vesicular nerve-substance; within, they contain also a number of white medullary or tubular filaments. These are connected with the anterior or motor columns of the crura cerebri; and, through them, with the corpora pyramidalia of the medulla oblongata.

The *thalami* (*optici*) are rounded masses lying posterior to the corpora striata, and partly inclosed between them. They are composed externally of white nerve-substance, which, within, is blended and laminated with gray vesicular neurine. All the nerves of sensation are more or less directly connected with the thalami, or with the corpora olivaria of the medulla oblongata which are continuous with them.

The *tænia semicircularis* is a narrow cord of white nerve-substance, between the thalamus and the corpus striatum. Beneath it is a vein, the *vena Galeni*, which ends in the choroid plexus.

When the anterior crura of the fornix are divided and it is thrown backwards, a delicate membranous network is seen, which is the *velum interpositum*. Removing this, we expose under it the *third ventricle*. The velum interpositum is a continuation of the pia mater.

The *choroid plexus*, on each side, is the lateral margin of the velum interpositum; it consists of a red fringe of tortuous arteries and veins. The two choroid plexuses meet at the foramen of Monro.

The *third ventricle* is a narrow, oblong fissure, roofed by the fornix, floored by the posterior perforated space, corpora albicantia, tuber cinereum, and crura cerebri. It contains three transverse commissures, *anterior*, *middle*, and *posterior*. The middle is also called the *soft* commissure.

The *iter ad quartum ventriculum*, or aqueduct of Sylvius, is a canal from the posterior part of the third ventricle under the tubercula quadrigemina into the fourth ventricle.

The *iter ad infundibulum* is a canal from the anterior part of the third ventricle downwards into the infundibulum.

Behind the posterior commissure, are the small rounded bodies called *tubercula quadrigemina*, or *nates and testes*. The former are larger and anterior. They are connected with the optic thalamus by a bundle of white nerve-filaments also communicating with the cerebellum, called the *processus e cerebello ad testes*. The tubercula are thus between the cerebrum and cerebellum, and almost equally connected with both.

The *valve of Vieussens*, or valve of the brain, is a thin plane of white neurine continuous with the lower margin of the testes, whence it extends as the roof of the fourth ventricle.

The *pineal gland* is a conical mass of gray nerve-substance, lying beneath the base of the fornix upon the nates of the tubercula quadrigemina. It is joined to the velum interpositum; and, by two *peduncles*, to the thalami and crura of the fornix. Without any reason, it has been imagined to be the special seat of the soul.

The *commissures* or connecting portions of the brain, composed of bands or bundles of white tubular nerve-filaments, are the *superior longitudinal*, above the corpus callosum within each hemisphere of the cerebrum, the *fornix* or *inferior longitudinal* commissure, the *corpus callosum*, and the *three transverse commissures* of the third ventricle.

The *pons Varolii* is the transverse commissure of the *cerebellum*. The valve of Vieussens and processus e cerebello ad testes make a *cerebro-cerebellar* connection or commissure.

Cerebellum.

The cerebellum is much smaller than the cerebrum, averaging in weight a little over five ounces in the adult. It is behind and below the cerebral hemispheres. It consists of a right and a left lateral hemisphere, divided from each other by a fissure; this being interrupted above by a ridge-like connection called the *median lobe* or *superior vermiform process;* and below, by the *inferior vermiform process*. A *horizontal fissure* also divides each hemisphere into an upper and a lower portion; and out of this fissure proceed several lesser ones.

The outer portion of the cerebellum consists of a large number of delicate layers or *lamellæ*, laid one upon another. Making a vertical *section* of it, we see a tree-like arrangement within, called the *arbor vitæ*; consisting of white nerve substance inclosed in vesicular neurine. In the trunk of the arbor vitæ is an irregular mass of vesicular nerve substance, the *corpus dentatum*. The proportion of gray nerve-matter is large in the cerebellum, making the whole of its exterior lamellar surface. The pia mater dips in between its layers.

The name of *peduncles* of the cerebellum is sometimes given to— 1, the *processus e cerebello ad testes*, previously described; 2, the *crura cerebelli*; 3, the *corpora restiformia*, extending to the medulla oblongata. The crura cerebelli radiate from the pons Varolii (great transverse commissure of the cerebellum) into all parts of the cerebellum.

The *fourth ventricle* is a cavity between the medulla oblongata in front and the cerebellum behind. It is somewhat triangular, narrowest above. Its floor is the medulla oblongata and pons Varolii; its roof, the valve of Vieussens and tubercula quadrigemina. It communicates with the third ventricle by the *iter e tertio ad quartum ventriculum*, or aqueduct of Sylvius. The longitudinal fissure in the floor of the fourth ventricle presents a pen-like form, called the *calamus scriptorius*.

The *pons Varolii*, or *tuber annulare*, already named, is a rounded mass of about an inch in diameter, resting by its convex surface upon the *clivis* or junction of the occiput and sphenoid bone. It is composed of white tubular filaments, nearly all transverse, blended with gray vesicular nerve-substance. The transverse fibres connect the hemispheres of the cerebellum. The longitudinal filaments are continuous with the corpora pyramidalia of the medulla oblongata behind, and, in front, with the crura cerebri.

Medulla Oblongata.

Being the connecting portion between the spinal cord and the brain, the lower and posterior boundary of the medulla oblongata is the foramen magnum occipitis. Its form is pyramidal; its length, to the pons Varolii, about an inch and a quarter.

Besides an anterior and a posterior longitudinal fissure, continuous with the fissures of the spinal cord, it is, by other *sulci* or furrows, subdivided into four portions; the *corpora pyramidalia*, *corpora olivaria*, *corpora restiformia*, and *posterior ganglia*.

The *corpora pyramidalia* are anterior. They consist of bundles of white tubular nerve-substance. A *decussation* or crossing over of a fasciculus of each pyramid occurs about three-fourths of an inch below the pons. Above, after penetrating the pons, the corpora pyramidalia expand, and, passing on through or forming part of the crura cerebri, diverge to form a large part of the cerebral hemispheres.

The *corpora olivaria* are two elliptical bodies, external to the pyramidalia. They are composed of a mixture of white and gray nerve-substance; having a covering of the white, then a mass of gray vesicular material (corpus dentatum), and within this a central white portion. They send fibres to the tubercula quadrigemina and thalami.

The *corpora restiformia* are the posterior and lateral rope-like prolongations of the antero-lateral and posterior columns of the spinal cord. Above, they pass into the cerebellum, as the crura cerebelli.

The *posterior ganglia* or *posterior pyramids* are smaller, and lie next to the posterior fissure. They are entirely of white nerve-substance, continuous with the posterior tracts of the spinal cord.

SPINAL CORD.

The *length* of the spinal cord from the foramen magnum occipitis to the *cauda equina* in the lumbar region, averages about eighteen inches. Its *width* is greatest in the upper cervical region; less in the middle dorsal; enlarged again in the lower dorsal; and thence diminishing gradually to a conical point opposite the second lumbar vertebra.

Spinal Membranes.

These are continuous with the *dura mater, arachnoid,* and *pia mater* of the brain. Adhering to the first cervical vertebra, the *spinal dura mater* is loose in the vertebral canal down to an attachment to the *os coccygis*. It invests, by processes, each of the spinal nerves to the intervertebral foramen, and surrounds the ganglion on the posterior root of each.

The *spinal arachnoid* contains, between it and the pia mater, the *cerebro-spinal* fluid; communicating with the subarachnoid space of the brain.

The *spinal pia mater* is more dense and fibrous, and less vascular, than the pia mater of the brain. Between it and the arachnoid, on each side of the cord, is a narrow band called the *denticulate ligament*. This, being attached to the dura mater by fifteen or twenty processes, detains the membranes in their position in relation to the cord.

Fissures of the Cord.

The *anterior* fissure is the widest, but extends only to one-third of the diameter of the spinal marrow. At its bottom is a thin layer of white nerve-substance, the *anterior commissure*.

The *posterior* fissure is deeper. At its bottom is a layer of gray nerve-substance. Both are lined by the pia mater.

On each side is a *lateral* fissure; somewhat back of the middle of the cord. This does not run the whole length of the cord. Anterior to this, and posterior to it, on each side, are lesser fissures, the *antero-lateral* and *postero-lateral sulci;* corresponding with the anterior and posterior roots of the spinal nerves.

Columns.

Each half of the cord may be described as consisting of two columns; *antero-lateral* and *posterior* or *postero-lateral* column. The antero-lateral is much the larger; but the difference is greatest in the cervical region, and least in the lumbar.

Structure of the Cord.

A transverse section shows the gray vesicular nerve-substance to be inclosed within the white medullary portion. The gray substance presents, in section, the form of two crescents, connected by a commissure. The white columns are also joined by the anterior commissure. The white substance is composed of longitudinal laminæ of tubular filaments, in contact by their inner portion with the gray matter of the cord.

Origin of the Spinal Nerves.

There are, of these nerves, thirty-one pair; eight cervical, twelve dorsal, five lumbar, and six sacral nerves, for each side. Each nerve has its *anterior* and *posterior roots*. The *anterior* roots arise from the antero-lateral column, and emerge through the anterior lateral sulcus. The *posterior* roots enter the postero-lateral sulcus, to connect with the posterior column.

All, or nearly all, the fibres of both roots proceed through the white columns into the gray substance. Those of both *decussate* freely from side to side, and also pass upwards and downwards, as well as diverge in all directions. The complete history of their terminations and connections remains yet to be finally traced.

CRANIAL OR CEPHALIC NERVES.

Though not in all respects a satisfactory arrangement, these are usually described as nine pair: 1st, olfactory; 2d, optic; 3d, motor oculi; 4th, pathetic; 5th, trifacial; 6th, abducens oculi; 7th, facial or portio dura, and auditory or portio mollis; 8th, glosso-pharyngeal, pneumogastric, and spinal accessory; 9th, hypoglossal.

Olfactory nerve.—Arising (to use the common language of anatomists) by three roots, from the anterior and middle parts of the base of the cerebrum, the first nerve proceeds forward as a flat band upon the under surface of the anterior lobe, not far from the longitudinal fissure. On the ethmoid bone, it expands into the *olfactory bulb;* from which pass, through the *cribriform plate* of the ethmoid, about twenty filaments, to be distributed to the mucous (Schneiderian) membrane of the nostril.

The olfactory nerve-trunk is soft, and contains gray matter in its interior.

Optic nerve.—The second nerve of each side unites with its fellow at the *optic chiasm* or commissure, within the skull, in front of the tuber cinereum. There a partial decussation occurs. Some fibres cross from the retina of one eyeball to that of the other; some from one side of the brain to the other; some from the eye on one side to the brain on the other; and some from each eye to the same side of the base of the brain.

Back of the commissure, the *optic tract* divides on each side into two bands, which continue to the *thalami,* the *corpora geniculata,* and the *tubercula quadrigemina.*

Anteriorly, the optic nerve of each side emerges by the optic foramen of the sphenoid bone, pierces the sclerotic and choroid coats of the eyeball a little to the nasal side of its centre, and is distributed to the retina. The *arteria centralis retinæ* perforates it, with corresponding veins.

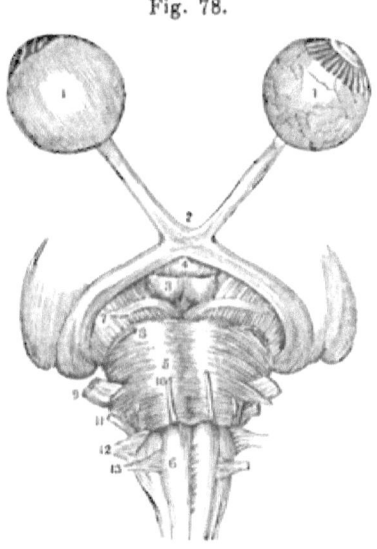

Fig. 78.

THE SECOND PAIR, OR OPTIC NERVES.—1, 1. Globe of the eye. 2. Chiasm of the optic nerves. 3. Corpora albicantia. 4. Infundibulum. 5. Pons Varolii 6. Medulla oblongata. 7. Third pair. 8. Fourth pair. 9. Fifth pair. 10 Sixth pair. 11. Seventh pair. 12. Eighth pair. 13. Ninth pair.

Motor oculi.—The third nerve originates in the crus cerebri in front of the pons Varolii. After receiving a few filaments from the cavernous plexus of the sympathetic, it divides into two branches, which enter the orbit through the sphenoidal fissure. It is finally distributed to all the muscles of the eyeball.

Pathetic nerve.—The fourth, sometimes called trochlear nerve, arises from the valve of Vieussens behind the tubercula quadrigemina. Winding around the crus cerebri, it passes into the orbit through the sphenoid fissure. It supplies only the superior oblique muscle.

Trifacial or fifth pair.—This, the largest of the cephalic nerves, arises, like the spinal nerves, by two roots; the posterior of which has a ganglion upon it. Both roots are connected with the medulla oblongata, through the pons Varolii. Near the apex of the petrous portion of the temporal bone, the posterior and larger root enters the *Casserian (semilunar) ganglion.* From this go off two great branches of the nerve, the *ophthalmic* and *superior maxillary;* which are therefore sensory nerves only. Below it, with fibres from both

Fig. 79.

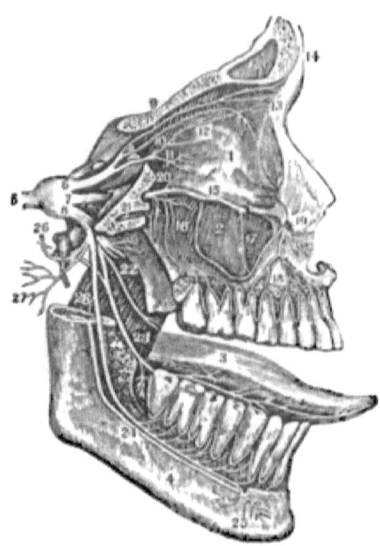

DISTRIBUTION OF THE FIFTH PAIR.—1. Orbit. 2. Antrum Highmorianum. 3. Tongue. 4. Lower jaw-bone. 5. Root of the fifth pair forming the ganglion of Gasser. 6. First branch of the fifth pair. 7. Second branch. 8. Third branch. 9. Frontal branch. 10. Lachrymal branch. 11. Nasal branch. 12. Internal nasal nerve. 13. External nasal nerve. 14. External and internal frontal nerve. 15. Infra-orbital nerve. 16. Posterior dental branches. 17. Middle dental branch. 18. Anterior dental nerve. 19. Terminating branches of the infra-orbital nerve. 20. Orbitar branch. 21. Pterygoid, or recurrent nerve. 22. Five anterior branches. 23. Lingual branch of the fifth. 24. Inferior dental nerve. 25. Its mental branches. 26. Superficial temporal nerve. 27. Auricular branches. 28. Mylo-hyoid branch.

roots, passes the *inferior maxillary,* which has sensory and motor filaments. The ophthalmic, or first branch of the fifth, goes out through the sphenoidal fissure; the second or superior maxillary, through the foramen rotundum of the sphenoid bone; the third, inferior maxillary, through the foramen ovale of the same bone.

Ophthalmic nerve.—The subdivisions of this are, the *lachrymal, frontal,* and *nasal* nerves. The *lachrymal* is smallest. It goes to the lachrymal gland, conjunctiva and tegument of the upper eyelid.

The *frontal* is larger. It divides into the *supra-trochlear* and *supra-orbital* branches. The first is distributed to the corrugator supercilii and occipito-frontalis muscles and the tegument of the forehead. The supra-orbital passes through the supra-orbital foramen, giving off filaments to the upper eyelid; then terminating in muscular, cutaneous, and pericranial branches, for the forehead and brow.

The *nasal* nerve leaves the orbit by the anterior ethmoidal foramen, enters the cavity of the cranium, goes over the cribriform plate of the ethmoid to the side of the crista galli, and there descends into the nose. Then it divides into the *external* and *internal* branches. The internal supplies the mucous membrane of the nostril, the external goes beneath the end of the nasal bone to be distributed with the facial nerve to the skin of the wing and tip of the nose.

Before this division, the nasal nerve gives off the *ganglionic*, two *long ciliary*, and *infra-trochlear* branches. The ganglionic enters the ciliary ganglion. The long ciliary go to the ciliary muscle and iris, with the short ciliary nerves from the ganglion. The infra-trochlear goes to the parts about the inner angle of the eye.

Superior maxillary nerve.—Its subdivisions are, 1. In the spheno-maxillary fossa : *orbital, spheno-palatine, posterior dental.* 2. In the infra-orbital canal : *anterior dental.* 3. On the face : *palpebral, nasal, labial.*

The *orbital* branch splits into the temporal and *malar*. The former passes through the malar bone into the temporal muscle and integument. The malar goes through a foramen in the malar bone to join the facial. The two spheno-palatine branches go to the ganglion of the same name. The posterior dental branches form a plexus with the anterior dental, from which filaments go to the molar and bicuspid teeth; supplying also the gums and buccinator muscle.

The *anterior dental* nerve enters a canal in the front wall of the antrum, and joins with the posterior dental. It sends fibres to the incisor, canine, and bicuspid teeth; some, also, to the tegument.

The *palpebral* branches go to the muscle, conjunctiva, and tegument of the lower eyelid.

The *nasal* branches supply the side of the nose, joining with filaments of the ophthalmic.

The *labial* branches are distributed to the skin and muscles of the upper lip and the mucous membrane of the mouth.

All these branches contribute, with some from the facial nerve, to make the *infra-orbital plexus*, just below the orbit.

Inferior maxillary nerve—After its exit from the foramen ovale, this nerve divides into a *smaller anterior* and a *larger posterior* trunk. The former gives off *masseteric, deep temporal, buccal,* and *pterygoid* branches, to the muscles of mastication. The *larger posterior* portion divides into the *auriculo-temporal, gustatory,* and *inferior dental* nerves.

The *auriculo-temporal* goes to join the temporal artery near the articulation of the lower jaw; thence upwards under the parotid gland, above which it divides into the anterior and posterior temporal branches. It has, first, the two *auricular*, the *articular* branches, two branches to the *meatus auditorius*, and *parotid* branches.

The *gustatory* or *lingual* nerve is deeply placed, and supplies the mucous membrane and papillæ of the tongue, anastomosing at the tip of the latter with the terminations of the hypoglossal nerve.

The *inferior dental* nerve goes down with the inferior dental artery to the dental foramen. Then, in the dental canal, beneath the teeth of the lower jaw, it passes forwards to the mental foramen; where it divides into the *incisor* and *mental* branches. The former supplies the incisor and canine tooth-pulps. The latter goes out at the mental foramen, and supplies the muscles, mucous membrane, and tegument of the lower lip. The inferior dental gives off, before this, the *mylo-hyoid* and *dental* branches. The latter go to the pulps of the molar and bicuspid teeth.

Ganglia connected with the fifth nerve.—These are: 1. The *ophthalmic, lenticular*, or *ciliary* ganglion; 2. *Spheno-palatine* ganglion. 3. *Otic* ganglion (of Arnold); 4. *Submaxillary* ganglion. Belonging to the sympathetic system, these will all be described in connection with it.

Abducens or sixth pair.—This nerve originates from the medulla oblongata, close to the pons Varolii. It enters the orbit of the eye through the sphenoidal fissure, and is distributed to the external rectus muscle of the eyeball.

Facial nerve.—This, the *portio dura* of the seventh pair, arises from the medulla oblongata, passes upon the crus cerebelli, and enters the internal auditory meatus with the *portio mollis* or auditory nerve. At the bottom of the meatus, it goes into and through the aqueduct of Fallopius of the petrous portion of the temporal bone. Then, emerging, from the stylo-mastoid foramen, it runs forward in the parotid gland, crossing the external carotid artery, to divide behind the ramus of the lower jaw into two primary branches, the *temporo-facial* and *cervico-facial*.

Within the aqueduct of Fallopius, this nerve gives off the *tympanic nerve* and the *chorda tympani*. At its exit from the stylo-mastoid foramen, the *posterior auricular, digastric*, and *stylo-hyoid*. The temporo-facial branch divides into the *temporal, malar*, and *infra-orbital nerves*. The cervico-facial, into the *buccal, supra-maxillary*, and *infra-maxillary*.

The *tympanic* branch supplies the stapedius and laxator tympani muscles.

The *chorda tympani* ascends in a canal parallel to the aqueduct of Fallopius, passes into and through the cavity of the tympanum, emerges from it near the Glaserian fissure, descends to meet the gustatory nerve, goes with it through the submaxillary gland, and terminates in the lingualis muscle.

The *posterior auricular* divides into the *auricular* and *occipital* branches; the latter being the larger.

The *stylo-hyoid* goes to the muscle of that name.

The *digastric* supplies the digastricus muscle; a filament goes through this to join the glosso-pharyngeal nerve.

Of the *temporo-facial* division of the seventh, the *temporal* branches join with branches of the fifth pair to supply the occipito-frontalis and orbicularis oculi muscles.

The *malar* branches supply the orbicularis oculi and corrugator supercilii muscle.

The *infra-orbital* branches are distributed, some deeply and others superficially, between the lower margin of the orbit and the mouth.

The *cervico-facial* division of the facial nerve sends *buccal* branches to the buccinator and orbicularis oris; *supra-maxillary* branches to the lower lip and chin; and *infra-maxillary* branches, of which some join the superficial cervical nerve from the cervical plexus, and others supply the platysma myoides and levator labii inferioris muscles.

The *auditory* nerve, *portio mollis* of the seventh pair, enters the meatus auditorius internus, and is distributed to all parts of the labyrinth or internal ear.

Eighth pair.—This is a threefold nerve, composed of the *glosso-pharyngeal, pneumogastric,* and *spinal accessory.*

Glosso-pharyngeal.—The origin of this is from the upper part of the medulla oblongata. Leaving the skull by the jugular foramen, it descends in front of the internal carotid artery, and arches on the side of the neck, to be finally distributed to the mucous membrane of the fauces, tonsils, and base of the tongue. While in the jugular foramen, this nerve has two enlargements, the *jugular* and *petrous* ganglia. The glosso-pharyngeal communicates by filaments with the pneumogastric, facial, and sympathetic nerves. A *tympanic* branch (nerve of Jacobson) goes off from the petrous ganglion, penetrates a canal of the temporal bone, enters the tympanum, and is there distributed, after dividing into three branches.

The divisions of the glosso-pharyngeal nerve are, the *carotid, pharyngeal, muscular, tonsillitic,* and *lingual.*

The *carotid* branches descend along the trunk of the internal carotid artery.

The *pharyngeal* branches form the pharyngeal plexus with filaments of the pneumogastric, superior laryngeal, and sympathetic nerves. From this plexus nerves pass through the muscular coat of the pharynx to supply the mucous membrane.

The *muscular* branches go principally to the stylo-pharyngeus muscle.

The *tonsillitic* branches form a sort of plexus around the tonsil.

The *lingual* branches are two. One goes to the base and the other to the side of the tongue. These are nerves of taste.

Pneumogastric or par vagum.—This part of the eighth pair arises from the medulla oblongata, and emerges from the cranium through the jugular foramen in the same sheath with the spinal accessory. In this foramen it presents an enlargement, called, sometimes, the *ganglion of the root* of the pneumogastric. Lower, it has the *inferior ganglion*, or ganglion of the trunk. It then descends within the sheath of the carotid; but has a different course on the two sides of the body. On the right side, it crosses the subclavian artery and goes down beside the trachea to the root of the lung, where it forms a sort of plexus. From this two branches go to the œsophagus, making with those of the other side the œsophageal plexus. These branches unite below into a nerve which runs down back of the œsophagus to be distributed upon the posterior surface of the stomach.

On the left side, the pneumogastric passes between the carotid and subclavian arteries, crosses the arch of the aorta, and descends behind the root of the lung and in front of the œsophagus, finally distributing branches over the anterior surface and lesser curvature of the stomach.

The *branches* of the pneumogastric are the *auricular, pharyngeal, superior laryngeal, recurrent laryngeal, cervical cardiac, thoracic cardiac, anterior* and *posterior pulmonary, œsophageal,* and *gastric* branches.

The *auricular* goes from the ganglion of the root, through the temporal bone, to the back of the ear.

The *pharyngeal* arises from the inferior ganglion, and is distributed to the muscles and mucous membrane of the pharynx; some filaments terminating, with some from the glosso-pharyngeal, upon the internal carotid artery.

The *superior laryngeal* goes from the inferior ganglion down by the side of the pharynx, and divides into the *external* and *internal laryngeal*. The external branch supplies the thyroid gland and crico-thyroid muscle. The internal passes to the mucous membrane of the larynx, and crico-arytenoid muscle. It communicates with the recurrent laryngeal.

The *inferior* or *recurrent laryngeal* winds around the subclavian artery on the right side, around the aorta on the left, ascends between the trachea and œsophagus, and is by its branches distributed to the muscles of the larynx, except the crico-thyroid. Cardiac, œsophageal, tracheal, and pharyngeal filaments go from it.

The *cervical cardiac* branches are two or three, which go to the great cardiac plexus, or to the cardiac branches of the sympathetic.

The *thoracic cardiac* branches arise lower, and have a similar distribution.

The *anterior pulmonary* branches are two or three small fasciculi, which join with sympathetic fibres, to make the anterior pulmonary plexus.

The *posterior pulmonary* are larger and more numerous; they form the posterior pulmonary plexus by union with filaments of the sympathetic. Both of these sets of ramifications attend the air-tubes in their distribution into the lungs.

The *œsophageal* and *gastric* branches have been already sufficiently described.

Spinal accessory.—This may be said to have two parts: the *spinal* portion and the portion *accessory* to the pneumogastric. The *spinal* part arises from the side of the spinal marrow, as low as the sixth cervical nerve, enters the cranium by the foramen magnum occipitis, again passes out by the jugular foramen, being there connected with the accessory portion; then, going behind the internal jugular vein, it descends obliquely behind the digastric and stylo-hyoid muscles to the sterno-cleido-mastoid muscles; finally, terminating in the trapezius muscle.

The *accessory* portion is smaller. It arises not far below the origin of the pneumogastric, in the medulla oblongata; goes out through the jugular foramen, communicating there with the par vagum, whose pharyngeal and superior laryngeal branches it accompanies in their distribution.

Hypoglossal nerve.—The ninth nerve of anatomists, more truly the twelfth of the series, originates in the medulla oblongata, and passes out through the anterior condyloid foramen of the occiput. Descending the neck, it winds around the occipital artery, crosses the external carotid, runs between the mylo-hyoid and hyoglossus muscles, to distribute branches to the whole of the tongue.

Its main branches are, the *descendens noni*, *thyro-hyoid*, and *muscular*.

The *descendens noni* is a slender branch, which goes across the sheath of the carotid vessels, making a loop with branches from the second and third cervical nerves. It sends thence filaments to the sterno-hyoid, sterno-thyroid, and omo-hyoid muscles; perhaps, also, to the phrenic and cardiac nerves.

The *thyro-hyoid* goes to the muscle of that name. The *muscular* branches supply the stylo-glossus, hyo-glossus, genio-hyoid, and genio-hyo-glossus muscles.

Functions of Cephalic Nerves.—So far as is yet ascertained, the following (from Gray) is a correct statement:—

Nerves of Special Sense.
Olfactory (1st),
Optic (2d),
Auditory (portio mollis of 7th),
Part of glosso-pharyngeal (of taste),
Lingual branch of 5th (of taste).

Nerves of Motion.
Motor oculi (3d),
Patheticus (4th),
Part of third branch of 5th,
Abducens oculi (6th),
Facial (portio dura of 7th),
Hypoglossal (9th).

Nerves of Common Sensation.
Greater portion of 5th,
Part of glosso-pharyngeal (of 8th).

Mixed Nerves.
Pneumogastric (part of 8th),
Spinal accessory (of 8th).

SPINAL NERVES.

Of these the roots have been described already, in their connection with the spinal marrow. Upon each *posterior* root, in the intervertebral foramen, outside of the dura mater, there is a *ganglion*; just beyond this, the two roots unite into one trunk, which soon again subdivides into *anterior* and *posterior* branches; each of which is furnished with filaments from both roots. The anterior branches are usually largest.

Eight Cervical Nerves.

The *first* leaves the spinal canal between the atlas and the occiput. It is called the *sub-occipital* nerve. It divides, as do *all* the spinal nerves, into anterior and posterior branches. The former joins the second cervical nerve. The latter goes to the recti and obliqui capitis and complexus muscles.

The *second* goes out between the atlas and the axis or second vertebra. Its anterior branch sends one fasciculus up to join the first nerve, and two down to connect with the third. Its posterior branch is larger, and goes to muscles of the back of the neck.

The anterior branch of the *third* is twice as large as that of the second; and so is the anterior branch of the fourth. Besides communicating with several other nerves, they unite at the cervical plexus. Their posterior branches go to the trapezius and other neighboring muscles and the integument.

Cervical plexus.—This is formed by the anterior branches of the first four cervical nerves. It lies in front of the first four vertebræ. Its branches are *superficial and deep*. Of the first, there are the ascending ones, viz., *superficialis colli*, *auricularis magnus*, and *occipitalis minor*; and *descending*, the *supra-clavicular*, subdividing into the *sternal, clavicular*, and *acromial* nerves. Deep branches of this plexus are, the *internal* ones, viz, the *communicating, muscular, communicans noni*, and *phrenic* nerves; and *external*, the

Fig. 80.

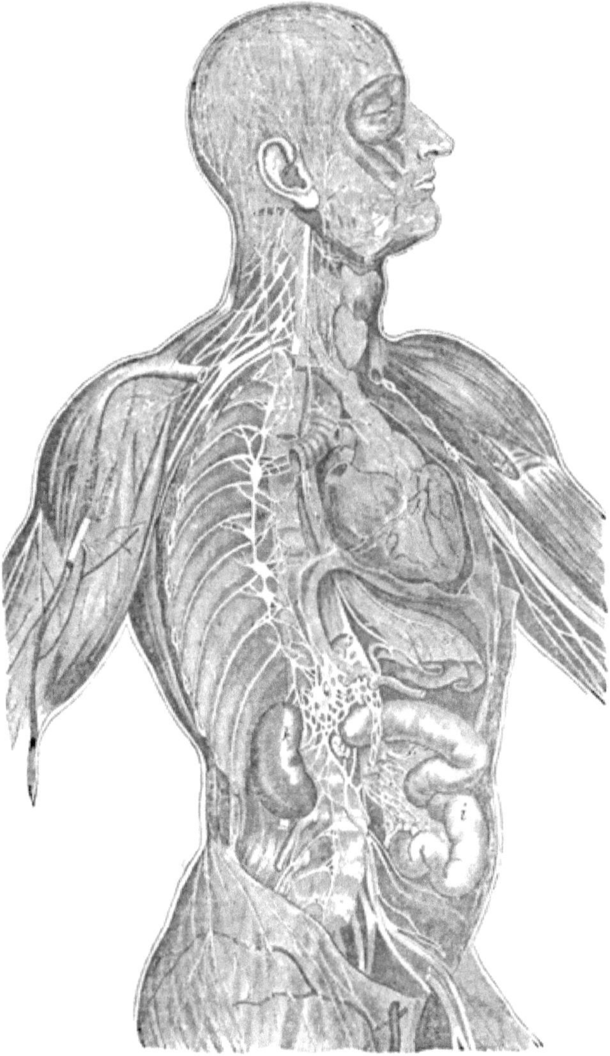

THE NERVES.

communicating and *muscular* nerves. These are distributed to the muscles and integument of the back of the head, and of the neck and chest. The *phrenic* nerve requires special description. Arising from the third and fourth cervical nerves, with a communicating branch from the fifth, it descends to the root of the neck, lying across the scalenus anticus muscle, passes between the subclavian artery and vein, and crosses the internal mammary artery as it enters the chest. The *right* phrenic nerve is shortest. It lies outside of the right vena innominata and the descending vena cava. The *left* phrenic crosses in front of the arch of the aorta to the root of the lung. Both are distributed to the diaphragm; sending filaments also to the pleura and pericardium. The right nerve communicates with the phrenic branches of the solar plexus; the left, with the phrenic plexus.

The Brachial Plexus.

Brachial plexus.—The anterior branches of the four cervical and the first dorsal nerves unite to form this. It extends from the lower portion of the neck to the axilla, and is quite wide. It communicates with the cervical plexus by a branch from the fourth to the fifth nerve. Its branches are—1. Above the clavicle: *communicating, muscular, posterior thoracic,* and *supra-scapular.* 2. Below the clavicle: to the chest, *anterior thoracic;* to the shoulder, *subscapular,* and *circumflex;* to the arm, forearm, and hand, *musculo-cutaneous, internal cutaneous, lesser internal cutaneous, median, ulnar,* and *musculo-spiral.*

The *communicating* goes to the phrenic. The *muscular* branches go to the scaleni, longus colli, rhomboideus, and subclavius muscles.

The *posterior thoracic* is a long branch, going through the scalenus medius muscle down to the bottom of the serratus magnus, to which it is distributed.

The *supra-scapular* nerve goes beneath the trapezius muscle to pass through a notch in

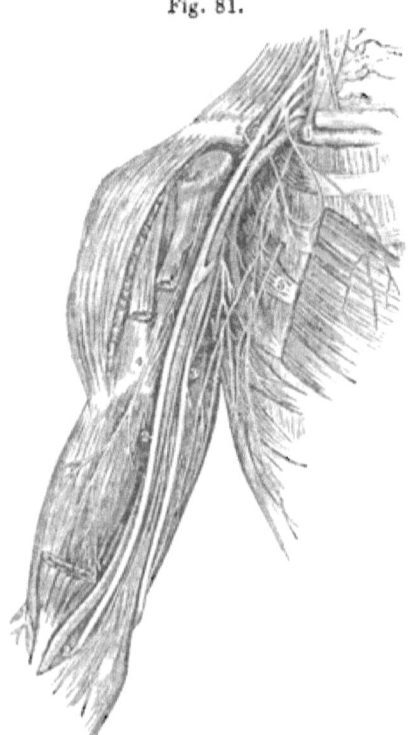

Fig. 81.

THE BRACHIAL PLEXUS.—1, 1. Scalenus anticus muscle. 2, 2. Median nerve. 3. Ulnar nerve. 4. Musculo-cutaneous nerve. 5. Thoracic nerves. 6. Phrenic nerve.

the upper border of the scapula, to supply the supra-spinatus and infra-spinatus muscles.

The *anterior thoracic* nerves are two, the *external* or superficial, and the internal or deep branch. They are distributed to the pectoralis muscles.

The *subscapular* nerves are three; they supply the subscapularis, teres major, and latissimus dorsi muscles.

The *circumflex* nerve goes down behind the axillary artery, and, below the subscapularis muscle, divides into the *upper* and *lower* branches. The upper winds around the neck of the humerus with the circumflex bloodvessels; supplying the deltoid muscle and the integument over it. The lower is distributed to the teres minor, deltoid, and triceps muscles, and integument over the last two.

The *musculo-cutaneous* nerve perforates the coraco-brachialis muscle, passes between the biceps and brachialis anticus, and, after sending off muscular filaments, becomes cutaneous on the outer side of the arm above the elbow; near the elbow-joint it subdivides into an *anterior* and a *posterior* branch. The anterior descends along the radial margin of the forearm, gets in front of the radial artery at the wrist, and goes with it to the back of the wrist. It receives a branch from the radial nerve. The posterior branch goes down back of the outer side of the forearm, to supply the skin of the lower part of the forearm; communicating also with the radial nerve, and with the external cutaneous branch of the musculo-spiral nerve.

The *internal cutaneous* nerve is a small branch of the brachial plexus. From its origin at the inner side of the brachial artery, it goes down along the arm to near its middle, when it emerges with the basilic vein, and divides into two cutaneous branches. Of these, the *anterior* branch goes usually in front of the median basilic vein; occasionally, behind it. Then it passes down on the anterior surface of the ulnar side of the forearm to the wrist. The *posterior* branch winds over the inner condyle of the humerus to the back of the forearm and is there distributed to the integument.

The *lesser internal cutaneous* nerve (nerve of Wrisberg) is the smallest branch of the brachial plexus. Passing through the axilla near the axillary vein, it goes with the brachial artery to the middle of the arm; then, perforating the fascia, it is distributed to the skin of the back of the arm.

Median Nerve.

The *median* is a more important nerve. Its two roots, from the brachial plexus, embrace the axillary artery, and then unite into one trunk. This descends the arm, at first outside, and then crossing to the inner side, of the brachial artery, to the bend of the elbow, over the brachialis anticus muscle. Then it passes between the two heads of the pronator radii teres muscle, and under the flexor sublimis digitorum, to become more superficial two inches above the wrist. Thence it passes under the annular ligament to the hand.

No branches go from the median nerve till it reaches the forearm. Then it gives off the *muscular, anterior interosseous,* and *palmar cutaneous* branches.

The *muscular* branches supply the superficial muscles in front of the forearm, except the flexor carpi ulnaris.

The *anterior interosseous* nerve goes to the deep muscles of the front of the forearm.

The *palmar cutaneous* nerve arises low down and divides above the annular ligament into an *outer* branch for the ball of the thumb, and an *inner* one for the palm of the hand.

The median nerve, reaching the palm, outside of the flexor tendons, divides into an *external* and an *internal* branch. The former supplies the muscles of the thumb and forefinger; the internal, the middle finger and part of the forefinger and third finger. Each of the five digital branches gives off a dorsal branch, which runs along the side of the back of a finger to its end. There it divides into a dorsal and a palmar branch, for the extremity of the finger.

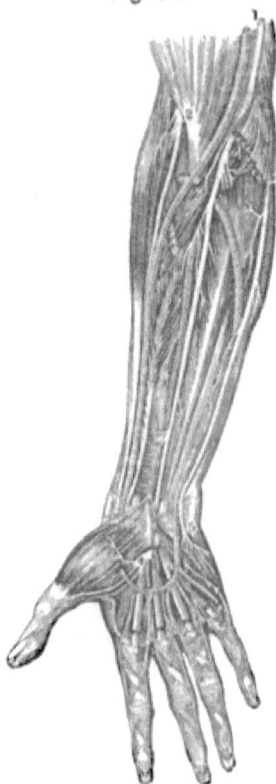

Fig. 82.

NERVES OF THE FRONT OF THE FOREARM.—1. Median nerve. 2. Musculo-spiral or radial. 3. Ulnar nerve. 4. Division of median nerve to the thumb, fore, middle, and radial side of ring finger. 5. Division of ulnar, to ulnar side of ring, and to little finger.

Ulnar Nerve.

This is behind the median nerve at their origin, and is smaller. Going on the inner side of the axillary and brachial artery, at the middle of the arm it runs obliquely across to descend between the olecranon process and the inner condyle of the humerus. It reaches the foramen between the two heads of the flexor carpi ulnaris muscle. Descending along the ulnar side of the forearm, it goes outside of the pisiform bone at the wrist, and, just below that bone divides into the *superficial* and *deep palmar* nerves. Besides these, its branches are, the *articular, muscular, cutaneous, dorsal cutaneous,* and *carpal articular.*

The upper *articular* branches are several small ones for the elbow-joint.

The two *muscular* branches pass off near the elbow to the flexor muscles.

The *cutaneous* arises about the middle of the forearm, and subdivides into superficial and deep cutaneous branches.

The *dorsal cutaneous* passes, from about two inches above the wrist, backwards beneath the flexor carpi ulnaris to the ulnar side of the wrist and the inner side of the little finger and adjoining sides of the little and the third or ring finger; communicating also with a branch of the radial nerve.

The lower *articular* filaments go to the wrist.

The *superficial palmar* terminal branch of the ulnar nerve supplies the skin on the inner side of the hand, and sends two digital branches; one to the ulnar side of the little finger, the other to the adjoining sides of the little and ring fingers.

The *deep palmar* branch follows the course of the deep palmar arterial arch beneath the flexor tendons; and sends filaments to the small muscles of the hand.

Musculo-Spiral Nerve.

The *musculo-spiral* nerve is the largest one that goes off from the brachial plexus. It passes behind the axillary artery and down in front of the tendons of the teres major and latissimus dorsi muscles; winds around the humerus with the superior profunda vessels, and then, on the outside of the arm, descends between the brachialis anticus and supinator radii longus to the front of the external condyle. There it subdivides into the *radial* and *posterior interosseous* nerves.

Its branches are, besides these, the *muscular* and *cutaneous*.

The *muscular* branches go to the triceps, anconeus, supinator longus, extensor carpi radialis longior, and brachialis anticus muscles.

The three *cutaneous* branches are, one *internal* and two *external*. They arise in or near the axillary space, and are distributed to the tegument, the lowest extending to the wrist.

Radial Nerve.

The *radial* nerve runs down on the front of the radial side of the forearm, two-thirds of its length, outside of the radial artery. Then it leaves that vessel, perforates the fascia outwards, and divides into an *external* and an *internal* branch. The former is small, and cutaneous in its distribution. The internal forms an arch on the back of the hand with the dorsal branch of the ulnar nerve. Then it gives off four digital nerves; one to the ulnar side of the thumb, one to the radial side of the forefinger, a third to the adjoining sides of the fore and middle fingers, the fourth to the adjoining sides of the middle and ring fingers.

The *posterior interosseous* nerve pierces the supinator radii brevis muscle, winds to the back of the forearm, and then passes down between the superficial and the deep layers of muscles to the middle of the forearm. Thence, over the interosseous ligament it reaches the back of the wrist; and, having a ganglion-like enlargement, gives off filaments from it to the wrist-joint. The branches of this nerve supply many of the muscles of the radial side and back of the forearm.

Twelve Dorsal Nerves.

The anterior branches of these, on each side, are the *intercostal* nerves. The posterior branches are smaller; they divide into *external* and *internal* ramifications. Their distribution is to the superficial and deep muscles of the back. Twelve pairs of *cutaneous* branches go off, six above from the internal, and six from the external posterior dorsal nerves.

Intercostal Nerves.

There are twelve on each side. The upper six go only to the chest (except the first); the lower six supply also the walls of the abdomen. The nerves of both sets go forward in the intercostal spaces below the artery and veins as far as the anterior terminations of the intercostal spaces. Those of the *upper* set, near the sternum, cross the internal mammary artery, penetrate the intercostal and pectoralis major muscles, and become the *anterior cutaneous* thoracic nerves. The *lower* ones, anteriorly, pass behind the costal cartilages, and between the internal oblique and transversalis muscles, to supply the rectus abdominis muscle; and afterwards become cutaneous. *Lateral cutaneous* nerves go off from the intercostal nerves, half way between the spine and the sternum.

The *first* intercostal nerve has no lateral cutaneous branch. That of the *second* is the *intercosto-humeral* nerve. Perforating the external intercostal muscle, it crosses the axilla to the inner side of the arm; there becoming cutaneous, and communicating with the cutaneous nerves of the arm. Sometimes the *third* intercostal nerve also gives off an intercostal humeral branch.

The *first dorsal* nerve has but a small intercostal branch; its anterior trunk mainly going into the brachial plexus.

The *last dorsal* nerve has a very large lateral cutaneous branch; reaching, in its distribution, as far as the surface over the hip-joint.

Five Lumbar Nerves.

The *anterior* branches of these, besides communicating with the sympathetic, send off muscular branches. Those of the *first four* lumbar nerves form the *lumbar plexus*. That of the *fifth*, joins the sacral plexus, in a trunk with the anterior branch of the sacral nerve (lumbo-sacral nerve).

The *posterior* lumbar branches subdivide into external and internal ; and are then distributed to the muscles of the lumbar region.

Lumbar Plexus.

This is located upon or in the psoas magnus muscle, near the transverse processes of the vertebræ, on each side. Its branches are, the *ilio-hypogastric, ilio-inguinal, genito-crural, external cutaneous, obturator, accessory obturator,* and *anterior crural* nerves.

Of these, the *ilio-hypogastric, ilio-inguinal,* and part of the *genito-crural* nerves are distributed to the lower abdominal walls. The rest of the *genito-crural* and the *external cutaneous, obturator, accessory obturator,* and *anterior crural* nerves, supply the fore part of the thigh and the inner side of the leg.

Anterior Crural Nerve.

This is the largest branch of the lumbar plexus. It sends muscular branches to the *iliacus internus* and *pectineus* muscles, and to all the muscles on the anterior part of the thigh ; also, cutaneous and articular branches.

In its course, the anterior crural nerve goes down between the psoas magnus and iliacus internus muscles, beneath Poupart's ligament, to the thigh ; there dividing into an anterior cutaneous and a posterior muscular portion. From the anterior part, it gives off the *middle cutaneous, internal cutaneous,* and *long saphenous* nerves, from the posterior part, *muscular* and articular branches.

The *long* or *internal saphenous* nerve is the largest of the cutaneous branches. It lies outside of the femoral artery from its passage beneath the sartorius muscle to the opening in the lower part of the adductor magnus. Thence it descends vertically along the inner side of the knee ; penetrates the deep fascia between the tendons of the sartorius and gracilis muscles, and, becoming subcutaneous, goes down the inner side of the leg with the internal saphenous vein. In the lower third of the leg it divides into two terminal branches. One goes along the tibia to the inner malleolus ; the other is distributed to the inner side of the foot, as far as the great toe.

Sacral and Coccygeal Nerves.

These are six on each side; the last one called coccygeal. The collection of their long roots together at the lower end of the spinal marrow makes the *cauda equina.* The lumbo-sacral nerve and the anterior branches of the three upper, and part of the fourth sacral nerve, form the *sacral plexus.*

Sacral Plexus.

This gives off from within the pelvis on each side the following branches: *muscular, superior gluteal, pudic, lesser sciatic,* and *great sciatic* nerves.

The *muscular* branches go to the pyriformis, obturator internus, gemelli, and quadratus femoris muscles.

The *superior gluteal* nerve passes with the gluteal bloodvessels through the great sacro-sciatic foramen; and then divides into a *superior* and an *inferior* branch. The former goes to the glutei muscles; the latter also to the tensor vaginæ femoris.

The *pudic* nerve leaves the pelvis through the great sacro-sciatic foramen, crosses the spine of the ischium, and then re-enters the pelvis by the lesser sacro-sciatic foramen. It goes with the pudic vessels upwards and forwards, and divides into the *perineal* and *dorsalis penis* nerve. A previous branch of it is the *inferior hemorrhoidal* nerve; which goes to the sphincter ani muscle and the integument around the anus.

The *perineal* nerve lies below the pudic artery, and divides into *cutaneous* and *muscular* branches; the *cutaneous*, into an anterior and a posterior branch.

The *dorsal nerve of the penis* runs with the pudic artery along the ramus of the ischium, and then penetrates the suspensory ligament of the penis and passes with the dorsal artery of the penis to the glans. In the female, the superior division of the pudic nerve goes to the clitoris; the inferior, to the external labia and the perineum.

The *small sciatic* nerve is distributed to the skin of the perineum, the back of the thigh, and the gluteus maximus muscle. Its branches are *muscular* or inferior gluteal, and *internal* and *ascending cutaneous* branches.

Great Sciatic Nerve.

This is the largest nerve in the body. Passing out through the great sacro-sciatic foramen it goes down between the tuberosity of the ischium and the trochanter major to the back of the thigh, and, at its lower third, divides into the *internal* and *external popliteal* nerves.

Before dividing, it sends off *articular* and *muscular* branches; the first to the hip-joint, the last to the flexor muscles on the back of the thigh.

The *internal popliteal* nerve passes down through the popliteal space, back of the knee, to the lower part of the popliteus muscle; then, with the artery, to become, under the arch of the soleus muscle, the *posterior tibial* nerve. Its branches, before this, are the *articular, muscular,* and *external* or *short saphenous* nerves.

The *articular* branches go to the knee-joint. The *muscular*, to the muscles of the calf of the leg. The *external* or *short saphenous* nerve descends between the heads of the gastrocnemius muscle, perforates the fascia at the middle of the leg, and, with a branch (*communicating peroneal* nerve) from the external popliteal nerve, goes down the outer margin of the tendo Achillis; then, with the external saphenous vein it winds around the external malleolus and is distributed to the skin on the outer side of the foot.

The *posterior tibial* nerve descends with the posterior tibial vessels to the space between the internal malleolus and the heel; there dividing into the *external* and *internal plantar* nerves. Above, it lies to the inner side of the artery; soon, however, it crosses it, and keeps to its outer side to the ankle. Its branches are *muscular* and *plantar cutaneous;* besides the two terminal branches, external and internal plantar. The *muscular* ones go to the deep muscles of the leg; the *plantar cutaneous*, to the heel and the sole of the foot.

Fig. 83.

POSTERIOR TIBIAL NERVE.

The *internal plantar* nerve goes with the internal plantar artery along the inner side of the foot. Opposite the bases of the metatarsal bones it gives off four terminal *digital* branches; communicating also with the external plantar nerve. Before these, it gives off *cutaneous muscular*, tarsal and metatarsal *articular* branches of the foot.

The *external plantar* nerve supplies the little toe and half of the fourth toe, as well as some of the muscles of the foot.

The *external popliteal* or *peroneal* nerve descends obliquely in the popliteal space near the margin of the biceps muscle to the fibula. An inch below the head of that bone it perforates the peroneus longus muscle and divides into the *anterior tibial* and *musculo-cutaneous* nerves. Before dividing, it gives off two *articular* branches to the knee, and two or three *cutaneous* branches to the back and outer side of the leg.

The *anterior tibial* nerve runs obliquely forwards under the extensor digitorum, joining the anterior tibial artery, on its outside, above the middle of the leg. Then descending with the artery to the front of the ankle, it divides into an *internal* and an *external* branch. Before this, it gives off some *muscular* branches to the leg.

The *internal* branch goes with the dorsal artery of the foot along the inner side of the latter, and divides into two branches for the adjoining sides of the great toe and the second toe.

The *external* or *tarsal* branch of the anterior tibial nerve goes outwards across the tarsus, and forms a ganglion-like enlargement; from which proceed branches to the extensor brevis of the toes and the tarsal and metatarsal articulations.

The *musculo-cutaneous* branch of the external popliteal or peroneal nerve passes forwards to penetrate the deep fascia at the lower third of the leg, in front and on the outer side. Then it divides into an *internal* and an *external* branch; having, first, given off some *muscular* and *cutaneous* branches.

Fig. 84.

ANTERIOR TIBIAL NERVE.—1. Musculo-cutaneous nerve. 2, 2. Anterior tibial nerve.

The *internal* of these branches goes in front of the ankle and along the dorsum of the foot to the great toe, and the adjacent sides of the second and third toes; also supplying the skin over and above them. It communicates with the internal saphenous nerve, as well as with the anterior tibial nerve.

The *external* terminal branch of the musculo-cutaneous nerve is larger. It runs along the outside of the dorsum of the foot, to the adjacent sides of the third, fourth, and fifth toes, and to the integument of the outer ankle, and outer side of the foot.

SYMPATHETIC NERVE.

This, the ganglionic system of physiologists, consists of a series of ganglia, connected together, and communicating with the spinal marrow and various organs and vessels, by cords of gray and white nerve filaments.

The ganglia of the two sides meet above, at the ganglion of Ribes, upon the anterior communicating artery of the brain. Below, they converge to the *ganglion impar* in front of the coccyx. The ganglia or centres of the sympathetic system are numerically as follows: four *cephalic;* three *cervical;* twelve *dorsal;* four *lumbar;* five *sacral;* and one *coccygeal* ganglion. Their branches are: 1. Communicating branches between the ganglia. 2. Branches connecting the ganglia with the cephalic or spinal nerves. 3. Branches distributed to the arteries, to the viscera, and to the thoracic, abdominal and pelvic gangia and plexuses.

Cephalic Ganglia.

Ophthalmic, ciliary, or *lenticular ganglion.*—This is a flattened vesicular mass of about the size of a pin's head, at the back of the orbit of the eye, between the optic nerve and the external rectus muscle. Its *communicating* branches are three: one with the nasal branch of the ophthalmic nerve, or first branch of the fifth pair. Another, with a branch of the third or motor oculi nerve. The other connects with the cavernous plexus of the sympathetic. Its branches of *distribution* are the *short ciliary* nerves, ten or twelve, going to the ciliary muscle and iris.

Spheno-palatine, or *Meckel's ganglion.*—This is the largest of those of the head. It is located in the spheno-maxillary fossa; and has a somewhat triangular shape. Its roots (as some anatomists call the communicating filaments) are, one from the facial (portio dura of seventh pair) through the vidian; one from the fifth; and one from the carotid plexus of the sympathetic, through the vidian nerve.

Its small *ascending* branches go into the orbit by the sphenomaxillary fissure, and supply its periosteum.

The *palatine* branches of this ganglion are the *anterior, middle*, and *posterior palatine nerve*. The *anterior* goes down through the posterior palatine canal, out by the posterior palatine foramen upon the hard palate, and passes forwards in a groove of the latter almost to the incisor teeth. In the canal, it sends off inferior nasal filaments, which go to ramify in the middle meatus and the turbinated bones.

The *middle* palatine nerve goes through the same canal as the above, to the posterior palatine foramen, giving off branches to the tonsil, soft palate, and uvula.

The *posterior* palatine nerve goes through the small posterior palatine canal, emerging behind the posterior palatine foramen; to be distributed to the soft palate, uvula, and tonsil.

The *internal* branches of Meckel's ganglion are the *anterior superior nasal* and the *naso-palatine* nerves. The former are four or five; they enter the nasal fossa through the spheno-palatine foramen, and supply its mucous membrane.

The *naso-palatine* nerve (nerve of Cotunnius) enters the nasal fossa with the above, goes inwards to the septum narium, then downwards and forwards along the septum to the anterior palatine foramen. It goes down to the roof of the mouth, where those of the right and left sides join, and supply the mucous membrane.

Posterior branches of the spheno-palatine ganglion are, the

Vidian nerve and the *pharyngeal* or pterygo-palatine nerve. The *Vidian* nerve runs through the Vidian canal to the foramen lacerum, and divides into the *large petrosal* and the *carotid* branches.

The *large petrosal* nerve passes beneath the ganglion of Casser in a groove of the petrous part of the temporal bone; then through the hiatus Fallopii into the aqueductus Fallopii, connecting with the ganglionic enlargement of the facial nerve.

The *carotid* branch of the Vidian enters the carotid canal of the temporal bone, outside of the artery, and joins the carotid plexus. A ganglionic enlargement upon it is called the carotid ganglion, or *ganglion of Laumonier*.

The *pharyngeal* nerve goes through the pterygo-palatine canal with the artery of that name, to the mucous membrane of the pharynx.

Otic ganglion (of Arnold).—This is a small, flattened, oval ganglion, located below the foramen ovale, upon the third branch of the fifth nerve. The cartilaginous portion of the Eustachian tube is at its internal side. It communicates with the third branch of the fifth pair, with the sympathetic plexus around the middle meningeal artery; and with the glosso-pharyngeal of the eighth, and the facial of the seventh pair, through the small petrosal nerve. Branches from this ganglion are distributed to the tensor tympani and tensor palati muscles.

Submaxillary ganglion.—This is small and circular, located near the posterior margin of the mylo-hyoid muscle. It is connected with the gustatory nerve, with the chorda tympani, and with the *nervi molles* of the sympathetic which surround the facial artery. Branches go from it, five or six in number, to the mucous membrane of the mouth, and to the submaxillary gland and its duct.

Cervical Ganglia.

These are three on each side: the *superior*, *middle*, and *inferior*, ganglia of the neck.

The *superior* is largest. It is opposite to the second and third, sometimes fourth cervical vertebræ. In front of it are the internal carotid artery, jugular vein, and glosso-pharyngeal nerve; behind it, the rectus capitis anticus major muscle; outside of it, the pneumogastric nerve. Its branches are *superior, inferior, external, internal,* and *anterior.*

The *superior* branch runs into the carotid canal, and divides into an *inner* and an *outer* branch, both distributed to the internal carotid artery. On the *inner*, the *cavernous plexus* is formed; on the *outer* branch, the *carotid* plexus.

The *inferior* branch of the first cervical ganglion connects with the second or middle ganglion.

The *external* branches are numerous. They communicate with the cephalic nerves, and with the four upper spinal nerves.

The *internal* ones are three : the *pharyngeal* and *laryngeal* branches, and the *superior cardiac* nerve. The last-named, on the right side, goes to the deep cardiac plexus; on the left side, to the superficial cardiac plexus.

The *anterior* branches of the same ganglion (nervi molles) form plexuses about the external carotid artery and its branches.

The *middle* cervical ganglion is the smallest of the three. It lies opposite to the fifth cervical vertebra, close to the inferior thyroid artery.

Its *superior* branches go up to the superior ganglion; its *inferior* ones, down to the inferior ganglion. *External* branches pass to the fifth and sixth spinal nerves. Its *internal* branches are the *thyroid* nerves, to the thyroid artery and gland, and the *middle cardiac* nerve. This last nerve goes to the deep cardiac plexus.

The *inferior* cervical ganglion lies near to the neck of the first rib. Its *superior* branches go up to the middle ganglion ; its *inferior* ones, down to the first thoracic ganglion. Among them is the *inferior cardiac* nerve, which joins the deep cardiac plexus. Some *external* branches connect with the seventh and eighth spinal nerves; others form a plexus around the vertebral artery.

Cardiac Plexuses.

These are the *superficial* cardiac plexus, the *deep* cardiac plexus, and the *anterior* and *posterior coronary* plexuses.

The *superficial cardiac* plexus is beneath the arch of the aorta, in front of the right pulmonary artery. It is supplied by the left superior cardiac nerve, the left, and sometimes the right, inferior cardiac branches of the pneumogastric, and connecting branches from the deep cardiac plexus. It sends filaments to form the anterior coronary plexus, and some to the left anterior pulmonary plexus.

The *great* or *deep cardiac* plexus lies in front of the bifurcation of the trachea, behind the arch of the aorta. It is formed by the cardiac nerves already described, and by the cardiac branches of the pneumogastric and recurrent laryngeal nerves. This plexus supplies the posterior coronary plexus and part of the anterior coronary plexus ; sending filaments also to the auricles of the heart and the pulmonary plexuses.

The *anterior coronary* plexus lies with the right coronary artery on the anterior surface of the heart.

The *posterior coronary* plexus embraces the branches of the coronary artery upon the back of the heart.

Beneath the endocardium Valentin and Robert Lee have demonstrated the existence of fine nervous ramifications, some within the substance of the heart, having many ganglia upon them.

Thoracic Ganglia.

These are usually twelve; placed on each side of the vertebral column, against the heads of the ribs. They have, each, two *external* branches, connecting with the dorsal spinal nerves. From the *upper six* ganglia, small *internal* branches go to the thoracic aorta and its branches; the third and fourth give filaments also to the pulmonary plexus.

Internal branches pass from the *six lower* thoracic ganglia to unite and form the *great splanchnic, lesser splanchnic* and the *renal splanchnic* nerves.

The *great splanchnic* nerve is whiter than the ordinary ganglionic nerves. It goes obliquely downwards and inwards to penetrate the diaphragm, and enter the semilunar ganglion.

The *lesser splanchnic* nerve also perforates the diaphragm, and connects with the cœliac plexus.

The *renal* or *smallest* splanchnic nerve goes from the lowest thoracic ganglion through the diaphragm to join the renal plexus, as well as the lower part of the cœliac plexus.

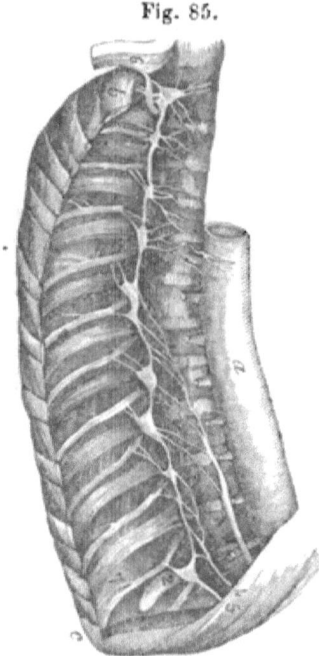

Fig. 85.

THORACIC GANGLIA.—*a.* Aorta. *b.* First rib. *c.* Eleventh rib. 1. First thoracic ganglion. 2. Last thoracic ganglion. 3. Great splanchnic nerve. 4. Lesser splanchnic nerve. 5. Renal splanchnic. 6. Part of brachial plexus.

Solar Plexus.

This name is given to a dense network of nerves and ganglia, behind the stomach, and front of the aorta. It is the termination of the great splanchnic nerves, part of the lesser splanchnic, and the right pneumogastric nerve. Branches from it form plexuses which surround the branches of the abdominal aorta.

The *semilunar ganglia*, one on each side, are the largest ganglia in the body. They lie very near to the supra-renal capsules and the cœliac axis artery.

The plexuses connected with or proceeding from the solar plexus are named as follows: *phrenic, cœliac, gastric, hepatic, splenic, supra-renal, renal, superior mesenteric, spermatic,* and *inferior mesenteric* plexuses. They accompany the arteries of corresponding names. The *aortic* plexus is also formed by branches of the solar and renal plexuses, receiving some filaments from the lumbar ganglia.

Lumbar Ganglia.

These are four, lying in front of the spinal column. *Superior* branches connect each with the ganglion above it; *inferior* ones with that next below; and *external* branches, with the lumbar spinal nerves. *Internal* branches go in part to the aortic plexus; some to form the *hypogastric* plexus over the promontory of the sacrum.

Pelvic Ganglia.

Of these there are four or five in front of the sacrum. They approach below, and unite in the *ganglion impar* on the front of the coccyx. The distribution of their branches closely resembles that of those of the lumbar ganglia.

The *hypogastric plexus* is formed by nerves from the two upper pelvic ganglia, the lumbar ganglia, and aortic plexus. Its branches are sent to the pelvic viscera.

The *inferior hypogastric* or *pelvic* plexus is an extension downwards of the above; supplied from the three or four lower pelvic ganglionic branches on each side, and some filaments directly from the ganglia. It lies by the side of the rectum and bladder in the male; by the rectum, bladder, and vagina in the female. From it branches go to all the pelvic viscera, with the branches of the iliac artery.

Back of it, from its branches is formed the *inferior hemorrhoidal* plexus. In front of it, the *vesical* plexus. Below it, the *prostatic* plexus. The nerves from this last are large, and pass in the male to the vesiculæ seminales, prostate, and erectile structure of the penis. The *large* and *small cavernous* nerves are the principal branches.

In the female, the *vaginal* plexus is below the pelvic or inferior hypogastric pelvis. It supplies the vagina.

From the *hypogastric* plexus pass off the nerves that supply the uterus. In the substance of that organ, as in that of the heart, Dr. Robert Lee has shown the existence of nervous filaments and ganglia.

CHAPTER X.
ORGANS OF SPECIAL SENSE.

THE EYE.

THE eyeball is nearly globular. Its coats are the *conjunctiva, sclerotic, choroid,* and *retina.* The transparent refracting media are the *cornea, aqueous humor, crystalline lens,* and *vitreous humor.*

Interior parts, connected with the choroid coat, are, also, the *iris, ciliary ligament* and *muscle,* and *ciliary processes.* Appendages of the eye are the eyelids, eyelashes, eyebrows, lachrymal gland, Meibomian glands and ducts; muscles, nerves, and blood-vessels.

The *conjunctiva* is a mucous membrane which lines the eyelids and is reflected over the front part of the sclerotic and the cornea.

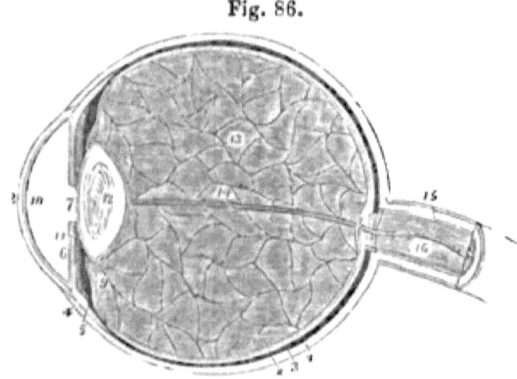

Fig. 86.

SECTION OF THE EYE.—1. Sclerotic coat. 2. Cornea. 3. Choroid coat. 4. Ciliary ligament. 5. Ciliary processes. 6. Iris. 7. Pupil. 8. Retina. 9. Canal of Petit. 10. Anterior chamber containing the aqueous humor. 11. Posterior chamber. 12. Lens inclosed in its capsule. 13. Vitreous humor in the hyaloid membrane. 14. Sheath of the hyaloid membrane. 15. Neurilemma of the optic nerve. 16. Arteria centralis retinæ.

Near the inner angle of the eye it forms a fold called *plica semilunaris.* The *caruncula lacrymalis* is at the inner side of this; it is a small, reddish, conical enlargement, consisting of follicles.

The *sclerotic,* or hard coat of the eye, is fibrous in structure. Externally, it is white; internally, brownish, and marked with grooves for the ciliary nerves. The muscles of the eyeball are inserted into it. The optic nerve perforates it. In front, the cornea is continuous with its edge; being set into it as a watch-glass in its case.

The *cornea* projects anteriorly, constituting about one-sixth of the ball of the eye. It is most prominent in early life. Its proper tissue is fibrous, covered in front and behind by elastic laminæ. No capillaries extend into the cornea.

The *choroid* coat is beneath the sclerotic. It has three layers. The *external* layer consists of the branches of the short ciliary arteries and, in greater number, the curved veins called *venæ vorticosæ*. Among the vessels are stellate pigment-cells. The *middle* layer (tunica Ruyschiana) is a plexus of fine capillaries from the short ciliary arteries. The *internal pigmentary* layer consists of six-sided pigment-cells, with nuclei and color-granules within them.

The *ciliary processes* are folds of the anterior margin of the choroid coat, making a circle around the edge of the crystalline lens, behind the iris. They are from sixty to eighty in number. Their vessels are larger than those of the choroid coat, and are mainly longitudinal.

The *iris* is a circular curtain, suspended behind the cornea in the aqueous humor; with a central aperture, the pupil. Its circumference is connected in part with the choroid, and, outside of that, by the ciliary ligament, with the sclerotic and cornea at their junction. It is composed of a fibrous stroma or central tissue, with commingled muscular fibres. These are the *radiating* muscular fibres, reaching to the circumference, and the *circular* ones, surrounding the pupil. Pigment-cells are also contained in it, giving the different colors to the eyes of different persons. Behind, the iris is purple; this surface is called the *uvea*.

The *ciliary ligament* is a narrow circle around the circumference of the iris, at the place of its union with the choroid, sclerotic, and cornea. A minute canal exists where it joins the sclerotic, the *sinus circularis iridis*.

The *ciliary muscle* is a circular band of unstriped fibres, longitudinal in direction, connecting the junction of the sclerotic and cornea with the choroid coat. Its action is upon the ciliary processes, and perhaps upon the lens.

The *retina* is the innermost coat of the eye. Outside of it is the choroid; within it, the vitreous humor. The optic nerve enters it a little to the inner side of the centre of the ball. Anteriorly it terminates a little behind the ciliary ligament, in a jagged edge, the *ora serrata*. Exactly in the centre of its posterior part, in the axis of the eye, is the *yellow spot of Sœmmering*, the point of most perfect vision. The *central artery of the retina* enters through the centre of the optic nerve. This point is destitute of vision.

The retina consists of three (Kölliker says four) layers: 1, external, *Jacob's membrane*, layer of *rods and cones;* 2, middle,

198   ANATOMY.

Fig. 87.

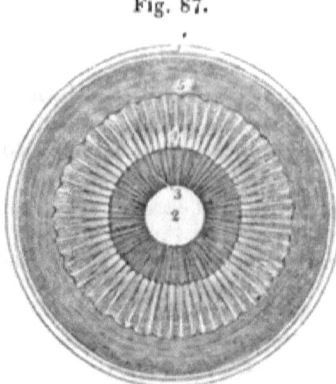

TRANSVERSE SECTION OF THE EYE.—1. Edge of sclerotic, choroid, and retina. 2. Pupil. 3. Iris. 4. Ciliary processes. 5. Border of retina.

granular layer; 3, internal layer; *expansion of the optic nerve* into a network, and arrangement of *nerve-cells.* A very delicate membrane, *membrana limitans,* separates the retina from the vitreous humor.

The *aqueous humor* is divided by the iris into an *anterior* and a *posterior chamber* The anterior is the larger. This humor is, as its name indicates, of a watery consistence.

The *crystalline lens* lies behind the posterior chamber of the aqueous humor, and in front of the vitreous humor. It is a double convex transparent body, inclosed in an elastic membranous capsule. It measures about one-third of an inch across, and one-fourth of an inch from before backwards. The posterior surface is the most convex. In structure, the lens is formed of a number of concentric laminæ, which may be separated by boiling or immersion in alcohol. By the same means a partition of the lens may be shown into three triangular *segments.* More minutely, the laminæ consist of parallel fibres with wavy margins. The *suspensory ligament* of the lens is a thin membrane between the anterior

Fig. 88.   Fig. 89.

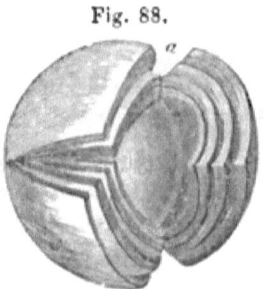

Crystalline lens divided.

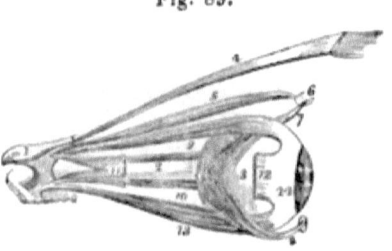

MUSCLES OF THE EYEBALL.—1. Fragment of the sphenoid bone. 2. Optic nerve. 3. Globe of the eye. 4. Levator palpebræ muscle. 5. Superior oblique muscle. 6. Its cartilaginous pulley. 7. Its reflected tendon. 8. Inferior oblique muscle. 9. Superior rectus muscle. 10. Internal rectus. 11. External rectus. 12. Extremity of the external rectus. 13. Inferior rectus muscle. 14. Sclerotic coat.

surface of the lens and the anterior margin of the retina. Posteriorly, this is separated from the hyaloid membrane by a space called the canal of Petit.

The *vitreous humor* forms about four-fifths of the whole eyeball. It is a transparent jelly-like material, inclosed in the *hyaloid membrane*.

The *arteries* of the eye are the long, short, and anterior ciliary arteries, and the central artery of the retina. The *nerves* of the eye are the optic, long ciliary, and short ciliary nerves.

Appendages of the Eye.

Each *eyelid* (palpebra) is composed of the following structures: skin, connective or cellular tissue, orbicularis oculi muscle, tarsal cartilage, Meibomian glands, and conjunctiva. The *tarsal cartilages* are two thin elongated strips of fibro-cartilage along the margin of the lids. The tensor tarsi muscle is attached to their inner ends. The *Meibomian glands* are about thirty in number for the upper lid; rather less in the lower. They are about as long as the width of the tarsal cartilages. Each has a minute duct opening on the margin of the eyelid.

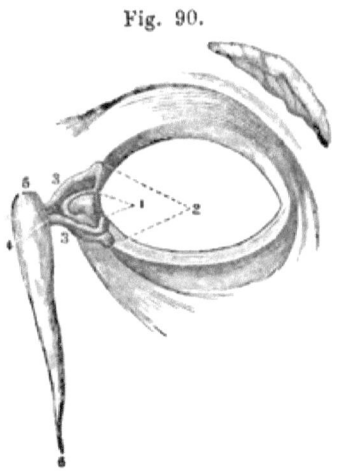

Fig. 90.

LACHRYMAL CANALS.—1. Puncta lacrymalia. 2. Cul-de-sac at the orbital end of the canal. 3. The course of each canal. 4, 5. The saccus lacrymalis. 6. Ductus ad nasum.

The *eyelashes* are short, thick, curved hairs, in a double, or sometimes triple row on each lid; the convexity of the upper ones being downwards, and of the lower ones upwards.

The *lachrymal gland* lies in a small fossa within the orbit, at its upper and outer, anterior part. It is oval, and about the size of an almond. Six or seven ducts carry tears from it to open through the conjunctiva, over the globe of the eye. Moisture being thus diffused over the ball, collects near the edges of the lids, and passing between the tarsi in the groove which they make when closed, or along the lower one when open, enters the *puncta lacrymalia*, near the corner of the eye. From these, two minute *canaliculi* convey the secretion (unless when, as in weeping, it is so profuse as to overflow) into the *lachrymal sac*. This is an enlargement at the upper part of the *nasal duct;* which duct or tube lies in a bony canal, three-quarters of an inch long, formed by the lachrymal (os unguis), superior maxillary, and inferior turbinated bones. It opens into the inferior meatus of the nose.

THE EAR.

External Ear.—The expanded part of the ear is the *pinna*, or *auricle;* the auditory opening (and canal) is the *external meatus.* The *pinna* consists of cartilage covered by skin. Its external prominent rim is the *helix.* Parallel with and in front of this, but dividing above, is the *antihelix.* Within the inclosure of the latter is a cavity of some size, the *concha.* In front of this, over the meatus, is the pointed projection, the *tragus;* and opposite to the latter, a smaller one, the *anti-tragus.* The softer lowest part of the ear is called the *lobule.*

The *external meatus* or auditory canal is about an inch and a quarter long, and directed forwards and inwards, slightly curved. At its end or bottom is the membrana tympani. The meatus is formed partly of cartilage and partly of bone, and is lined with a thin skin. Near the orifice are hairs and sebaceous glands; and, further in, the *ceruminous* glands, which secrete the ear-wax.

Middle ear, or tympanum.—Being within the petrous portion of the temporal bone, the cavity of the tympanum is separated from the external meatus by the *membrana tympani.* It communicates with the pharynx by the Eustachian tube. Within it are the *ossicles* or small bones of the ear; *malleus, incus, orbicular,* and *stapes.*

The *malleus* or hammer-bone has a head, neck, handle or *manubrium,* and two processes. The head is the oval upper extremity, which connects with the incus. The manubrium is vertical in position, and is attached along its margin to the membrana tympani. The *processus gracilis* is long and delicate; it gives origin to the laxator tympani muscle. The *processus brevis* or short process gives origin to the tensor tympani muscle.

The *incus* or anvil-bone has an irregular four-sided body, and a long and short process. The body joins with the malleus. The long process connects with the os orbiculare.

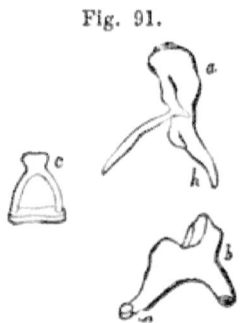

Fig. 91.

OSSICLES OF THE EAR.—*a.* Malleus, its head. *h.* Handle of malleus. *b.* Incus. *o.* Orbiculare. *c.* Stapes.

The *stapes* or stirrup-bone has very much the shape of a stirrup. By its head it is attached to the orbiculare; by its foot to the *fenestra ovalis* of the vestibule. To its neck is connected the stapedius muscle.

The *orbiculare* is a very small round bone, between the long process of the incus and the head of the stapes. These bones are all united by ligaments, admitting of slight movement.

Besides the *fenestra ovalis*, there is another aperture, the *fenestra rotunda*, from the tympanum into the cochlea of the internal ear. The latter is closed by a membrane.

The *muscles* of the tympanum have been already named; *tensor tympani*, *laxator tympani*, and *stapedius*. The first of the three is the largest. It draws the membrana tympani inwards, making it more tense. The *laxator* reverses this action.

The *chorda tympani* nerve leaves the facial to enter the tympanum, and crosses its cavity to an opening near the Glaserian fissure. Several other nerves enter and communicate in the tympanum.

Internal ear, or *labyrinth*.—This is composed of the *vestibule*, *semi-circular canals*, and *cochlea*.

The vestibule is the middle portion. On its outer wall are the fenestra ovalis and fenestra rotunda. In its inner wall is the foramen of the *aqueduct of the vestibule*, going to the back part of the petrous portion of the temporal bone. In front is a large opening, communicating with the cochlea; *apertura scalæ vestibuli cochleæ*. Behind, five orifices open from the vestibule into the semi-circular canals.

The three *semi-circular canals* are above and behind the vestibule. They are unequal in length and different in direction; two being vertical and one horizontal.

The *cochlea* is shaped somewhat like a snail-shell. It is anterior to the vestibule. It consists of the *modiolus* or *columella*, which is its central axis, and a spiral canal wound around this, with the *lamina spiralis* (partly osseous and partly cartilaginous), contained within the canal. The spiral canal has two turns and a half. The interior is divided into two equal scalæ or staircase like passages (*scala tympani* and *scala vestibuli*, by the delicate *lamina spiralis;* upon which the nerve filaments of the auditory nerve are spread out. The terminations of that nerve are also distributed to the vestibule and semi-circular canals.

The whole inner surface of the bony labyrinth is lined by a fibro-serous periosteal tissue. In the vestibule and semi-circular canals, it separates the osseous from the membranous labyrinth; its fibrous coat being attached to the bone, and its free serous layer to the membranous interior structure. In the cochlea, it covers both surfaces of the osseous zone of the lamina spiralis and

Fig. 92.

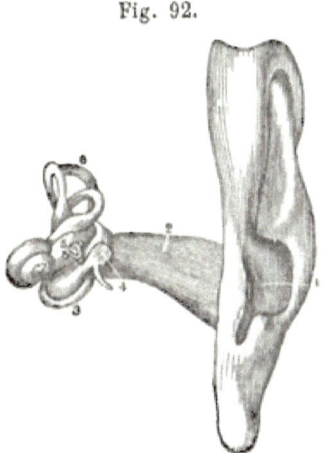

VIEW OF THE EAR.—1. The opening into the ear at the bottom of the concha. 2. Meatus auditorius externus. 3. Membrana tympani. 4. Malleus. 5. Stapes. 6. Labyrinth.

then forms the membranous zone (*zona Valsalvæ*), or continuation of the same. The *liquor Cotunnii* or *perilymph* is the fluid secreted within the periosteal double layer thus described.

The *membranous labyrinth* is the internal duplicate of the vestibule and semi-circular canals. The vestibular portion of it is formed into two sacs, the *utricle* and the *saccule*. The *endolymph*, or *liquor Scarpæ*, is the fluid contained within the labyrinth. The *otoliths* or *otoconia* are two small round collections of crystals of carbonate of lime, surrounded by fibrous tissue, in the vestibule.

THE NOSE.

Five cartilages, with the bony structures (nasal bones and processes of the superior maxillary bones) make up the framework of this organ; two upper and two lower lateral cartilages, and the nasal *septum*.

The *nasal fossæ* or cavities of the nostrils are lined by the *pituitary* or *Schneiderian* mucous membrane. It is covered by epithelium, tessellated at the upper part and near the aperture of the nares, but ciliated through the rest of its extent. The olfactory nerve is distributed to it. Mucous glands abound in it. The passages of the nares or nostrils are divided in front into the *superior*, *middle*, and *inferior meatus*. The *posterior nares* open into the pharynx.

Besides the olfactory, a number of nerves reach the nasal mucous membrane; principally branches of the fifth pair, already described.

CHAPTER XI.

ANATOMY OF HERNIA.

Inguinal Hernia (see Fig. 67).—Dissecting off the skin and superficial fascia from the groin, the lower portion of the *external oblique* muscle is exposed. The diverging fibres of this muscle, just above and outside of the pubes, give passage to the *spermatic cord;* in the female, the round ligament. This natural opening is the *external abdominal ring*. It is, rather, a triangular aperture; the inner column of which is that part of the tendon of the muscle which goes to the symphysis pubis; and its outer column, a portion or reflection of *Poupart's ligament; i. e*, the tendinous margin of the external oblique muscle, extending from the anterior superior spine of the ilium to the pubes. *Direct* or *ventral* hernia occurs immediately through the external ring; carrying over it the common tendon of the internal oblique and transversalis.

The *inguinal canal* leads from the external ring, downwards, forwards, and inwards, about an inch and three-fourths, to the *internal* abdominal ring. Along it the spermatic cord passes to and through the *internal* abdominal ring. This is an opening in the *fascia transversalis* through which the cord enters, or, more properly, *emerges* from the abdomen, on its course to the testicle.

The fascia gives a fibrous investment to the cord which continues to the testis; this is the *fascia propria*. As the *cremaster muscle* covers the cord, as a "carrying down" of fibres of the internal oblique, a protruding knuckle of intestine, in inguinal hernia, has the following coverings: *skin*, *superficial fascia* (*intercolumnar fascia*, from the external ring), *cremaster muscle*, *fascia propria*, and *peritoneal coat* or *sac*. The epigastric artery lies on the *inside* of oblique hernia, *i. e.*, that through the inguinal canal. In *direct* or *ventral* hernia it is *outside* of the hernial tumor.

Femoral Hernia.—This occurs beneath Poupart's ligament, at the place of transit of the femoral vessels from the groin. Between the vessels and the pubes is a space through which the intestine escapes. It then protrudes farther by the *saphenous opening*, by which the internal saphenous vein joins the femoral vein. Over that opening is the *cribriform fascia;* so called from its having a number of small perforations. The curved margin, by which, at the saphenous opening, the fascia lata, the thigh doubles or dips in, is sometimes called the *falciform border*.

The *crural arch* is between Poupart's ligament and the pelvis. Under it, besides the iliacus internus and psoas muscles, pass the anterior crural nerve, femoral artery, and femoral vein. To the inner side of the latter is the crural ring.

Gimbernat's ligament is a triangular reflection and expansion of the outer column of the external ring; that is, of the insertion of the external oblique muscle, attached to the *linea pectinea* of the crest of the pubes; and directing a concave border towards the femoral vessels. The part of this border turned most directly to the vessels is *Hey's ligament*. The vessels, as they escape from the pelvis, receive a funnel-shaped fascial envelope—the *crural canal* or sheath of the *femoral vessels*.

A femoral hernia, therefore, has, as its coverings, the *skin, superficial fascia, cribriform fascia, sheath of the vessels* (*septum crurale*, of fibrous tissue from the margin of the crural ring), and *peritoneal sac*.

When, in the descent of the testicle in the process of development, the intestine enters the canal with it, the peritoneum over it not being obliterated, it forms *congenital inguinal hernia*.

Umbilical and other forms of hernia require no special anatomical description here. The student is, however, advised not to be satisfied, in regard to the surgical anatomy of hernia, without repeated *demonstrations* or *dissections*.

CHAPTER XII.
THE PERINEUM.

THIS region lies between the tuberosities of the ischia; with the arch of the pubes in front, and the coccyx behind it. In and

Fig. 93.

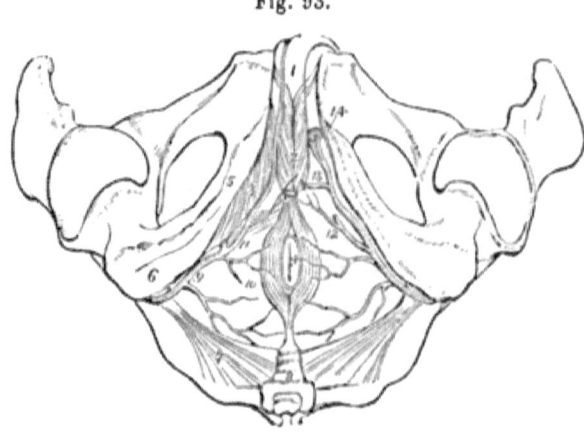

ARTERIES OF THE PERINEUM.—1. Penis. 2. Acceleratores urinæ muscles. 3. Erector penis. 4. Anus. 5. Ramus of ischium and pubes. 6. Tuberosity of ischium. 7. Lesser sacro-sciatic ligament. 8. Coccyx. 9. Internal pudic artery. 10. Inferior hemorrhoidal branches. 11. Superficialis perinei artery. 12. The same. 13. Artery of the bulb. 14. Terminal branches of the pudic artery.

beneath the two layers of the superficial perineal fascia, are several vessels and nerves.

The *inferior* or *external hemorrhoidal artery* is a branch of the internal pudic artery, leaving it, behind the tuberosity of the ischium, to go to the levator and sphincter ani.

Farther forward, the internal pudic gives off the *superficial perineal* artery. Curving upwards to the ramus of the pubes, it runs obliquely forwards to terminate in the scrotum.

THE PERINEUM.

The *transversalis perinei* artery goes off from this at a right angle, parallel with the transversus perinei muscle, to the sphincter ani.

The *artery of the bulb* leaves the pudic to enter the corpus spongiosum at its posterior margin. *Sometimes* it is a branch of the superficial perineal artery.

The *internal pudic artery* is an important branch of the internal iliac. It goes out by the great sacro-sciatic foramen, crosses the spine of the ilium, re-enters the pelvis through the lesser sacro-sciatic foramen, and, reaching the ramus of the ischium, an inch in front of its tuberosity, runs beneath its edge to the symphysis pubis. There, as the *arteria dorsalis penis*, it goes on to end at the glans.

Nerves of the perineal region are the internal pudic nerve, perineal cutaneous nerves, and their branches.

Muscles of the same part are, besides the *sphincter* and *levator ani*, the *transversus perinei, erector penis*, and *acceleratores urinæ* muscles.

The *transversus perinei* muscle is a small bundle of fibres, passing across from the tuberosity of the ischium to the middle line, when it meets its fellow. The *erector penis* goes from the tuberosity and ramus of the penis. The *acceleratores urinæ* arise from the perineal centre, and, by diverging fibres, surround the basal portion of the penis.

The *triangular ligament* is a portion of the deep perineal fascia, occupying the space under the arch of the pubes. Between its two layers are Cowper's glands. Through it the membranous portion of the urethra passes. The external pudic arteries and arteries of the bulb are, for part of their course, included within it.

A MANUAL

OF

PHYSIOLOGY.

CONTENTS.

	PAGE
DEFINITIONS	213

PART I.—GENERAL PHYSIOLOGY.

CHAPTER I.
ORGANIC MATTER.

Organic Products 215
Organizable Principles 215
    Albumen, 215—Gelatin, 215—Ostein, 215—Chondrin, 215—Keratin, 215—Neurin, 215—Fibrin, 216—Musculin, 216—Globulin, 216—Casein, 216—Pulmonin, 216—Mucosin, 216—Hæmatin, 216—Melanin, 216—Fatty Principles, 216.

CHAPTER II.
ORGANIC FORCES.

Vital Force 217

CHAPTER III.
ORGANIC FORMS.

Blood 218
Chyle 221
Lymph 221
Elementary Solid Forms 221
    Cells 222
    Fibres and Membranes 225
Tissues 226
    Connective, 226—Fibrous, 226—Elastic, 226—Cartilaginous, 226 Osseous, 226—Dermoid, 227—Corneous, 227—Fatty, 228—Mucous, 228—Serous, 228—Glandular, 228—Parenchymatous, 229—Muscular, 229—Nervous, 230.

CHAPTER IV.
ORGANIC FUNCTIONS.

	PAGE
Comparison of Animals and Plants	231

PART II.—SPECIAL OR FUNCTIONAL PHYSIOLOGY.
CHAPTER I.
ALIMENTATION.

Mastication and Insalivation	233
Deglutition	233
Gastric Digestion	234
Intestinal Digestion	236
Absorption	236
Assimilation	238
Nutrition	239

CHAPTER II.
CIRCULATION.

Action of the Heart	240
Arteries	243
Capillaries	245
Veins	246
Route of the Circulation	246

CHAPTER III.
RESPIRATION.

Movements	247
Quantity of Air Changed	248
Changes of the Air Breathed	249
Changes produced by Respiration in the Blood	250
Animal Heat	250

CHAPTER IV.
EXCRETION.

Secretion of Bile	253
Secretion of Urine	256
Excretion by the Bowels	259
The Skin	259

CHAPTER V.
REPRODUCTION.

	PAGE
General Considerations	260
Male Organs of Generation	262

CHAPTER VI.
MUSCULAR ACTION.

Voluntary Muscles	266
Mixed Muscles	267

CHAPTER VII.
FUNCTIONS OF THE NERVOUS SYSTEM.

GENERAL CONSIDERATIONS	267
REFLEX ACTION	270
GANGLIONIC NERVOUS APPARATUS	272
SPINAL MARROW	273
ENCEPHALON OR BRAIN	275
Medulla Oblongata	276
Cerebellum	277
Sensorial Ganglia	278
Cephalic Nerves	279
Cerebral Hemispheres	282
SLEEP	284
ORGANS OF SENSATION	285
Touch	285
Taste	286
Smell	287
Hearing	288
Vision	291

CHAPTER VIII.
THE VOICE.

CHAPTER IX.
DEVELOPMENT.

CONCEPTION	301
THE EMBRYO	301
Amnion	301

		PAGE
Allantois	. . . .	304
Decidua	. . . .	304
Chorion	. . . .	305
Placenta	. . . .	306
Umbilical Cord	. . . .	307
FŒTAL GROWTH	. . . .	307
CHANGES AT BIRTH	. . . .	309
INFANCY AND ADOLESCENCE	. . . .	309

PHYSIOLOGY.

DEFINITIONS.

PHYSIOLOGY is the science of the functions of living beings; including, in its widest acceptation, the study of all the changes which they undergo. There may be, therefore, *vegetable* and *animal* physiology; also, *human* and *comparative* physiology. *Biology* is a word now getting into use, meaning the whole science of life. *Pathology* is the physiology of the body and its organs in a state or states of disease; it is fundamental to the scientific practice of medicine.

PART I.
GENERAL PHYSIOLOGY.

This considers the *materials, forces,* and *forms* of organized bodies.

CHAPTER I.
ORGANIC MATTER.

THE *matter* of which plants and animals are, or have been, composed, is called, from its being or having been present in their *organs* or instrumental parts, *organic matter*. All other substances, with properties not affected by the presence of life, are inorganic. A distinction is perceptible and important between *organizable* matter and that which *has been* organized, but is no more capable of active function or new formation; for the last, the term *postorganic* would be convenient, although it is not usual.

Between the organic and inorganic materials, differences exist—

1st, in complexity of composition ; 2d, in instability ; 3dly, in the forms which they tend to assume, especially under the influence of life. Of the whole number of elements in nature supposed by chemists to be simple or undecomposable, scarcely twenty are found taking part in the composition of plants or animals. In mineral and other inorganic bodies, binary compounds are not rare, and ternary ones common ; while, in organic substances, four, five, or a still larger number of elements are more often combined ; with, also, a large number of *equivalents* of each. From this complexity of composition results great *instability;* shown by the rapid decay or putrefaction to which vegetable and animal structures are liable after their death. This complexity is greatest in *animal* bodies; most of all in the highest animals.

Carbon, hydrogen, oxygen, and nitrogen are the most nearly universal elements in organic matter. With them occur sulphur, phosphorus, calcium, iron, potassium, sodium, chlorine, silicon, fluorine, and some others, in variable quantities. Animal tissues, except fat (and some very few other partial exceptions among the lowest animals), always have carbon, hydrogen, oxygen, and nitrogen ; vegetable substances may consist of the first three of these, without nitrogen ; although the latter is also frequently present in plants.

Remembering that *water* is the most abundant of all substances in organized structures, as, for instance, in our own bodies, where it acts as a constituent of both solids and fluids, and as a vehicle of circulation and transmission, we may enumerate the most important other *organizable* proximate (or compound radical) elements, as follows :—

Nitrogenous :

Albumen,	Musculin,
Gelatin,	Globulin,
Ostein,	Casein,
Neurin,	Pulmonin,
Fibrin,	Mucosin.

Also, *pigmentary* or coloring principles, as

Hæmatin, of the blood,   Melanin, of the skin, iris, hair, &c.

Non-nitrogenous—oleaginous or fatty :

| Olein, | Stearin, |
| Margarin, | Cerebrin.[1] |

Saline substances in the blood furnish some materials for the organization of certain tissues, besides being essential, apparently, to the vital properties of the blood itself. Chlorides of potassium

---

[1] Liebreich asserts, instead of cerebrin and cerebric acid, the existence in the brain-substance of *protagon;* for which he gives the formula, C_{232}, H_{240}, N_4, P, O_{44}.

and sodium, and carbonates, phosphates, and sulphates of potassa, soda and lime, seem to be the most abundant and important of the salts of the blood.

Organic Products.

Unorganizable (post-organic) compound substances found in the blood and in various secretions and excretions, in process of removal from the body, are, chiefly, as follows :—

Nitrogenous:

Ptyalin,	Taurin,
Pepsin,	Taurocholic acid,
Pancreatin,	Glycocholic acid,
Creatin,	Urea,
Creatinin,	Uric acid.

Also, the pigmentary matters of the bile (biliverdin, cholepyrrhin) and urine (urosacin, uroxanthin.)

Non-nitrogenous:

Lactin,	Cholesterin,
Lactic acid,	Excretin,
Glycogen,	Stercorin.

Besides a number of *saline* substances (urates, sulphates, phosphates, &c.), found in the urine, bile, perspiration, tears, &c.

The full history of these belongs to organic chemistry. A few words may find place here concerning the *organizable* proximate elements.

Organizable Principles.

Albumen is found in blood and lymph, and, though not quite identically one substance, in the white of eggs. Its main peculiarity is its coagulation by heat (160° Fahrenheit); it is coagulated also by strong alcohol, tannin, mineral acids, ferrocyanide of potassium in an acid solution, and salts of lead, mercury, and copper.

Gelatin is present in all the hard and elastic tissues of the body; as cartilages, ligaments, tendons, membranes, &c. It appears, however, that the gelatinous constituents of these tissues undergo some alteration in the common process of extraction by long boiling. The fact that pure gelatin, so obtained, is not capable of sustaining life, when used alone as food, is thus explained.

Ostein is the animal matter of bone. It differs in some chemical reactions from cartilage-gelatin (chondrin). Nails and hairs contain *keratin*.

Neurin is a complex substance, presenting two varieties; gray or ash-colored nerve-substance, found in the cellular or vesicular structure of ganglia, and white neurin, of the tubular filaments, of nerves and commissures.

Fibrin is the spontaneously coagulable ingredient of the blood. The name is given to it because of its forming, in its coagulation, a fibrillary solidification, imitating a low form of tissue.

Musculin, myosin, or syntonin, was long supposed to be identical with fibrin. It is the special constituent of **muscular** tissue.

Globulin is the substance of the blood-corpuscles.

Casein abounds most in milk, of which it makes about forty parts in a thousand. It is coagulated by feeble acids; as by lactic acid, in the souring of milk. It is highly nutritious. Condensed, it becomes cheese.

Pulmonin is the name sometimes applied to the peculiar substance of lung tissue. Verdeil discovered in this tissue an acid substance, *pneumic acid*, to which importance has been ascribed in the detachment of carbonic acid from the blood in respiration.

Mucosin, of mucus, secreted by mucous membranes, is probably the substance of the membrane slightly altered.

The *pigments* of the blood, skin, iris, and hair contain carbon, hydrogen, oxygen, and nitrogen; sometimes other elements. *Hæmatin*, of the red blood-corpuscle, is notable for the amount of its iron (7 per cent.). *Melanin* is a convenient name for the dark coloring matter of solid parts.

Non-nitrogenous tissue-forming substances are the fats; *olein, margarin, stearin*, and *cerebrin*. The first three are similar in composition. Each consists of a *fatty acid*, oleic, margaric, stearic acid, combined with a base (oxide of glyceryl). Stearin is solid up to 130° or 140° Fahr.; margarin melts at about 120°; olein is liquid down to 25°. Human fat consists of margarin and olein, with very little stearin; deposited in cells. Cerebrin is a more complex substance, found in the brain, associated with phosphorus. Some chemists name, as contained in brain-substance, a nitrogenous acid called cerebric acid, and glycero-phosphoric acid, syntonin, olein, cholesterin, &c.

CHAPTER II.

ORGANIC FORCES.

REASON exists for designating by a special name that agency in organized bodies, *i. e.*, plants and animals, which gives them the characters of living beings; and the best name for it is *vital force*, or life-force. *Nerve-force* may, similarly, designate that which is peculiarly the attribute of ganglia and nerves; generated, accumulated, and reflected by ganglia, and transmitted by nerves. Animals only, not plants, possess this.

The ordinary *cosmic* forces (*i. e.*, forces common to all nature), heat, light, electricity, chemical attraction, gravitation, all affect organized beings. They are not unfrequently generated or transformed by vital processes; and have importance in various functions.

The doctrine of the correlation and convertibility of the forces of nature is indispensable, now, in physiology, as it is in physics. By it we mean, that heat, light, electricity, &c., are (not substances, but) different *modes of movement;* and that one kind of motion may be transmuted into another, by a change of conditions. This is true of life-force and nerve-force; these being dependent upon heat, light, &c., for their sustenance, and being sometimes, conversely, transmuted into those forces. Of the last change, the luminosity of the fire-fly is probably an example.

Vital Force.

Life-force ought to be studied, then, like the other forces of nature. By exclusion, we find, that, after most functions of the animal or vegetable organs have been explained by reference to chemical, mechanical, or other ordinary physical laws, something is still left. That is, *formation, growth, development;* the construction, from a formless liquid (blood, sap), of definitely formed structures, going through a series of changes for a definite period. We call the cause of this adaptive formation and change *life*. When it ceases, *death* is characterized by the loss of all that was peculiar to the being, and the return of its materials to the inorganic (through the post-organic) state.

We may enumerate the main facts established concerning vital force, as follows: 1. It is common to *animals* and *plants*. 2. It never originates except from parentage; *omne vivum ex vivo*. 3. Its action is essentially *formative* and *reparative*. 4. In the living body, it *controls* and *directs* the other physical forces, as chemical affinity, etc., modifying their results. 5. It acts *expansively*, from centres outwards; as shown by the production of rounded forms, cells, &c. 6. Sometimes it may be *transmuted* into other forces during life, and is altogether so at death. 7. Other forces, *especially heat, sustain* it, or are converted into it. 8. Sometimes it may be *suspended* for a time; as in the winter torpor of certain animals. 9. It is always *definitely limited* in duration under any particular form; that is, each individual can live only for a certain time, longer or shorter according to its species. 10. Life-force may *vary in degree*, in the same body at different times, and in the different parts of the same organism. This last proposition affords the best foundation for rational pathology. Yet it would be a serious error to suppose that *all* disease consists merely in diminished vitality, general or local.

CHAPTER III.
ORGANIC FORMS.

LIQUIDS and solids together make up every organized body which has active functions. The liquids in plants are the sap, and sometimes special juices; in animals, the *blood, lymph, chyle,* and various secretions. The solids are the *organs,* composed of various *tissues;* and these, of elementary forms, viz., *cells, fibres, membranes, tubes.*

The Blood.

As seen under the microscope, the blood consists of a colorless liquid (liquor sanguinis) in which float the red and the white or colorless corpuscles; from fifty to five hundred of the red, to one of the colorless in human blood. Of the former, the diameter is about $\frac{1}{3300}$th of an inch; of the white corpuscles, $\frac{1}{2500}$th. The latter are nucleated; the red corpuscles in man are not. The shape of the red corpuscles is disk-like or car-wheel-like; *i. e.,* circular, flattened, and concave at the middle. Carbonic acid and some other gases, when absorbed, swell the corpuscles into a more globular form; oxygen widens and flattens them. The difference in color between arterial (oxygenated) and venous blood is by some ascribed to this change of shape. The biconcave corpuscles concentrate the light which they reflect, giving a brilliant effect; while the diffusion of rays reflected from the more convex surfaces of the corpuscles of venous blood produces a dull purple color. Another view is, that blood-corpuscles contain a special principle, *cruorin,* the color of which is altered by the absorption and loss of oxygen.

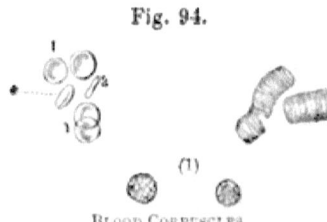

Fig. 94.

BLOOD CORPUSCLES.

Human blood has a salt taste, a peculiar odor, alkaline reaction to test paper, and, while in the living bloodvessels, a temperature of 100°. When drawn from the body, in about ten minutes it begins to *coagulate* or clot; the total separation of the coagulum from the liquid *serum* taking place gradually, and requiring many hours. In the *clot* is the fibrin of the blood, and its corpuscles; in the *serum,* albumen, the salts, and water. A *buffy coat* is sometimes seen on the surface of the clot, when the red corpuscles sink with unusal rapidity. In like cases, as of inflammatory disease, the coagulum may present a *cupped* top.

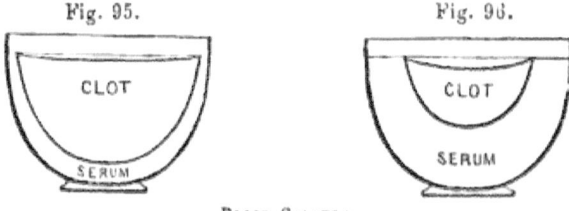

Fig. 95.   Fig. 96.

BLOOD COAGULA.

The cause of the coagulation of blood out of the body, or even within it if the circulation be arrested (as within the sac of an aneurism) is not known, further than that it depends upon the presence of fibrin. Richardson's theory, that it was owing to the escape of ammonia, whose presence in the blood kept the fibrin liquid, is open to many objections, and has now been abandoned by its proposer. The simplest view is, that *vitality* maintains the fluidity of the fibrin of the blood, and that when dead it becomes solid; as heat makes many things liquid, which congeal when they cool. The analogy is legitimate.

Coagulation is retarded by cold, and favored by rest, free access of air, and the multiplication of points of contact. In the act of death, or shortly before it, clots sometimes form in the heart and

Fig. 97.

BLOOD-CRYSTALS.

obstruct its valves. When blood has been at rest for a considerable time, *blood-crystals*, of hæmatoidin or hæmato-crystalline often form in it, of various shapes.

The *amount* of blood in a human body is not exactly ascertainable. It may be estimated at from fifteen to twenty pounds. Its *composition* is given by Dr. Kirkes, on the basis of numerous analyses by different observers, as follows:—

Average proportions of principal constituents in 1000 parts:—

Water	784
Red corpuscles	131
Albumen	70
Saline matters	6.03
Extractive and fatty matters	6.77
Fibrin	2.2
	1000.00

It is remarkable that the blood never contains any gelatin. Its potassium is contained chiefly in the corpuscles; chloride of sodium in the liquor sanguinis or plasma.

Average proportions of all the constituents of the blood in 1000 parts:—

Water		784
Albumen		70
Fibrin		2.2
Red corpuscles:—		
globulin		123.5
hæmatin		7.5
Fatty matter:—		
cholesterin	0.08	
cerebrin	0.4	
serolin	0.02	1.3
oleic and margaric acids		
volatile and odorous fatty acids		
fat containing phosphorus		
Inorganic salts:—		
chloride of sodium		3.6
chloride of potassium		0.36
tribasic phosphate of soda		0.2
carbonate of soda		0.84
sulphate of soda		0.28
phosphates of lime and magnesia		0.25
oxide and phosphate of iron		0.5
Extractive matter, with salivary matter, urea, creatin, creatinin, lactic acid, biliary coloring matter, gases, and accidental substances		5.47
		1000.00

The *gases* of the blood are, ordinarily, oxygen, nitrogen, and carbonic acid gas.

The *development* of the blood is an obscure subject. There appear to be two sets of blood corpuscles; the first peculiar to fœtal life, originating as primary cells in the *vascular layer* of the embryo; the others afterwards, in modes not demonstrated. Probably they, as well as the *plasma* or liquor sanguinis, are formed of materials furnished, and vital influence supplied by the mesenteric glands, to the chyle which passes through them. Chyle corpuscles and lymph corpuscles are, in appearance, identical with the colorless corpuscles of the blood. Whether the red blood corpuscles result from modification of these, or are generated independently, has not yet been rendered certain.

The *uses* of the blood are, to give nourishment and vital stimulation to all parts of the body; nutrition by the material which transudes from its plasma through the walls of the capillaries, and stimulation also by the oxygen it conveys. It is likely that the red corpuscles are the principal oxygen carriers, and that the carbonic acid is mainly absorbed by the liquid part (whose salts facilitate its absorption), but partly by the corpuscles. The constant movement of properly aërated blood is essential to life; and its momentary failure is made known by the cessation of functional activity in the great organs. Thus it is in *syncope*, or fainting; when the heart ceases to send fresh blood to the brain, unconsciousness results.

Chyle.

This is the fluid taken up by the *lacteals* or absorbent vessels of the small intestine. After digestion of food it is milky in character, from the amount of oily matter it contains (lacteals, from *lac*, milk). Passing through the mesenteric glands, the amount of fibrin and of colorless corpuscles and molecules increases. All the lacteals empty into the thoracic duct; and this terminates at the junction of the left subclavian and internal jugular vein, in the upper part of the left side of the thorax or chest. The obvious purpose of the chyle is to replenish the blood with nutritious, especially fatty, material.

Lymph.

The *lymphatics*, or common absorbents, take up this, as the effused or transuded plasma of the blood, in different parts of the body, not all consumed. Almost all organs of the body have lymphatic vessels permeating their substance. These vessels all pass through lymphatic *glands*, whose precise action is not understood. *Assimilation* is probably their function; that is, preparing their contained fluid for use in nutrition, by rendering it more like the tissues. This function is believed to be shared by the liver, spleen, and, in very early life, the thymus and thyroid glands.

Elementary Solid Forms.

These are *cells*, *nuclei*, *fibres*, *membranes*, and *tubes*. All of them are rounded; none angular. This is an important difference between *organic* and *inorganic* forms; the latter being very often angular. A crystal is a typical example of the inorganic; a cell of organic form.

Fig. 98.

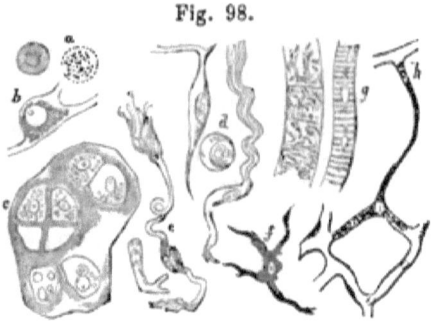

NUCLEATED CELLS.—*a*. Blood corpuscles. *b*. Nerve cells. *c*. Cartilage cells. *d*. Connective tissue cells. *e*. Elastic fibre cells. *f*. Pigment cells. *g*. Muscular fibres. *h*. Capillary vessels.

The importance of the cell in morphology (science of forms), vegetable and animal, is very great. Prominent in physiology and pathology, from Schleiden and Schwann (the first cell-discoverers 1836–38) to Virchow, has been the cell-theory; according to which *all* activity, for development and functional performance, belongs to cells only. *Omnis cellula e cellulâ* is Virchow's maxim, in his *Cellular Pathology*.

All physiologists, however, do not admit the *exclusive* activity of cells. Dr. Beale asserts two states of organic matter, in the tissues and organs of the living body; *germinal* matter (mostly, but not universally or necessarily, contained in cells), and *formed* material, which has changed from the active to the passive state, from which it becomes effete and excrementitious. The latter is usually farthest from the central and more vital portions of cells or other forms.

Dr. J. H. Bennett advocates *molecular* physiology; believing that activity for development and function resides in the *molecules* or particles, within or without organic cells and other formative elements. Probably there is some truth in all these views; while each is too exclusive. Life is probably the endowment of the whole germ from which the organism is evolved; and pervades all parts, though not with equal intensity or degree, according to their uses and functions.

Organic Cells.

Cells are, when first originated, nearly globular (spheroid) in shape. Mutual compression often makes them polygonal (many-sided), and especially hexagonal (six-sided). By development and transmutation, some cells are converted into fibres, and others into tubes. Some remain as cells until disintegrated and destroyed. Some, as blood cells, chyle and lymph corpuscles, and spermatozoa, float or move in liquids. Others become fixed as parts of firm tissues; as, for example, those of which the prismatic columns of tooth-enamel are constructed.

Powers ascribed to organic cells, under different circumstances, are various. 1. *Selection.*—Each cell has capacity to select and absorb from the common reservoir, the blood, just such materials as are appropriate to its place and action; fat in the adipose tissue, neurin in the ganglia, urea in the kidney, &c. 2. *Elaboration.*—Some cells only possess this power, to modify or elaborate material selected. Thus, in the liver, bile is partly a product of such an action in the cells of the organ; and so is milk in those of the mammary gland. 3. *Simple absorption* is effected by some cells, as those which cover the *villi* or minute projecting tufts of the lacteals, through which chyle is taken up from the small intestine. 4. *Elimination,* or secretion and excretion, is performed by others, as those of the kidney, sweat-glands, &c. 5. *Aeration* of the blood is accomplished by the vesicles of the lungs, which, however, are larger than ordinary cells, so called. 6. *Conveyance of oxygen* is attributed to the red cells or corpuscles of the blood. Some micrologists deny the character of

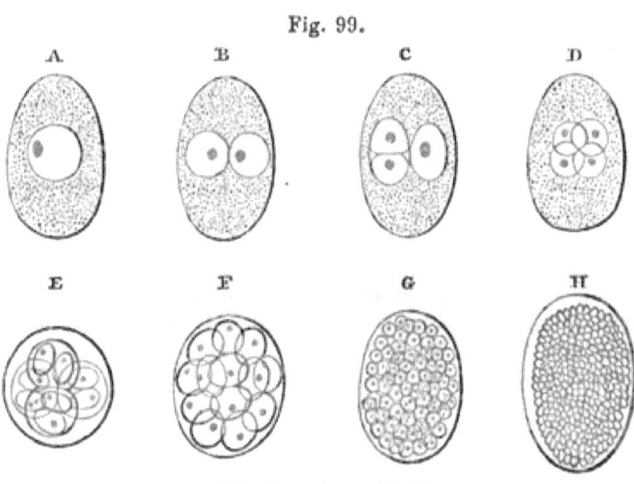

Fig. 99.

MULTIPLICATION OF CELLS.

true cells, with a cell wall, to these corpuscles. 7. *Change of form,* producing linear *contraction,* belongs exclusively as a function to muscle cells. 8. *Sensation, thought*, and *emotion*, so far as they are physiologically related, have for their instruments gray nerve-cells in the brain. 9. Lastly, cells may *reproduce* other cells— by subdivision or proliferation, *i. e.*, new cells being formed from them.

Most cells are *nucleated;* that is, have within the cell wall a more minute body, often itself cellular. The action of chemical substances, as acetic acid, is not the same upon the cell-wall and the nucleus. Cells may contain several nuclei. Sometimes nuclei become separately developed. The *contents* of different cells vary much more than their walls; but the *causes* of their selective and other peculiar powers are beyond special explanation. A plausible supposition as to the origination of some cells has been founded upon an experiment of Ascherson; viz., on dropping very small drops of oil into a solution of albumen, each droplet surrounds itself with an albuminous pellicle or covering, resembling closely an organic cell.

Partial explanation, only, of the transmission of fluids through cells and other elements of tissue, is found in *osmosis*, or endosmosis and exosmosis. These terms are applied to the transition of fluids through animal membranes; so that, two different fluids being on opposite sides of a membrane, an exchange takes place between the two. For example, if a fresh piece of membrane be stretched between a solution of salt and a quantity of pure water, after a while the solution will become more dilute, and the water on the other side will taste of salt. *More* of the pure water, however, has passed through, so that the quantity of the salt solution is increased, and that on the other side is lessened. The more abundant imbibition is called *endosmosis;* the lesser, *exosmosis*. Instead of salt, sugar, gum, albumen, &c., may be employed.

Conditions affecting osmosis are, 1. The freshness of the membrane. 2. Extent of contact between it and the liquids. The greater this is, the more rapid the flow. 3. The nature of the materials used. Dutrochet found that, through ox-bladder, water would pass into a solution of albumen or of sugar with a force more than twice as great as into one of gum, and three times as great as into a solution of gelatin. 4. The position of the membrane; *e. g.*, with mucous membrane, as to which of the liquids the mucous surface is presented. 5. Temperature. Endosmosis is most active at a moderately elevated heat. 6. Pressure affects it considerably. The greater the pressure, other things being equal, the more rapid the imbibition.

Sometimes, as when water and albumen are used, endosmosis occurs without any returning exosmosis. Generally, the stronger current is from the less dense to the more dense liquid. In the case of alcohol and water, it is, as an exception, the reverse. It appears that the force of osmosis does not depend upon the degree of attraction of the liquids for each other, but upon the attraction of the membrane for the liquids, by which it takes them into, and passes them through its substance. The movement of a fluid in a continuous current, as it occurs constantly during life, always favors endosmosis. So does the extent of contact produced by the minute ramification of the capillaries. *Albumen*, under ordinary circumstances, as already stated or implied, is not capable of endosmosis or exosmosis itself, though water will endosmose into it.

Dialysis is a name given by Graham to a process of transition of materials in solution through a permeable septum, by which substances mixed together may become separated from each other. He distinguishes bodies into *colloids* and *crystalloids;* the former, comprising organic matters and a few inorganic ones, not passing through animal membrane used as a *dialyser*, while crystalloids do, freely. This property or process no doubt has its influence in the living body, although its bearings are not yet fully discerned.

Fibres and Membranes.

It is almost certain that *fibres* are *directly* organized from plasma, in the white fibrous and yellow elastic tissues. Of simple *membranes* so formed, probably the only examples are, the capsule of the lens of the eye, the layer at the back of the cornea, and the "basement membranes" of mucous tissue, upon which cell-layers are formed.

In connection with the latter, allusion may be made to the variety of forms of such cells as compose these layers. Three kinds of epithelial or pavement-cells are found: *tessellated, cylindrical,* and *ciliated* epithelium. The first are rounded at first but become flattened and polygonal. They exist in the cuticle or scarf-skin, conjunctiva of the eye, the mouth, pharynx, œsophagus, vagina, serous and synovial membranes, many gland-ducts, and bloodvessels and lymphatics. *Cylindrical* or *conical* epithelial cells are found from the cardiac orifice of the stomach along the alimentary canal to the anus, and lining the gland-ducts communicating with the stomach and intestines; also, in the greater portion of the male urino-genital apparatus, and in the female urinary passages. *Ciliated* cells are so called from the presence on their summits of extremely minute *cilia*, or lash-like processes; which, during life and

Fig. 100.

PAVEMENT EPITHELIUM.

Fig. 101.

CILIATED EPITHELIUM.

for a time after death, are incessantly in waving motion; compared, when viewed under a microscope, to the undulations in the wind of a field of grain. The result of their movement is usually to produce a current (in a tube, for example) in one direction. Such cells are found, in the human body, lining the nasal cavity, except the strictly olfactory portion; in the frontal sinuses, lachrymal canal, and mucous surface of the lids of the eye; the upper part of the pharynx, soft palate, Eustachian tube, and middle ear or tympanum; all the respiratory tract from the glottis to the terminating ramules of the bronchial tubes; in the epididymis of the testicle; and in the female generative organs, from the neck of the uterus to the fimbriated extremities of the Fallopian tubes.

Tissues.

As all the tissues are compounded of the few elementary forms already mentioned, with more or less modification of the contents of cells and tubes, and of structureless intervening substance, their classification must be somewhat arbitrary. The clearest and most convenient arrangement of them is the following:—

Connective,	Fatty,
Fibrous,	Mucous,
Elastic,	Serous,
Cartilaginous,	Glandular,
Osseous,	Parenchymatous,
Dermoid,	Muscular,
Corneous,	Nervous.

To these might be added, perhaps, *tubular* tissue; that of the capillary bloodvessels, lacteals, and lymphatics.

Connective tissue is also called areolar or cellular tissue. It is the most abundant of all, and is the *packing* tissue of the body; being placed between muscular fibres and nearly all other closely contiguous yet separable parts. Its structure is essentially fibro-cellular. Connective-tissue *corpuscles* (irregularly stellated) are recognized in it by Virchow.

Fibrous tissue is white, tough and flexible. It exists in ligaments, tendons, periosteum, and certain membranes, as the dura mater of the brain, and the outer layer of the pericardium; and in the outer coat of the arteries.

Elastic tissue is yellowish in color, and microscopically more *tangled* in structure than the white fibrous tissue. Some of the ligaments of the spinal column possess it, and it is united with muscular tissue in the middle coat of arteries. It contains *elastin*, which differs somewhat from gelatin.

Cartilage is met with in many places in the body; between the vertebræ of the spine, between the ends of bones, at the joints connecting the ribs with the sternum, making the flexible portions of the nose and ears, and the edges of the eyelids. It is also the basis of formation of the bones.

Osseous or bony tissue is compounded of ostein or bone-cartilage and mineral matter, chiefly phosphate of lime (see *Anatomy*). A

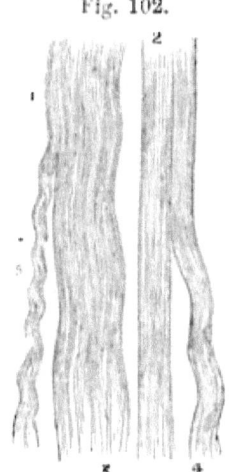

Fig. 102.

WHITE FIBROUS TISSUE.

Fig. 103.

YELLOW FIBROUS TISSUE.

modification of it is seen in the *dentine* of the teeth; which also present two other peculiar substances, the *cementum* which covers the fangs, and *enamel*, the covering of the crowns of the teeth.

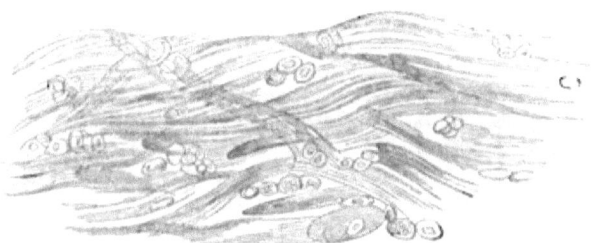

Fig. 104.

FIBROUS CARTILAGE. FROM THE SYMPHYSIS PUBIS. MAGNIFIED.

Dermoid tissue, or skin, is intermediate between, or composed of, fibrous and areolar or connective tissue (see Anatomy).

Corneous or horny tissue is represented in man by the nails; the

Fig. 105.

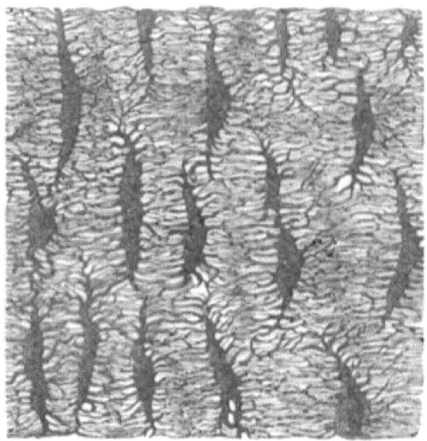

Osseous Tissue.

hairs are not far removed from it, although of a somewhat special tubular structure.

Fatty tissue is formed by the distribution of drops of semi-fluid oleaginous matter in cellular spaces of connective tissue. Fat lies under the skin, and gives roundness to the face, trunk, and limbs; besides furnishing cushions to prevent pressure on other parts, and, by its non-conducting power, to protect the body from undue loss of heat. Cushions of fatty deposit are also found behind the eyeball, around the heart, kidneys, and other parts. Fat is absorbed when waste of the body exceeds the supply of food; seeming to afford fuel for the "combustion" which generates animal heat.

Mucous tissue lines all those cavities of the body which communicate with the exterior; the orbit of the eye, the mouth, nostrils, throat, alimentary canal, bladder, vagina, uterus, &c. It consists of a basement membrane, on which is a layer of epithelial cells, already described.

Fig. 106.

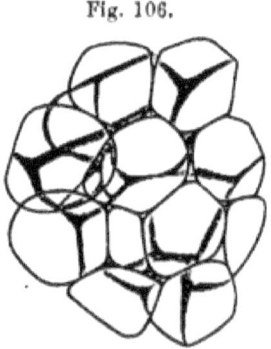

Fat Vesicles.

Serous tissue *envelopes* the organs contained in the great cavities of the body; as the lungs, abdominal organs, &c. It is in those situations always *double*, to lessen friction. It is thin, and very smooth, covered with epithelium, and moistened with serum or serosity.

Glandular structure is not identical in all the glands; but in each consists of *cells*, clustered or conglomerated together variously. The differences belong to descriptive anatomy.

Parenchymatous tissue is spoken of as existing in the liver, kidneys, lungs, and other large organs. The analogy between the lungs and secreting glands is close; but the air vesicles are larger, and open freely into the bronchial ramifications.

Muscular tissue is of two kinds. That of the voluntary muscles (of locomotion, &c.) is red, and composed of *fibres;* each fibre of *fibrils,* and, as shown by the microscope, each of these of *cells,* (sarcous elements) end to end.[1] *Striæ* or circular marks indicate the line of separation between the cells of the fibrillæ of a fibre, bound in its sheath or *sarcolemma.* The contraction of this red, striated or striped muscular tissue, occurring by the widening and shortening of the fibrillary cells, is quick, limited, and short in duration. In the heart only is this striped tissue altogether beyond the influence of the will; although it is almost entirely involuntary in the pharynx except at its upper part, and is rather controlled by emotion than by will in the muscles of expression in the face.

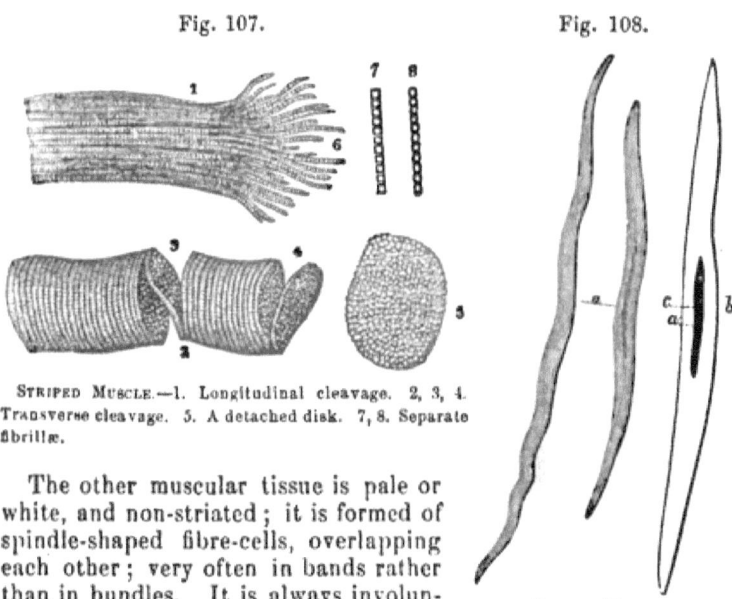

Fig. 107.   Fig. 108.

STRIPED MUSCLE.—1. Longitudinal cleavage. 2, 3, 4. Transverse cleavage. 5. A detached disk. 7, 8. Separate fibrillæ.

SMOOTH MUSCLE.

The other muscular tissue is pale or white, and non-striated; it is formed of spindle-shaped fibre-cells, overlapping each other; very often in bands rather than in bundles. It is always involuntary. Examples of it are in the muscular coat of the stomach and intestines,

---

[1] The truly cellular nature of these minute muscular elements is now denied by several physiologists.

20

230 PHYSIOLOGY.

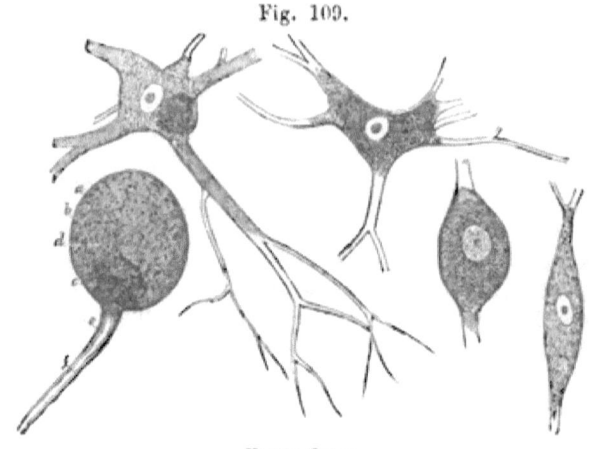

Fig. 109.

NERVE CELLS.

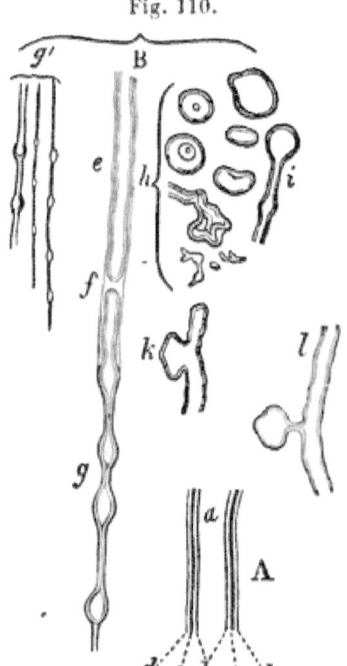

Fig. 110.

and in that of the bladder; the uterus, especially during pregnancy; in gland ducts, and the middle coat of arteries.

Nervous tissue is also of two kinds; the *gray vesicular* and the *white tubular;* the first in the ganglia, spinal marrow, and brain, though not constituting the whole of these. The first, formed of nucleated nerve-cells, many of them with processes (bipolar, multi-polar), is active, cumulative or reflective in function; the other, white tubular matter of the nerves and commissures, is simply capable of transmission and communication. Some of the nerves, especially of the "sympathetic" or ganglionic system, are gray and gelatinous; but still unlike the gray or cineritious nerve-tissue of the ganglionic centres.

NERVE FILAMENTS.—A. Diagram of nerve-tubule. *a.* Axis cylinder. *b.* Inner border of white substance. *c, c.* Outer border of same. *d, d.* Tubular membrane. B. Tubular fibres. *e.* In natural state. *f.* Under pressure. *g, g'.* Varicose fibres.

CHAPTER IV.
ORGANIC FUNCTIONS.

Two classes of functions (meaning by function action, operation) are performed by organs or apparatus in the animal body. One class is of actions common, in nature though not in method, to plants and animals. These are the functions connected with the *nutrition* of the organism; *organic* or *vegetative* functions. Such are absorption, assimilation, circulation of fluid, aeration, secretion, and reproduction. Others are entirely peculiar to animals; as sensation and spontaneous locomotion. These are *animal* functions, or functions of *relation*.

Comparison of Animals and Plants.

The differences between animals and plants are several. Plants are without most of the organs which the movements and endowments of animals require; as the stomach, liver, heart, etc. The *chemical composition* of the tissues of plants is simpler. Plants require *inorganic* matter for their *food;* as, carbonic acid, ammonia, potassa, etc.; animals must have *organic* matter; as, what we call our vegetable and animal food. Plants may thus be said to *prepare* organizable material for animals. Animals *elaborate* it further for their own substance, and then, by various actions, *restore* it again to the inorganic world. Water, though an inorganic substance, is a common vehicle for both. So, we find *water, carbonic acid,* and *ammonia* to be typical inorganic materials absorbed by plants through their leaves and roots, to be assimilated in the sap, and organized into stem, leaves, flowers, etc. Again water, carbonic acid, and ammonia are common and representative results of life-processes in animals, *after* organization has made them effete and thrown them out in excretions and exhalations.

Animals and plants both require *aeration*. But their action upon the air is different, even opposite. During the daytime, plants absorb carbonic acid from the air, and give out oxygen. Animals absorb oxygen, and return carbonic acid.

Sensation and locomotion are altogether animal functions. Yet, apparent exceptions to this exist; as, in the shrinking of the sensitive plant when touched, the closing together of the leaflets or lobes at the base of the leaf of the Venus' fly-trap, rhythmic movements of the stems of plants called oscillatoriæ, and actual locomotion of the zoospores (germinative seed-like particles) of Algæ. These are difficult of precise explanation; but it is evident that they are not the same in nature as the sensibility and truly spontaneous locomotion and other varied movements of animals. In their lowest and simplest forms, animals and plants approach each other very closely; so much so that doubt exists, in certain instances, to which kingdom to refer some of them. Some naturalists (Cassin and Wilson) have proposed, therefore, a third intermediate kingdom of *primalia*.

The *animal* functions, sensation, and spontaneous motion, are called functions of *relation*, because they bring the body into relation with the external world, and its own different parts with each other.

Man, as an animal, resembles in general structure the other animals of the class *Mammalia*, *i. e.*, those who suckle their young.

Fig. 111.

HAND OF MAN AND OF ORANG.

But he has some peculiarities, which, apart from his higher mental and spiritual endowments, separate him, even, from the apes and other *quadrumana*, which resemble him most. These are, in brief, the erect posture, curves of the spine, width and capacity of the pelvis, depth of the socket of the hip-joint, long legs and short arms, wide and strong knee-joint, firmly arched foot, backward-projecting heel, prominent chin, even rows of teeth, absence of intermaxillary bone in the upper jaw in the mature skeleton, head balanced *equally* upon the spine, large head and brain, speech, laughter, and tears.

PART II.
SPECIAL OR FUNCTIONAL PHYSIOLOGY.

CHAPTER I.
ALIMENTATION.

THIS comprehends, after *prehension*, or the taking of food, *mastication, insalivation, deglutition, digestion, absorption, assimilation,* and, as the final result of all, *nutrition.*

Mastication and Insalivation.

These are accomplished together; by the muscles of the jaws, which make the teeth divide the food, acting at the same time with those of the tongue, to mix the saliva with it. The temporal, masseter, and pterygoid muscles are those of mastication.

The *salivary glands* are the *parotid, submaxillary,* and *sublingual.* Different opinions exist as to the respective actions of their secretions. The *mucus* of the mouth, from numerous follicles, is added to them in mastication. It seems to be shown, that the mixed fluid will, out of the body, act upon starch, converting it first into dextrin, and then into sugar. Several experimenters have found saliva from the different glands to act in the same manner. As this effect is prevented by the presence of acid, it appears to be entirely arrested in the stomach. Dr. Dalton concludes that the saliva does not, therefore, digest starch; but that its solution is effected in the small intestine by the pancreatic juice. Dr. A. Flint, Jr., asserts the more generally held opinion, that at least a considerable part of the starch of food is changed into sugar by the saliva. This fluid is peculiar in containing, though not invariably, *sulpho-cyanogen.* Its active principle, ptyalin, or salivin, is a nitrogenous substance, analogous in composition and catalytic agency, to *diastase* of the seeds of plants; in which the change from starch to sugar is effected at a certain stage of germinal development.

Deglutition.

When the "bolus" of masticated food is forced by the muscles of the tongue and palate through the fauces and over the epiglottis into the pharynx, the constrictor muscles of the latter carry it downward; the contraction of the œsophagus upon it conveys it through the cardiac orifice of the stomach into that organ. An indispensable part of the process of deglutition is the adjustment of the *epiglottis*, as a lid to the larynx, which is by it protected

Fig. 112.

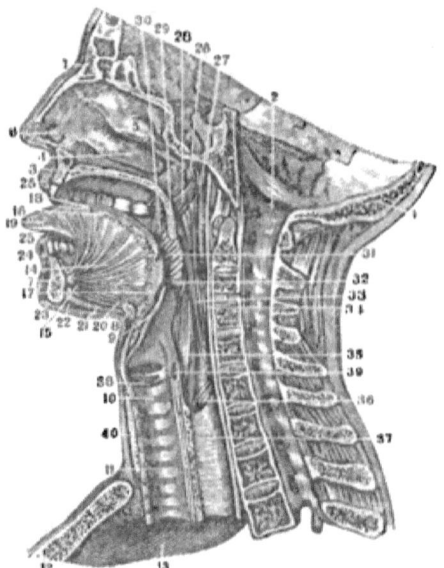

DEGLUTITION —1, 1. Section of head. 2. Spinal canal. 3. Hard palate. 9. Epiglottis. 11. Trachea. 30-34. Pharyngeal muscles. 35-37. Œsophagus.

from the entrance of what is swallowed; which must pass over the glottis, as the windpipe lies in front of the pharynx. If one breathes (as in laughing) at the moment of swallowing, *choking* ensues, from a morsel or a drop of liquid going "the wrong way." The irritability of the larynx is itself protective, by the violent spasmodic efforts produced, expelling the intruding substance. When we swallow, the pharynx and larynx are *raised up*, by the stylo-pharyngeus and superior constrictor muscles.

Gastric Digestion.

Entering the stomach, the food is still kept in motion; being carried slowly around by a sort of churning movement of the muscular fibres of the stomach. Thus the *gastric juice* is thoroughly mixed with it. This fluid consists of an acid solution containing *pepsin*. The acid is either chlorohydric (muriatic) or lactic acid; sometimes an acid salt of phosphoric acid. Dr. Dunglison found chlorohydric acid in Alexis St. Martin's[1] stomach, in 1833. The same subject was experimented upon by Profs. R. E. Rogers and F. G. Smith in 1856; these observers concluding that the principal agent of digestion was lactic acid.

Fig. 113.

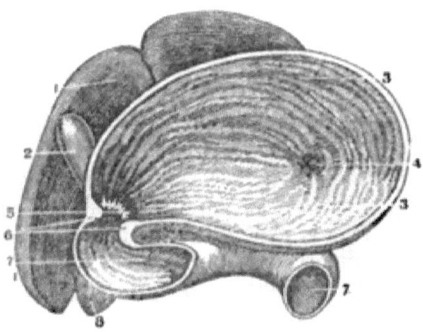

THE STOMACH.

Pepsin is a nitrogenous body, considered to be analogous in its mode of action to the fermenting principle of yeast, or to the

Fig. 114.

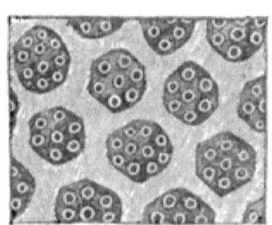

MUCOUS MEMBRANE OF THE STOMACH, MAGNIFIED.

Fig. 115.

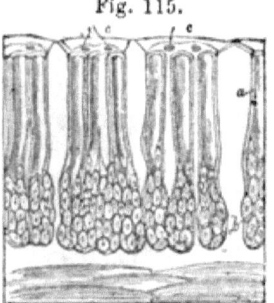

PERPENDICULAR SECTION OF THE SAME — *a*. Neck of a gastric tubule. *b*. Fundus. *c*. Orifices of tubules. *m*. Muscular coat.

diastase of plants, already alluded to. Being complex, it is prone to chemical change; and, by contact, institutes the *same kind* of chemical movement among the particles of food with which it is mixed. The gastric juice is believed to digest especially *nitrogenous* food; as the lean of meat, the gluten of bread, and the casein of milk. The products of this digestion have been, by Lehmann, called *peptones;* that of albuminoid food, *albuminose.*

[1] A patient of Dr. Beaumont of Ohio, whose stomach was wounded by the bursting of a gun, leaving a permanent opening or fistula.

Gastric juice is not present in any quantity in the stomach when it is empty; but begins to be secreted by its glands as soon as food is taken. The amounts of the digestive fluids secreted in a man in 24 hours are thus stated by Dr. Dalton:—

Saliva	2.880 pounds.
Gastric juice	14.000 "
Bile	2.420 "
Pancreatic juice	1.872 "
	21.172 "

The result of gastric digestion is called *chyme*. Besides the action of the gastric juice upon nitrogenous matters, no doubt some easily soluble substances are ready for absorption as soon as they enter the stomach. This is probably the case with sugar and dilute alcohol. *Strong* alcohol, as raw spirits, irritates the stomach, and interferes with secretion and absorption. Many dissolved medicines also are absorbed at once, into the capillary bloodvessels of the stomach.

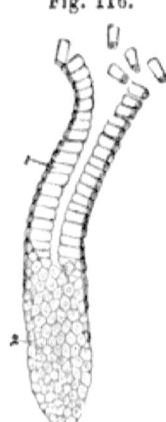

Fig. 116.

FOLLICLES OF PIG'S STOMACH.

The *pylorus* is a muscular valve constricting the left end of the stomach, so as to prevent the passage of undigested food. When reduced to chyme, it is allowed to pass into the duodenum.

Intestinal Digestion.

Bile and *pancreatic* secretion are poured into the duodenum by the ducts of the liver (and gall bladder) and pancreas. It is a common opinion that both continue and complete the digestion of food; especially *fatty* food. In two modes is this believed to take place. One by the alkaline material (soda, potassa) of the two secretions *saponifying* the fat, that is, making a soap by combining with the fatty acid, oleic, margaric, or stearic, etc. Soap is soluble in water, and thus absorbable; as fat or oil is not. Also, an *emulsification* or suspension takes place, like that made by mixing oil first with gum Arabic and then with water; so that in the state of minute subdivision, almost identical with solution, the oil may be absorbed. Dr. Dalton, however, does not admit that the bile takes part in digestion, although acknowledging that very little of it is excreted from the bowels, and that the larger amount of it not passing out must have *some* important function to perform before its re-absorption from the intestinal canal.

Absorption.

The pancreatic juice, containing the organic agent *pancreatin*, is considered by Dr. Dalton and others to *emulsify* fatty materials of food. By this process, and the continued action of the gastric juice derived from the stomach, *chyle* is formed. This is absorbed by the *villi* or minute velvety tufts, of the terminations of the *lacteal* vessels. Each villus contains the loop-like beginning of a lacteal tube; and each is covered by a layer of epithelial cells. It is uncertain whether these cells fill by absorption and then burst into the interior of the villus, or whether they simply *transmit* the chyle; probably the latter. Lacteals receive their name from the milky appearance of chyle, which is especially marked after a meal. All of these vessels pass through mesenteric glands to empty into the thoracic duct. Besides lacteal absorption, the *bloodvessels* of the small intestine, like those of the stomach, absorb the products of digestion. The veins of the upper portion of the alimentary canal empty into the *portal vein*. This

Fig. 117.

VILLI OF INTESTINE.

Fig. 118.

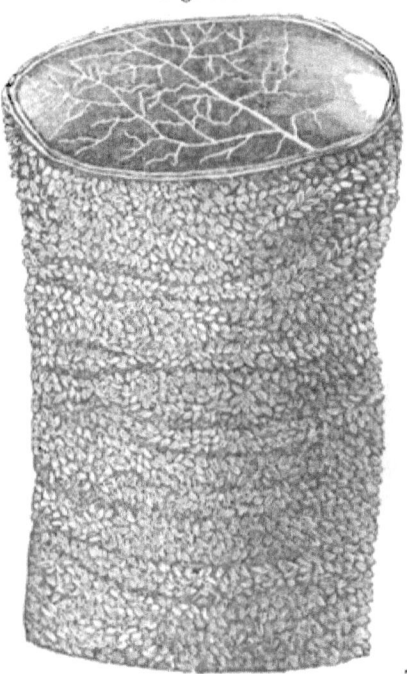

PIECE OF ILEUM.

goes to the liver, subdividing into capillaries as it enters that organ. Probably the secretion of the *glands of the small intestine* (succus entericus) may have some digestive action; but it has not yet been demonstrated. *Undigested* food is, as refuse, conveyed, by the peristaltic action of the intestinal tube, through the ileo-colic valve to the large intestine, to be excreted as a part of the feces.

Assimilation.

With good reason, this is ascribed as a principal function to the liver; through which so much blood, enriched, after eating, by the materials digested and absorbed, passes, entering by the portal vein. Exactly what is done in this process we cannot explain. After death, Bernard and others have found a *saccharine* substance, glucose, liver-sugar. Pavy asserts this to be a post-mortem educt. A *sugar-producing* substance, at least, glycogen (hepatin, liver-dextrin), must be admitted to be formed naturally in the liver. Its after destination is doubtful, except that, by the hepatic vein, it goes on toward the general circulation. Some believe that it acts as a "fuel for combustion," for animal heat, being "burned off" in the lungs. Dr. McDonnel has proposed the view that, in assimilation of blood brought by the portal vein, glycogen combines with nitrogenous materials of the food, to make plastic material for tissue.

Besides the liver, the *mesenteric glands* are almost certainly assimilating organs. Chyle is obviously altered by its passage through them. In fœtal and infantile life the *thymus* and *thyroid* glands probably have the same use. The *lymphatic* glands are supposed to restore to the lymph reabsorbed by the lymphatics (as the surplus of nutrition) all over the body, some qualities necessary for its farther utility. Gray ascribes to the *spleen* the office of regulating the quality of the blood, by producing new blood corpuscles when they are deficient, and destroying a portion of them when they are excessive. It may be remarked that this theory is not certainly established. More probable is the opinion that the spleen is a *diverticulum* or reservoir of blood, receiving it especially from the stomach when that organ is function-

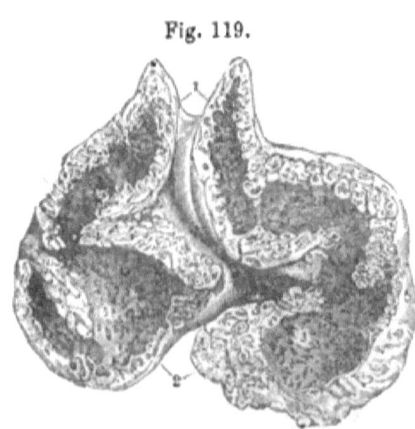

Fig. 119.

THYMUS.

NUTRITION. 239

ally inactive. A case has been reported (*Med. Times and Gazette,* Dec. 7, 1867) in which the spleen was removed entirely; yet the woman recovered and seemed to have good health.

Nutrition.

This term may, of course, be attached to *all* that concerns the alimentation of the body. Physiologically, however, it is applied

Fig. 120.

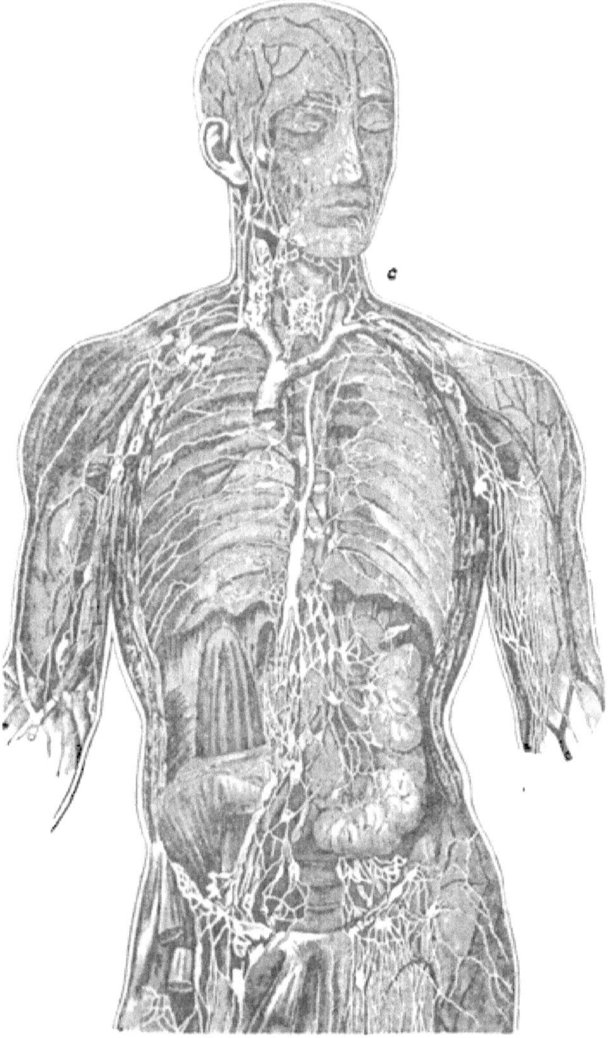

THE LYMPHATICS.—*a.* Receptaculum chyli, commencing thoracic duct. *c.* Descent of the latter to its termination *r.* Innominato vein.

more especially to the direct appropriation of plasma of the blood to the building up of the tissues. Nutrition, in this sense, comprises four processes: *formation, development, growth,* and *repair.* The first two of these are predominant in embryonic life. Growth, as well as development, goes on from conception *in utero* to maturity. After that, *repair* is the only result of nutrition; repair of tissue destroyed in active or passive waste, or by disease or injury. Construction and destruction are going on together throughout life.

In the formation and repair of organs, the *selective* power of cells, or their nuclei, is manifest. *Harmony* of action is also observed in the construction of contiguous or related parts, as though there were a purposive combination among them. Thus, the eye and its orbit, the brain and the skull, are proportioned to each other. Most wonderful is the exact *symmetry* of the two sides of the body and of most of the organs. When this is interfered with, during gestation and development, deformity results; as *spina bifida* from imperfect union of the two halves of the spinal column, or hare-lip from a similar want of closure near the middle of the face.

The conditions necessary to the healthy nutrition of any part of the body are, 1, a sufficient supply of blood; 2, good quality of the blood; 3, supply of nerve-force; 4, functional exercise; 5, due intervals of repose.

CHAPTER II.

CIRCULATION.

The distribution of the blood throughout the body is effected by the *heart, arteries, capillaries,* and *veins;* a continuous closed system, with no outlet, except by transudation through the walls of the capillaries.

Action of the Heart.

The *heart* in man is double, as though two hearts were placed side by side. In fœtal life they communicate directly; but after birth indirectly only. One-half of the heart, the right, receives venous blood from the body and propels it to the lungs; the other, the left, receives arterial blood from the lungs, and, through the aorta and its branches, sends it all over the body. The right half might, therefore, be called the *respiratory* heart, and the left the *systemic.*

Cavities.—Each of these halves of the heart has two cavities; an auricle and a ventricle. The first is a receiving, and the last a propelling cavity. The right auricle receives venous blood from the venæ cavæ, and pushes it on into the right ventricle. This then propels it, through the pulmonary artery, to the lungs. The left auricle receives blood from the lungs, and transfers it to the left ventricle, which then propels it out by the aorta.

The size of the heart is about that of the closed fist. Anatomists state that it continues to grow later in life than any other organ. Each of the ventricles will hold about three ounces; each of the auricles, rather less. The walls of the ventricles are much thicker than those of the auricles; those of the left being thickest. The force of contraction of the left ventricle is estimated by Valentin at $\frac{1}{50}$ of the weight of the whole body; of the right, half as much. The latter has to send blood only through the lungs; the former, through the whole body.

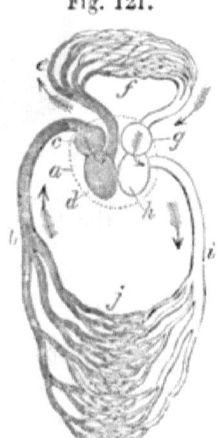

Fig. 121.

DIAGRAM OF THE CIRCULATION.

The tissue of the heart (inclosed in the pericardial sac) is *muscular;* of red muscular fibres, spirally arranged; a sort of double or returning spiral. When it contracts, that is, during the *systole* or contraction of the ventricles, the heart elongates (as shown by Dr. Pennock, of Philadelphia), and is twisted forwards so as to strike the left side below the nipple. This constitutes the *impulse.* The *dilatation* of the cavities of the heart appears to result from elasticity only; it has almost no appreciable suction power.

Valves.—Between each auricle and the corresponding ventricle there is a membranous and muscular valve; the tricuspid valve for the right side, the mitral for the left. (See *Anatomy.*) After the auricle contracts, the contraction of the ventricle follows; and, with this, the muscular columns of the auriculo-ventricular valve close it against the return of the blood. When the ventricles have contracted, the rebound of the arteries pushes out the pocket-like semilunar valves of the aorta and pulmonary artery, so as to close them together.

The *cause* of the heart's action is, probably, the contractility of its muscular tissue, under the stimulation of oxygenated blood. Brown-Séquard's theory of its being due to the action of carbonic acid in the blood is untenable. Rhythmic (*i. e.*, regularly alternating or successive) contraction is the general, indeed the universal law of healthy muscular tissue; as has been proved lately by M. Marey's experiments, even when it seems to be continuous.

The form of the heart and the arrangement of its fibres, are such as to give a *magnitude* to its *alternation* of action and repose, such as is only seen in the body elsewhere in the muscular movements of respiration. Although minute ganglia are discoverable in the

Fig. 122.

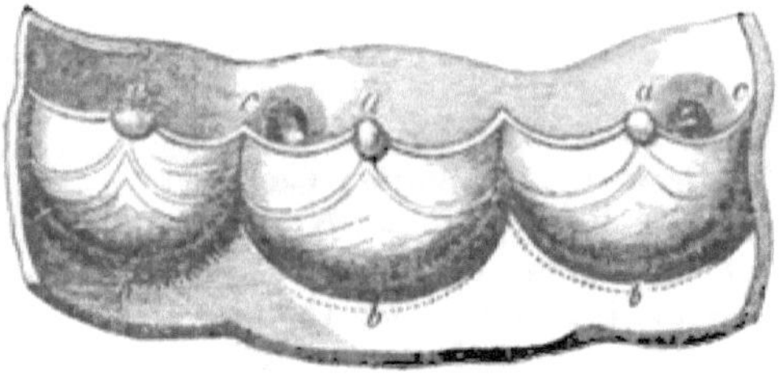

SEMILUNAR VALVES.

tissue of the heart, and branches of the pneumogastric nerve go to it, so that it is under the influence of the nervous system, and is often much affected by its condition (as in emotion), yet this influence seems to be modifying rather than essential.

Sounds.—Placing the ear over the heart, we hear two sounds—lub-dup—the first longest and loudest. If we divide the whole time of the two sounds and the following pause into four equal parts, the first sound, and the interval between it and the second, will occupy two of these, or half of the whole time of the rhythm; the second sound, nearly one part or one-fourth of the whole; and the pause a little more than one-fourth. The first sound occurs with the *systole* or contraction of the ventricles; the second, with their *diastole* or dilatation.

The *causes* of the first sound are, 1, the *closing, with vibration*, of the *auriculo-ventricular valves;* 2, the impulse against the wall of the chest; 3, the rush of blood into the vessels; 4, the friction of the muscular fibres of the heart against each other. The cause of the second sound is, the flapping together of the semilunar valves of the aorta and pulmonary artery, with the arterial rebound during the diastole of the ventricles. During the *first* sound, the ventricles are contracting, and the auriculo-ventricular valves are closed; the semilunar arterial valves are open. During the *second* sound, the ventricles are dilating, the auricles contracting, the mitral and tricuspid valves are open, and the semilunar valves of the arteries closed.

The heart contracts, in an adult, from 70 to 75 times in a minute, while in health and at rest. Its average rate is:—

At birth, times in a minute	.	.	.	.	.	140 to 120
First year " " "	.	.	.	.	.	120 " 115
Second year, times in a minute	.	.	.	.	115 " 100
Third year " " "	.	.	.	.	.	100 " 90
Seventh year " " "	.	.	.	.	.	90 " 85
Fourteenth year " " "	.	.	.	.	85 " 80
Middle life " " "	.	.	.	.	.	75 " 70
Old age " " "	.	.	.	.	.	70 " 50

In very advanced age, however, sometimes it quickens greatly. Dr. Guy found that the pulse was most rapid in the standing

Fig. 123.

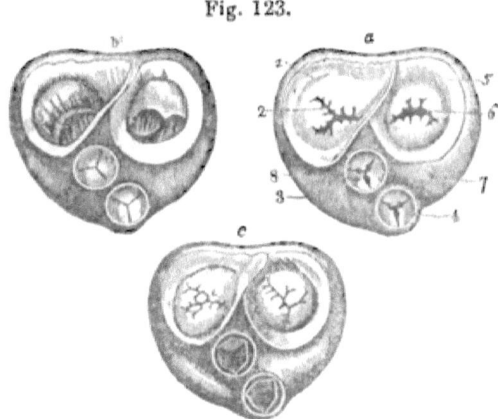

VEINS OF THE BASE OF THE HEART.—*a*, 1. Right auricle. 2. Tricuspid valve. 3. Right ventricle. 4. Pulmonary artery. 5. Left auricle. 6. Mitral valve. 7. Left ventricle. 8. Aorta. *b* shows the mitral and bicuspid valves open, and the arterial valves closed. *c* shows the opposite; as during the *systole* of the ventricles.

posture; next sitting, slower lying down. This depends on the muscular effort made in supporting the body. In the female it is a little more rapid than in the male of the same age. In disease it is much more often accelerated than retarded. Great debility is nearly always accompanied by acceleration as well as feebleness of the pulse at the wrist, and of the impulse of the heart. Guy found the pulse somewhat more rapid in the morning than in the evening during health. In disease, it is commonly most rapid in the evening.

Ordinarily, there is a nearly constant relation between the pulse and the frequency of the respiratory movements; there being one pulsation of the heart to three or four respirations.

The Arteries.

All arteries, except the largest, having (unstriped) muscular tissue in their middle coat—the smallest arteries the most—this must have some influence upon the flow of the blood through them after the reception of it from the heart. The arteries are found empty after death; as, in their last contraction, they force the blood into the less resistant veins.

The common view among physiologists at the present time is, that the sole office of the muscularity of the arteries is, to *limit*, by *resistance*, the amount of blood passing through them. That is, as Virchow has expressed it, the more healthy and vigorous the action of an artery, the less blood goes through it. This opinion is founded upon some experiments of the Webers; who found that the intestinal canal, heart, and arteries, when powerfully acted upon by galvanism, were thrown into a state of rigid or *tonic* contraction. They thence concluded, that, though the intestinal tube has peristaltic contraction, and the heart an alternating impulse, the arteries have normally only a power to become rigid with a certain force when blood is forced into them.[1] The *elasticity* of the arteries must have much to do with the change of the flow of blood, gradually, from an intermittent to a steady stream, such as we find passing from the capillaries to the veins, and through the latter to the heart. If we admitted an active propulsive power in the arteries, supplementary (as Sir Charles Bell held) to that of the heart, the pulse at the wrist, or elsewhere, would be explained by these forces combined. On the current view, however, the pulsation of arteries is due entirely to the impelling action of the heart, driving the blood through them. At the same time, it is well understood that the regulation of the varying supply of blood to different parts must depend mainly upon the condition of the arteries; as the heart acts impartially towards all, having one trunk only, the aorta, to give out its supplies. In the growth of the deer's horn in the spring, the *rutting* or periodical genital excitement of many animals, the development of the uterus during gestation, and of the mammary gland before and during lactation, there is an unusual flow of blood through those organs. Under other circumstances, as in Bernard's experiment of dividing the sympathetic nerve in the neck, arteries are *dilated* passively, by paralysis of the muscular coat. The flushing of the skin upon a blow or friction, or under the stimulation of mustard, ammonia, etc., seems to show a *reflex* action of the bloodvessels, under nervous influence.

On the fact that the *vaso-motor* nerves, *i. e.*, those which go to the bloodvessels, are all derived from the ganglia of the sympathetic system, certain speculations concerning the action of cold

---

[1] See *Transactions of the American Medical Association*, 1856, for an argument by the present writer (Prize Essay on the Arterial Circulation) in opposition to this view, and advocating the existence of an *actively propelling* power in the arteries. Also, *American Journal of Medical Sciences*, July, 1868, p. 288.

and heat upon the circulation, through the ganglia, are founded; making the basis of the "ice-bag" and "hot-water-bag" practice of Dr. John Chapman. Neither the practice nor the theory is as yet established; nor does the one necessarily depend upon the other.

The Capillaries.

Having but a single coat, without muscularity, these intermediate vessels can have no office but to subdivide, in fine networks,

Fig. 124.

 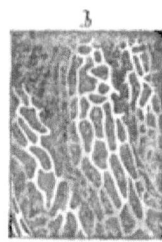

CAPILLARIES.—*a*. Capillaries of the papillæ of the skin of a finger. *b*. Capillaries of villi of small intestines.

the blood; so that it may afford nutriment, by transudation of the plasma through their walls; or, in the glands, allow secretion; or, in the lungs, expose the blood to the air. Capillaries contract only by elasticity; so as, after being dilated, to return, on the withdrawal of pressure, to their ordinary dimensions.

Yet, two powers have been pointed out as, in the capillary region, contributing to the movement of the blood. One of these, as shown by Dr. Draper, is common to plants, animals, and some materials of an inorganic nature; viz., capillary action; that is, the attraction of the walls of fine tubes for liquid in which they are immersed, varying with the smallness of the tube and the nature of the materials used.

The other agency, first pointed out by Dr. Draper, is also present both in animals and plants. It is

Fig. 125.

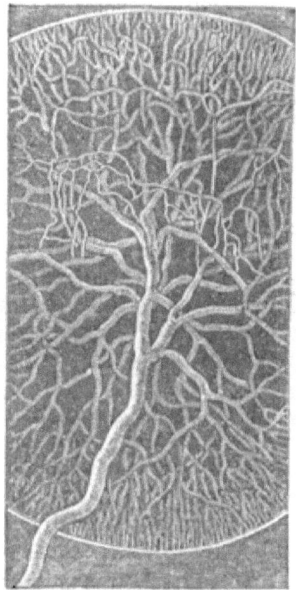

CAPILLARIES OF A TOOTH.

the attraction which the tissues of the organism have for the nutritive materials circulating in its vessels. This "vital or nutritive affinity" is a *vis à fronte*; which, as it constantly takes, in nutrition, particles from the blood in the capillaries, must diminish pressure in resistance, and favor the onward flow.

The Veins.

Valves along the course of nearly all the veins, opening only towards the heart, economize the power used in returning the blood through them (from the capillaries) to the heart. The pressure of the muscles during exercise contributes to the same end. So does *inspiration* tend to promote the return of the blood through the venæ cavæ to the heart; since the lifting of the ribs lessens the pressure upon the heart's surface, or, in other words, exerts some "suction" power upon it. The *larger* veins have an appreciable amount of muscular tissue; the smaller ones, none. It is natural to suppose that this is because the largest, being nearest the end of the round of circulation at the heart, require the most power to complete the circuit. The *velocity* of the blood-movement is greater in the arteries than in the veins. The *capacity* of the venous system is about three times as great as that of the arterial system; and the angles at which branches join the veins are much larger than those at which branches leave the arterial trunks. These facts account for the greater slowness of the venous current.

By experiments with chemical reagents introduced into the veins, it has been shown that the blood passes round its course in *less than half a minute*, in some instances, and in others in about a minute. It is not equally rapid at all times, nor even through all the different organs of the same body.

Route of the Circulation.

Although perhaps not necessary in this place, we may recapitulate the round of the circulation, as follows: Beginning at the aorta, the blood is distributed by its branches to all parts of the body. The small arteries terminating finally in capillaries, these, in various networks, subdivide the blood for the supply of the different organs. Then the capillaries unite to form veins, and these, larger ones, till finally all combine to end in the ascending and descending *venæ cavæ*. These empty into the right auricle. This pours its blood through the tricuspid valve into the right ventricle; which, by the pulmonary artery and its branches, sends it to the lungs. Thence, by the four pulmonary veins, the blood is brought to the left auricle. Through the mitral valve, it is passed on into the left ventricle; whence it again is thrown into the aorta. The *portal* circulation of the liver is, as has been already explained, a deviation from this simple course of the general system; since the portal vein, instead of going to empty its contents directly into the venæ cavæ, breaks up into capillaries to enter the liver; whose blood is then collected by the hepatic vein, by which it is conveyed to the vena cava.

The discovery of the course of the circulation was made by Dr. William Harvey, about 1619.

CHAPTER III.
RESPIRATION.

This function has for its purpose the aeration of the blood. It is accomplished by the exposure of the venous blood, brought from the right half of the heart, to the air received into the air-vesicles of the lungs. The immense number of these vesicles (about six hundred millions), provides a very large expansion of surface. Air and blood both periodically enter and pass through the lungs; although the blood is entirely confined within the capillaries. The heart sends a new supply of venous blood with every systole; the lungs receive a fresh quantity of air with each inhalation.

Movements of Respiration.

These are, *inspiration* and *expiration*. The first is accomplished by expanding the chest, so as to take pressure from the outside of the lungs, while the mouth or nostrils are open to allow the entrance of air. It is precisely the action of filling a pair of bellows with air by drawing apart the handles. This expansion of the chest is effected in two ways: 1, elevation of the ribs by the intercostal muscles; 2, depression of the diaphragm by its own contraction.

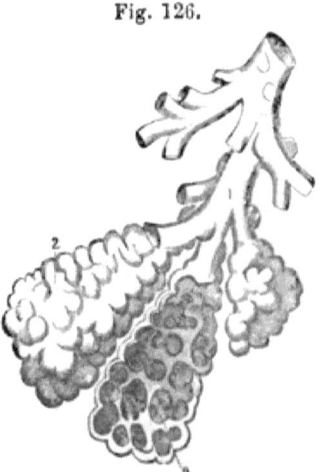

Fig. 126.

DIAGRAM OF AIR-CELLS.—1. Small bronchial tube. 2. Vesicular portion of lobule. 3. The same, laid open.

The intercostal muscles, internal and external, have their fibres crossing in opposite directions. This obliquity adds to the extent of their action in lifting the ribs; as every muscle shortens, in its contraction, about one-third of its length, therefore the longer it is, the greater the distance through which it draws what it moves. Breathing chiefly by action of the intercostal muscles in lifting the ribs is called *costal* respiration; by the diaphragm mainly, *abdominal* respiration. The latter is observed in young children; the costal type, in women. Abdominal respiration is so called, because, when the diaphragm descends, it forces out the organs of the abdomen perceptibly.

In violent efforts of inspiration, as in asthma or croup, accessory muscles of respiration assist the intercostals and diaphragm. The principal of these are the *levatores costarum*. But other muscles may contribute aid—even those of the neck; and sometimes the nostrils are forcibly dilated in the struggle for air.

Ordinary *expiration* follows the cessation of the muscular act of inspiration, not requiring any positive effort of muscular contraction. The weight of the ribs causes them to fall; the elasticity of the diaphragm makes it ascend; the same property in the lungs induces their contraction and the expulsion of the air. The elasticity of the costal cartilages also assists. *Forced* expiration, however, as in blowing hard, involves (besides the accessory action of the *sterno-costalis* muscle) compression upward of the diaphragm, by the superficial *muscles of the abdomen* (external and internal oblique, transversalis, and rectus) pressing in the contents of the cavity under them.

Quantity of Air Changed.

With each breath, a man changes about twenty-five cubic inches of air. By forced expiration, one can expel a much larger amount. Still a quantity will remain in the lungs, which cannot be driven out. This *residual* air varies probably from forty to two hundred and sixty cubic inches.

After a deep inspiration, a healthy man, five feet seven inches in height, can, on the average, expel from his lungs two hundred and twenty-five cubic inches of air. This was called, by Mr. Hutchinson, the *vital capacity*. For every inch above the height just named, the capacity of the chest increases about eight cubic inches; and for every inch below, it is diminished in the same proportion. Less regular correspondence exists in regard to weight. Mostly it does not vary much with weight under one hundred and sixty-one pounds. Over that, each additional pound of weight brings a cubic inch of diminution in vital capacity, so called. With age, this increases from fifteen to thirty-five years, at the rate of five cubic inches per year. Then it diminishes, one and a half cubic inches each year, to sixty-five years. Bourgery states that women have but half the breathing capacity of men of the same age.

The *number* of respirations usual in an adult in health is from fourteen to eighteen in a minute. The *force* of ordinary inspiration is calculated to be equal to a weight of two hundred pounds lifted.

Changes of the Air Breathed.

Common air consists of about seventy-nine volumes of nitrogen, and twenty-one of oxygen; with about four parts in ten thousand of carbonic acid, a variable amount of watery vapor, and some non-essential gases and particles.

After passing through the lungs, a portion of air has become *warmer;* its *oxygen* is *diminished,* its *carbonic acid* and *watery vapor increased.* The increase of carbonic acid may be shown by the milky turbidness produced by breathing into clear lime-water. The vapor of water in the breath is made known in the open air on a cold day, by the cloud condensing near the nostrils or mouth.

For every thousand volumes of carbonic acid exhaled, Valentin and Brunner assert that over one thousand one hundred and seventy-four volumes of oxygen gas are absorbed into the blood. In an hour, 1583.6 cubic inches of oxygen are absorbed on the average. Pettenkofer has shown that more oxygen is absorbed at night than during the day.

Of carbonic acid, about 1345.3 cubic inches are given out every hour, containing one hundred and seventy-three grains of carbon; or eight ounces of carbon in twenty-four hours. The amount of this exhalation varies, however, with *age, sex, temperature,* and *purity* of the air.

From eight to thirty years of age, in males, the amount of carbonic acid exhaled increases; from thirty to forty it is nearly the same; after that time it diminishes gradually. In females, it is less than in males of the same age; it increases from eight years till puberty, and then remains stationary throughout the menstrual and child-bearing period of life.

The *faster* one breathes, the *less* is the proportionate amount of carbonic acid exhaled. As to *temperature,* between $38°$ and $75°$ Fahr., every rise of $10°$ is attended by a lessening in the carbonic acid given out, of two cubic inches per minute.

An atmosphere containing five or six per cent. of carbonic acid gas is not capable of long sustaining life. Ten per cent. may produce immediate danger. While pure carbonic acid is irrespirable, its dilution causes it to be tolerated by the breathing organs. Hence the peril of life in some deep wells, brewers' vats, and rooms in which charcoal is allowed to burn without ventilation.

The use of *food* increases the amount of carbonic acid in the air expired. *Alcoholic* drinks diminish it. *Exercise* increases it. Sleep diminishes it.

The amount of *watery vapor* exhaled from the lungs is, on the average, a pint in twenty-four hours.

Changes Produced by Respiration in the Blood.

The *color* of the blood is altered in the lungs, from dark crimson or purple to bright scarlet. The blood is also 1° or 2° *warmer;*[1] it contains *more oxygen, less carbonic acid,* and more fibrin. The introduction of oxygen gas, and the elimination of carbonic acid gas, are, as already observed, the two great purposes of respiration. Venous blood is that which has been, by various influences during its flow, rendered unfit for the support of vital energy. Arterial blood has been revivified by its purification and oxygenation.

When these changes are prevented, as in strangulation, drowning, or *asphyxia* by irrespirable gases, the dark blood is unable to maintain the vitality of the nerve-centres; and the blood ceases even to flow through the vessels. Drowning occurs, therefore, not from any directly injurious effect of the water in the lungs, but from the simple exclusion of air. So in some of the deaths from inhalation of chloroform, the cause probably has been the deficient admixture of air with the anæsthetic. That substance is, however, capable of causing fatal arrest of respiration, apparently by its toxic influence upon the *medulla oblongata,* the nerve-centre of respiration.

Recovery from drowning seldom occurs when the individual has been submerged as long as five minutes. Rare instances are narrated, in which it has been fifteen minutes. Even practised pearl-divers can seldom stay under water for one whole minute at a time.

Animal Heat.

In the armpit, or under the tongue, the temperature of the adult human body is, in health, 98°.4 or 98°.5 Fahr. The heat of the blood is 100° to 103°. Children have a temperature two or three degrees higher. In disease, especially in scarlet fever and yellow fever, it has been known to reach 108°, and, it is said, 112°. Cholera, pernicious fever and *cyanosis* are attended by depression of temperature. In cholera, it has gone down during life to 77°. Other parts of the body are cooler than the armpit; the sole of the foot does not average in health above 90°. During sleep, the heat of the body goes down about 1½ degrees. It is highest early in the morning, fluctuates through the day, and is lowest about midnight. Exercise elevates it considerably; eating does so to a less extent. The reaction following a cold bath may raise it one degree or more.

As we are constantly giving off heat by radiation, conduction, and evaporation to surrounding bodies, it must be supplied by processes going on within the system. The explanation long held

---

[1] This is not admitted by all observers.

by physiologists to be most probable is, that our animal heat is produced by *slow combustion;* that is, the union of oxygen with the carbon, hydrogen, nitrogen, sulphur and other elements of the blood and tissues, giving out heat less rapidly, but in the same quantity, as when wood, coal, oil or other fuel is burned in the air. Liebig has asserted, on calculation, that the amount of carbon and hydrogen shown to unite with oxygen in the body is sufficient to account for all its animal heat. Warm-blooded animals always breathe a great deal of air (birds, for example), and consume a great deal of carbonaceous food. Whether materials of food are ever "burned off" from the blood in the generation of heat, without entering first into the tissues, is not certain; probably it is so. In cold climates, Arctic explorers have found the demand for fatty (carbohydrogenous) food to be very much greater than in warm or temperate regions.

It must be understood, however, that the "combustion" of materials in the body is not, like that of wood or coal, a simple process of direct conversion of carbon, by oxidation, into carbonic acid, and of hydrogen into water. Step by step combinations are formed, of which the *last results*, only, are these familiar substances.

Prof. Dalton regards animal heat as the result of a chemical, but not strictly of a combustive process. His language is, in part, as follows:[1] "The numerous combinations and decompositions which follow each other incessantly during the nutritive process, result in the production of an internal or vital heat, which is present in both animals and vegetables, and which varies in amount in different species, in the same individual at different times, and even in different parts and organs of the same body."

The *nervous* system has a considerable though unexplained influence over animal heat. This is shown by the coldness following great shocks to the nervous centres, the loss of temperature in paralyzed limbs, and the occasional increase of temperature under nervous excitement.

The power of resisting the depressing action of exposure to cold is greatest in adolescence; least in infancy and old age. Clothing, by its non-conducting property, *retains* heat, that is, prevents or retards its loss; but it does not *make* us warm, in a positive sense.

[1] Treatise on Physiology, 4th edition, p. 247.

CHAPTER IV.

EXCRETION.

BESIDES carbonic acid, which must be thrown out from the blood, other substances, results of chemical changes in the different parts of the living body, have to be removed from it. No particle seems to remain *permanently* in the form and condition into which it is organized; but each passes from the organic to the *effete* or post-organic state; when, if retained, it will be obstructive and injurious to the system. Poisonous and even fatal effects may result from the retention in the blood of excrementitious matter; as, in *uræmia*, when the action of the kidneys is suppressed; *cholæmia*, when the liver fails to secrete bile, etc. *Toxæmia* is blood-poisoning from any cause. This is prevented very often, even when deleterious agents have been taken into the blood, by the emunctories or excretory organs eliminating it.

Excretion is always a secretory process; but secretion is not always excretion. The former term is applicable whenever anything is, by glandular or follicular action (*i. e.*, by the selective power of cells), separated from the blood. The latter, excretion, occurs only when the material removed is altogether *waste*, and cannot be used for any purpose connected with the organism. Milk, for instance, is a secretion, but not an excretion; because it is available, and is produced, for the nourishment of offspring. Urine and feces are entirely excretory. Bile is only partly so.

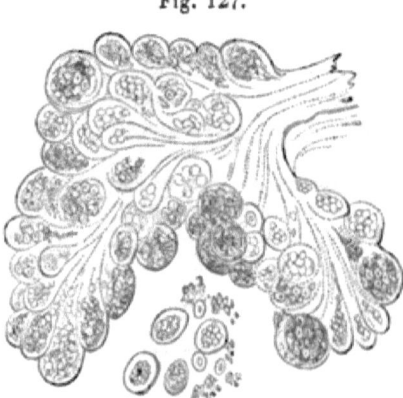

Fig. 127.

BRUNNER'S GLAND, MAGNIFIED.

Secretion and excretion being, however, so nearly alike in nature, we may, without impropriety, enumerate together their most definite products, as follows:—

Ptyalin,
Pepsin,
Pancreatin,
Creatin,
Creatinin,
Lactic Acid,
Lactin,
Butyrin,

Cholesterin,
Taurocholic Acid,
Glycocholic Acid,
Glycogen,
Excretin,
Stercorin,
Urea,
Uric Acid.

Pigments, as biliary coloring matter (biliverdin, biliphein, or cholepyrrhin) and coloring matter of the urine (urosacin, uroxanthin).

Also, excretory *salts;* as urates, phosphates, sulphates, &c., dissolved in water.

The most important excrementitious substances of the body are thus stated by Dalton :—

1. Carbonic Acid . . . . . CO_2.
2. Urea . . . . . . $C_2H_4N_2O_2$.
3. Creatin . . . . . . $C_8H_9N_3O_4$.
4. Creatinin . . . . . $C_8H_7N_3O_2$.
5. Urate of Soda . . . . $NaO,C_5HN_2O_2+HO$.
6. Urate of Potassa . . . . $KO,C_5HN_2O_2$.
7. Urate of Ammonia . . . $NH_4O,2C_5HN_2O_2+HO$.

The organs which are altogether excretory, in the human economy, are, the kidneys and the large intestine. Partly so, are the lungs, liver, and skin. Having considered already the functional action of the lungs, we may now briefly attend to that of the liver, kidneys, bowels, and skin.

Secretion of Bile.

Only the liver, of all the glands of the human body, is supplied with venous as well as arterial blood. Although the main purpose of this is, probably, the assimilation of crude blood coming from the digestive organs, it is not possible to say whether the bile is mainly produced from the blood of the hepatic artery or from that of the portal vein. That vein is supplied by branches from the stomach, spleen, pancreas, and small intestine. Entering the liver by two main branches, the portal vein subdivides and ramifies into the *interlobular veins*. These, as well as the minute branches of the hepatic artery, make the capillary networks, which surround the *acini* or lobules of the liver. From the centre of each of these lobules or "islets," goes off a ramule (intra-lobular vein) contributing

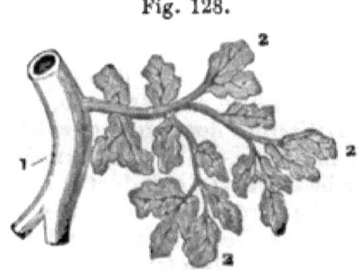

Fig. 128.

LOBULE OF LIVER.

to the hepatic vein. From the same acini also pass off the tubules which carry bile, and which by combining make finally the biliary or hepatic duct. Among the capillary meshes of the acini, and inclosed within each, so as to be in direct contact with the biliary tubules, lie the *secreting hepatic cells.* These take from the blood materials from which they elaborate the bile.

Leaving the liver, the bile commonly goes backward through the gall-duct to the *gall-bladder,* where it is held in reserve, to be

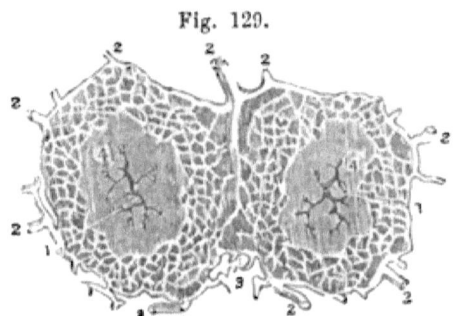

Fig. 129.

LOBULE OF LIVER.

forced out, by the *ductus communis choledochus,* into the duodenum, from time to time.

Human bile is yellowish-brown in color, and of a peculiar acrid or bitter taste. Its reaction to test paper is disguised by its bleaching litmus; but it is probably neutral when fresh, tending to alkalinity on keeping. It makes a lather-like foam when shaken in a tube. Nearly two and a half pounds of bile are estimated to be secreted by an adult in twenty-four hours.

Characteristic *ingredients* of bile are, *biliverdin* (coloring principle), *cholesterin, glyco-cholate* and *tauro-cholate* of *soda;* also, chloride of sodium, oleate, margarate, and stearate of soda and potassa, carbonate and phosphate of soda and potassa, and phosphates of lime and magnesia.

Biliverdin does not pre-exist in the blood. It must be formed in the liver. After its formation, it may be re-absorbed, when, for instance, the gall-duct is obstructed by gall-stones, and then it may be thrown out from the blood into the skin (jaundice) and tissues and secretions generally. It is a nitrogenous substance.

Cholesterin is a fat-like non-nitrogenous crystallizable substance, distinguished from the fats by not making soap with alkalies. It is not formed in the liver, but reaches it in the blood, being derived apparently from the waste of tissue in the brain and other parts of the nervous system, and from the spleen. Cholesterin is, according to the investigations of Prof. A. Flint, Jr., changed into other substances (stercorin, excretin) in the intestinal canal; not being found in the feces.

Bilin or *biliary resin* consists chiefly of *glyco-cholate* and *tauro-cholate of soda*. The former of these crystallizes readily; the latter with difficulty, if at all. They are distinguished also by the fact that the first is precipitated by acetate of lead, while the other is not. Both are nitrogenous; but tauro-cholic acid is peculiar in containing sulphur. These substances are *formed in the liver*.

Pettenkofer's test for bile is believed to be the best. It consists in mixing with the liquid to be examined a little cane sugar, and then adding sulphuric acid, drop by drop. A red color appears, changing gradually to lake, and finally opaque purple.

Biliary *coloring* matter, but not the resinous salts of the bile, is tested by nitric acid; which produces a green color with it.

Fig. 130.

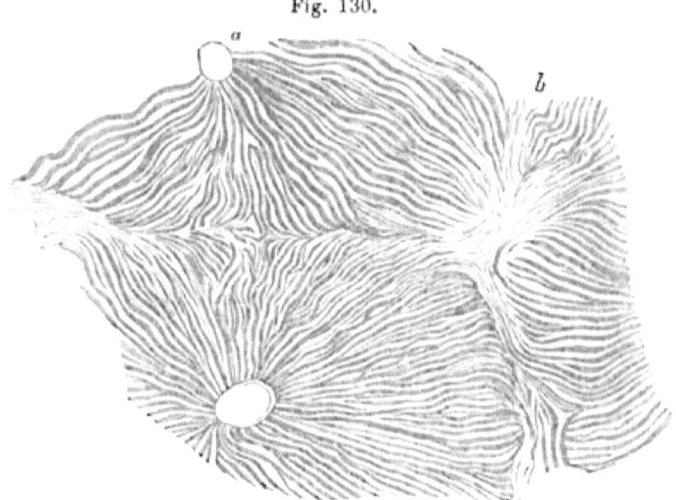

SECTION OF LIVER OF THE HORSE.

Uses of the Bile.—Most physiologists ascribe to the bile a share with the pancreatic secretion in the digestion of the fat of our food. It is usually secreted in largest amount not long after a meal. Nearly all of the biliary substances proper are reabsorbed from the intestine. Experiment shows that it is necessary to health and even to the life of an animal, not only that the bile should be secreted and discharged, but that it should be passed *into the alimentary canal*. All these facts combine to prove that it partakes in the completion of the digestive process. Against this, Dr. Dalton urges that experiments with bile out of the body have not succeeded in showing that it has any positive reaction with either albumen, starch, or fat, at a temperature of 100°.

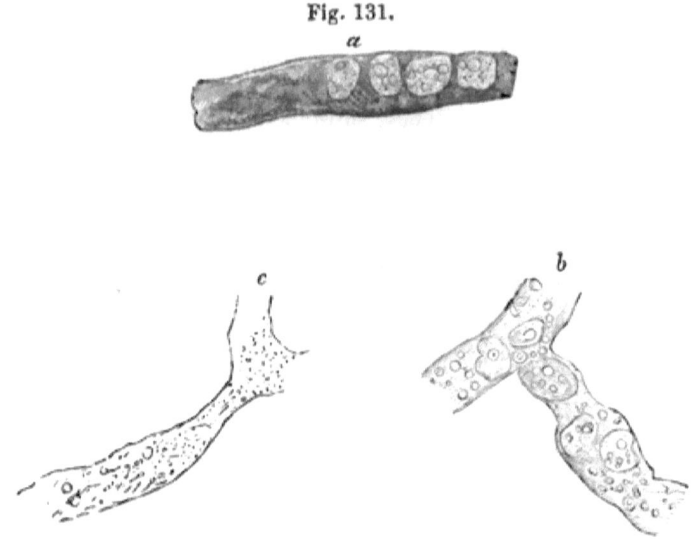

Fig. 131.

BILE-DUCT AND CELLS.

Probably the bile also acts as the natural "peristaltic persuader," or stimulant of muscular contraction in the intestine. By its proneness to alkalinity, it may neutralize excess of acidity in the bowels; and, by its antiseptic quality, retard putrefactive changes in the refuse of blood.

Secretion of Urine.

There is no doubt that it is in the *cortical* portion of the kidney that urine is secreted. There are the *cells* of the organ, in close relation to the beginnings of the uriniferous tubules, which then collect, in conical bundles, to end at the pelvis of the kidney. Capillary bloodvessels surround these cells; each minute tubule also begins in a capsule, which embraces a Malpighian corpuscle, or tuft of capillaries. Very possibly there may be an actual expression or filtration of a portion of the water and salts of the blood, from the Malpighian tufts, into the tubules, through the inclosing capsule. Besides this, however, there is a true *secretion*, or selective separation, of matters from the blood, by the cortical renal *cells*.

Urine is entirely excrementitious; serving, after it leaves the kidney, no functional purpose. Its ingredients are all taken from the blood; not manufactured, although perhaps somewhat modi-

SECRETION OF URINE.

fied, in the kidney.[1] The average daily amount in an adult is from thirty-two to thirty-five fluidounces. Its normal specific gravity (water being 1000) is 1024. Its quantity and character, however, both vary, even in health; and, greatly, in disease. Diabetes mellitus is marked by saccharine urine, which is very heavy; up to 1060 or 1070. Hysterical patients often have very abundant urine, pellucid and light; 1006 or 1005. *Albuminuria* is the presence of albumen in the urine. This occurs transiently in a number of diseases; permanently, in Bright's disease of the kidney.

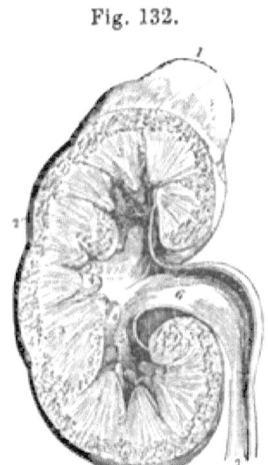

Fig. 132.

SECTION OF KIDNEY.

Diurnal variations take place in the urine in health. Dalton found that passed on rising in the morning to be dense, highly colored, and of acid reaction. During the forenoon, pale, light in weight, and neutral or slightly alkaline. In the afternoon and evening it becomes again dense, dark colored, and strongly acid.

The following are the constituents of the urine:—

Water	938
Urea	30
Creatin	1.25
Creatinin	1.50
Urate of soda, " " potassa, " " ammonia	1.80
Coloring matter and mucus	.30
Biphosphate of soda, Phosphate of soda, " " potassa, " " magnesia, " " lime	12.45
Chlorides of sodium and potassium	7.80
Sulphates of soda and potassa	6.90
	100.00

Urea is a soluble, crystallizable, neutral, nitrogenous substance; of which the daily average passed by an adult in the urine is from 400 to 600 grains. It is increased by exercise and by highly animalized food. Out of the body, decomposition converts it into carbonate of ammonia.

Creatin is a crystallizable, neutral, nitrogenous substance, origi-

---

[1] Zalesky asserts that the kidneys change creatin into urea.

nating in the muscular tissue as a result of its waste. Being absorbed into the blood, it is thrown out in the urine.

Creatinin contains two equivalents less of water than creatin. It is slightly alkaline. Muscular tissue yields it also. Probably creatin is converted into creatinin; as the latter substance is most abundant in the urine, and the former in the muscles.

Fig. 133.

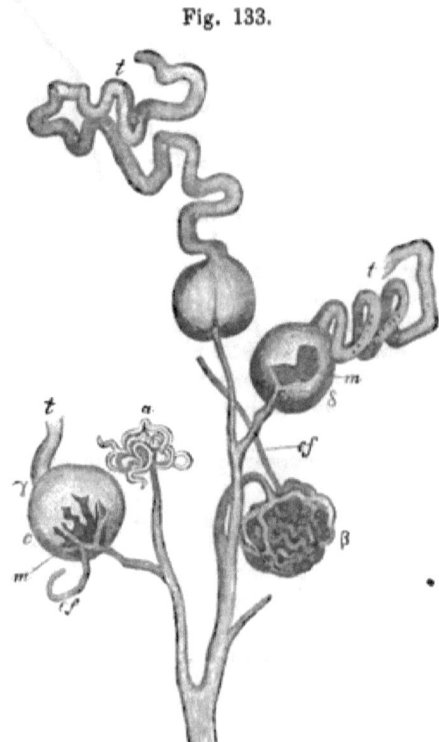

STRUCTURE OF KIDNEY.—*a.* Arterial branch. *b.* Malpighian tuft. *ef.* Efferent vessel. *m, m.* Capsule. *t.* Uriniferous tubule.

Urates, or salts of uric or lithic acid, are soluble and crystallizable salts, containing nitrogen. Urate of soda is the most abundant. They result from the waste or disintegration of the nitrogenous tissues. The *rate of metamorphosis* of tissue, therefore, can be approximately estimated by determining the amount of urea, urates, &c., passed.

The *coloring matter* of the urine, *urosacin*, is usually dissolved in the water of the secretion. Sometimes it is thrown down with other deposits as uric acid or the urates; making the "lateritious" or brickdust sediment.

Various medicinal and other substances pass from the blood into the kidneys, are thrown out by the urine, and give color, odor, or other properties to it.

Excretion by the Bowels.

In man, the large intestine has only an excretory function. The *feces* consist, 1st, of materials of food, not perfectly changed and rendered assimilable by digestion, from their nature or from excess in amount; 2d, of the secretions of the glands of the large intestine, viz., effete matter taken from the blood. The necessity of the regular action of the bowels for health is evident from this double nature of the material passed. Even when no food is taken, as in illness, *some* discharge, though it may be reduced in quantity, is required. In the feces, excretin, stercorin, ammonio-magnesian phosphate, and other salts, have been found along with remnants of undigested food.

The Skin.

Two important uses, besides secretion, evidently belong to the skin; *protection* of the organs beneath it, and the reception and conveyance of *sensation*. Two kinds of secreting glands are found in it; the *sudoriparous* or sweat-glands, and the *sebaceous* glands. The former are most abundant; on the palm of the hand, for instance, 2700 to the square inch. Each sweat-gland is a tubular coil, lined with epithelium, lying just beneath the skin. Its duct penetrates the skin, ending at the cuticle with an oblique valve-like opening. Altogether, nearly two pounds of perspiration pass off from the body of an adult in twenty-four hours. Its composition is as follows:—

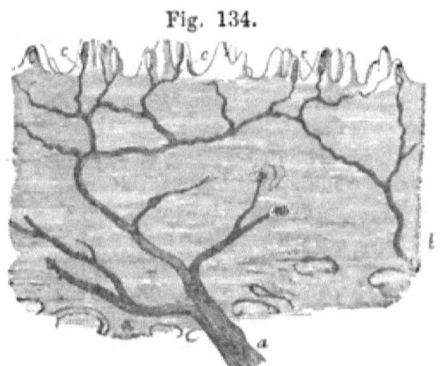

Fig. 134.

Section of Skin.

Water	995.00
Animal matters, with lime	.10
Sulphates, with substances soluble in water	1.05
Chlorides of sodium and potassium, and spirit-extract	2.40
Acetic acid, acetates, lactates, with alcohol extract	1.45
	1000.00

The *sebaceous* glands abound especially upon parts of the skin covered with hair. Their secretion is unctuous, and maintains the suppleness of the skin and hair. In the *external meatus* of the ear, the *ceruminous* glands discharge a matter of a peculiar consistence and odor, whose purpose seems to be to exclude insects from the ear.

Insensible perspiration is an exhalation of moisture from the whole surface of the skin. By its evaporation and that of the *sweat*, the heat of the body is moderated, under exercise, in summer weather or tropical climates. Thus, in a dry air-bath, the temperature of 250° can be readily sustained; in vapor, 150° would be dangerous. Chabert, the Fire-king, is said to have entered safely an oven heated to 600°.

CHAPTER V.

REPRODUCTION.

General Considerations

For the indefinite continuance of species of organized beings by the reproduction of individuals, the essential condition is the union of two oppositely polar or "sexual" cells; the *germ-cell* and the *sperm-cell*. In all the higher animals, as in man, these are always the products of different bodies, having concomitant sexual peculiarities. Among lower forms, animal and vegetable, true *hermaphrodism* is sometimes met with; *i. e.*, the existence of both sexes in the same individual; as in the tapeworm. Still, even if the uniting cells do not exhibit any palpable differences, the principle of *duality* seems to be universal in reproduction.

Apparent exceptions seem to exist to this law, in several instances. The propagation of plants from cuttings (as the grape-vine), or from "eyes" of tuberous roots (as the potato) is certainly not a *dual* process. If, then, it be infinite in its possible extent of multiplication, it must be exceptional. But we do not know that it is so. Degeneration of the potato and other plants under that method has been noticed. If this *multiplication by division* be exhaustible, it is really only the separate growth of dividual parts of the unit of organization from which they came. The tree grows in its cuttings; and, although their life is prolonged beyond that of the branches which are not planted out, it is still limited; only seed-life is perpetually renewable.

Other seeming exceptions occur in "parthenogenesis," or reproduction without impregnation, and the "alternation of generations" of certain animals (medusæ, salpæ, &c.), whose offspring are quite different from themselves. Careful examination has, however, shown that, while it may sometimes be deferred for several generations, sexual union does at intervals always occur. This is not necessarily in the bodies of the animals; as, in the case of fishes, the *spawn* and *milt* meet in the water outside of both parents.

Parasites within the cavities of the human and other animal bodies were once a serious puzzle to physiologists. It is now well understood, however, that all of them must be, and can be shown to be, derived from other, like or unlike, forms whose germs enter the body in food or drink, through the skin, or by being variously deposited as eggs. So, the tapeworm comes from the *cysticercus*, swallowed while very small, in food.

Some facts have often suggested the idea of "spontaneous generation;" that is, of the springing up of life in previously inanimate organic matter. Vegetation and animal life do certainly appear often, on the surface of decaying liquids and solids, without visible sources of origination. Any infusion of organic matter, left exposed to the air for some days in warm weather, will display under the microscope a number of animalcules and minute but definite vegetable growths. These, at least those whose motions give the idea of animality, are called *infusoria*. Do they ever begin to exist without previous germs?

Nearly all physiologists and naturalists are now agreed that the atmosphere and common water, which "teem with life," are always the sources of such development, by the ordinary methods of reproduction from parentage. Some experiments of Prof. Jeffries Wyman would seem to have settled this point fully. Having, with the greatest care, prevented all air except what had been

Fig. 135.

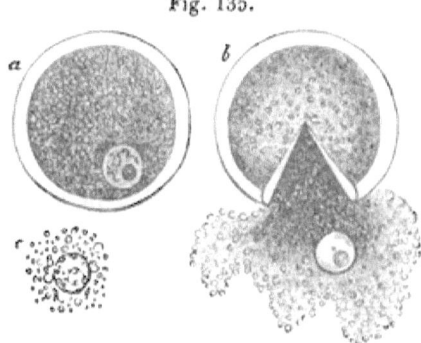

UNFERTILIZED OVUM.—*a*. Vitelline membrane, inclosing the yelk and germinal spot. *b*. The germ-cell, burst. *c*. The germinal vesicle, surrounded by granular matter.

exposed to high heat and disorganized by sulphuric acid, from reaching a preparation of organic matter, he found the number of infusoria and vegetations produced to be greatly diminished. Some appeared, even after four hours' boiling of the materials. After *five* hours' boiling, however, *none appeared*. This shows, first, that the resistance of some of these minute germs to the destructive action of high temperature is greater than had been supposed; and, also, that a certain degree and continuance of such exposure will destroy *all* living particles; after which none are spontaneously produced. *Omne animal ex ovo*, Harvey's dictum, is then verified.

Of woman, the *ovary* is the primary organ of reproduction. In it the ova are produced, and, once a month, an ovum is thrown off, by the Fallopian tube, into the uterus. If not impregnated by sexual intercourse, it is then carried out by the (mucous and

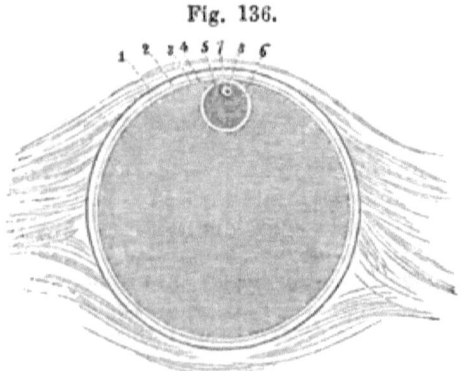

Fig. 136.

GRAAFIAN VESICLE AND OVUM.—1. Stroma of ovary. 2, 3. Tunics of Graafian vesicle. 4. Cavity of vesicle. 5. Yelk-sac. 6. Yelk. 7. Germinal vesicle. 8. Germinal spot.

hemorrhagic) *menstrual discharge* from the uterus. This, the womb, is the organ of *gestation;* i. e., the retention of the embryo during the term of fœtal life, till it is *viable* apart from the mother.

Male Organs of Generation.

The essential organs of reproduction in man are the *testicles*. In the seminal fluid are multitudes of *spermatozoa*, on which its generative potency depends. These are microscopically minute bodies, $\frac{1}{600}$ of an inch long, with a triangular head and elongated tapering tail; moving incessantly while their vitality continues. This motion, suggesting the idea of animality, gave rise to their name; but they are well understood to be *cell-filaments*, with a motility like that of some other reproductive forms, vegetable as well as animal, but not animalcular themselves. Perhaps each spermatozoon may be a ciliated cell, with but one cilium, the tail.

When the spermatozoa become dry, or are subject to extreme

heat or cold, or to disorganizing agents of any kind, they cease moving directly. Otherwise, it is probable that (in the genital organs of the female, for instance, after coitus) they may sometimes retain their vitality for hours, or even days.

Kölliker's account of the formation of the spermatozoa is as follows. At and after puberty, there are formed in the seminiferous tubes of the testicle certain vesicles; each containing from one to twenty nuclei, with nucleoli in them. In these vesicles, probably from the nuclei, the spermatozoa are developed in bundles. Then the vesicle gives way and disappears, and the spermatozoa are set free in the ducts, with a very small amount of fluid. This mingling of the spermatozoa occurs in the *rete testis* and head of the epididymis.

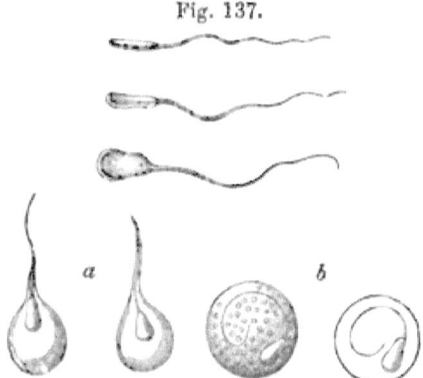

Fig. 137.

SPERMATOZOA.—*a*. Spermatozoa of the squirrel. *b*. Spermatozoa of the dog; two inclosed in the sperm-cells, and three free.

Passing through these and the *vas deferens*, a glairy mucus is added, and the material is accumulated in the *vesiculæ seminales*. When the sexual orgasm takes place, contraction of surrounding muscular fibres expels the semen from the vesiculæ seminales into the urethra. There it receives the secretions of the prostate gland, the glands of Cowper, and the mucous follicles of the urethra; all of which are excited together, by the act of coition. Entrance of the seminal fluid into the uterus is necessary for impregnation. Sometimes certainly, perhaps usually, a portion of it passes through a Fallopian tube to an ovary.

Periodicity in reproduction is observed in many animals. In the human female, monthly ovulation occurs, with the menstrual hemorrhage. Difficulty exists in ascertaining whether any regular periodicity is normal in the male. Observation makes it not improbable that a special proclivity to seminal secretion and discharge exists, in continent persons, about once in two or three weeks.

Like the mammary glands in the female, it is known that the testicles may, in the absence of excitation, remain inactive, so as for long periods to be free from discharge, without inconvenience. These organs, however, like others in the body, are excited to secretion by mental and emotional, as well as by physical stimulation.

CHAPTER VI.
MUSCULAR ACTION.

Some account has been given, on a previous page, of the two sorts of muscular tissue; the red, striated, usually voluntary, and the white, non-striated, always involuntary muscle; the former in the limbs, trunk, and face, as well as in the heart; the latter, in the alimentary canal, uterus, bladder, middle coat of arteries, gland-ducts, etc. Most of the physiological facts to be stated are true of both kinds, although the voluntary muscle-fibres are always the quicker in contraction.

No change of bulk, but only of shape, occurs when a muscle shortens; as it expands laterally at the same time. This may be seen and felt in many muscles; as, *e. g.*, in the biceps muscle of the arm.

Several *theories* concerning the source or nature of muscular power have been proposed. Haller, long ago, showed that contractility belongs to the muscular tissue itself; not *depending* on the nervous system, though ordinarily called into action under nervous influence. *Electricity* has by many been thought to afford the means of explaining muscular power; the analogy being closest, perhaps, to electro-magnetism, as that used in the telegraph. Matteucci's and Dubois Raymond's experiments are considered by Dr. Radcliffe to sustain an electrical theory.

Dr. B. W. Richardson has lately urged with emphasis the importance of the direct relation between *caloric* (heat) and muscular power.

Chemical change, similar to that which generates animal heat, all agree in believing to be either a cause or an accompaniment of muscular action. Many physiologists have supposed the change to be disintegration or waste of the *muscle itself*; that the consumption of the muscular tissue might be the source of the power. Some late experiments seem to contradict this view; especially those of Fick and Wislicenus; who found that, in a day's journey, climbing one of the Alps, there was no decided increase in the amount of nitrogenous waste (measured by the urinary solids) beyond that of repose. It would seem to be the *non-nitrogenous* material of the blood, supplied by food, that is consumed for the production of muscular power.

Only the *contraction* of any muscle is active. Its dilatation is produced sometimes by elasticity, often by opposing muscles. Almost every muscle in the body has its antagonist. So there are, for the limbs, the flexors and extensors; for the fingers, adductors and abductors; at the anus, the sphincter and levator ani, etc. All muscles have, during life, a continued slight *passive* contraction. Since the opposing groups of muscles are not exactly equal in power, the position of parts of the body when at rest is determined by the preponderance of one or another set; as in the flexion of the fingers during sleep.

After death, for a certain time the muscles will contract under the excitation of galvanic electricity. The signal of the loss of this irritability is the coming on of *rigor mortis*, the stiffening of death. This is not a vital contraction at all, but rather a physical change, the first result of the death of the muscle. After it follow relaxation and decomposition.

Rigor mortis may begin at any time from ten minutes to six or seven hours after death; usually it is an hour or two at least. Sudden death from violence in full health is followed by *late* rigidity, continued long. After protracted exhausting disease, it is apt to occur soon and to be short in duration. Death by lightning has been observed to be without any rigor mortis.

This stiffening affects the involuntary as well as the voluntary muscles. It begins in the left ventricle, and ends in the right auricle; the other muscular organs, including the arteries, contracting between these extremes in time.

Classifying the muscles of the body according to their method of action, they may be designated as *voluntary*, *involuntary*, and *mixed*. Purely voluntary are all the muscles of the surface of the trunk and of the extremities.

Voluntary Muscles.

The action of these, by their tendinous attachments, is, in most cases, upon the bones. These, as mechanical instruments of locomotion, may be divided into levers of the *first*, *second*, and *third* kinds. In the first, the *fulcrum* or fixed point is between the power applied and the weight or resistance. In the second, the *weight* or resistance is between the power and the fulcrum. In the third, the *power* is applied between the fulcrum and the weight or resistance.

Of the first kind of lever, an example is the movement forward, or backward, of the head upon the spine, by the muscles of the front or back of the neck.

Of the second kind, an instance is, raising the body upon the toes, by the action of the muscles of the calf of the leg.

Of the third kind, the action of the *biceps flexor cubiti* is the best example. In this, the power is applied at the insertion of the biceps tendon into the radius, below the elbow; the fulcrum is at the elbow-joint, and the weight is that of the forearm and hand.

Fig. 138.

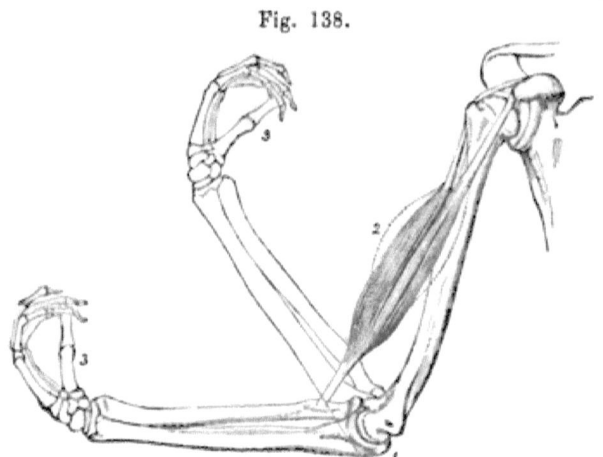

THE BICEPS MUSCLE.—1. The fulcrum. 2. The power. 3, 3. The weight.

This muscle affords an illustration of the fact that some muscles are arranged at a less advantage of *power* than might be given by a different insertion. If the biceps were inserted at the wrist, it would lift ten times as much. But, then, inconvenience in bulk and loss of beauty and grace would follow. The law of mechanics also applies, that what is gained in power is lost in velocity, and *vice versâ*.

Pulleys for special direction of muscular action exist in several parts of the body; as in the course of the *superior oblique* muscle of the eye, the *digastric* of the neck and lower jaw, etc.

The amount of effort made by the muscles is perceived by what is called the *muscular sense*. Its employment is exemplified in *weighing* anything in the hand, in balancing one's self (à la Blondin) on a tight rope, and in skating. The latter exercise is guided almost entirely by the muscular sense.

Involuntary muscular action has been considered sufficiently, in connection with the organs of nutrition, secretion, &c., whose functions it subserves.

Mixed Muscles.

Of this character are the muscles of the pharynx, of the respiratory apparatus, and of the face.

When we *swallow*, the will has control only over the beginning of the process; after the morsel gets fairly within the grasp of the constrictor muscles, its descent cannot be arrested.

Breathing is ordinarily involuntary, continuing during sleep. But we have the power to hold the breath for some seconds, as well as to modulate it for vocal utterance.

The muscles of *expression*, in the face, though yielding readily to volition, are most naturally controlled by *emotion*. Strong feeling involuntarily exhibits itself in the countenance. Efforts at the expression, by the face, of counterfeit emotion, when no feeling exists, seldom are successful. Actors and orators personate and convey emotion best, by throwing themselves for the time into the character or feeling required; so that it then expresses itself, naturally and effectively.

Many muscular movements which at first are performed entirely under direction of the will, by frequent repetition become habitual, and then involuntary. Walking is an act of this kind. Soldiers, on long marches, have, it is said, sometimes fallen asleep on the road, yet continued marching. An expert musician is said to have finished playing a piece on the piano, begun while awake, after going to sleep.

CHAPTER VII.
FUNCTIONS OF THE NERVOUS SYSTEM.

GENERAL CONSIDERATIONS.

THE purpose of the nervous apparatus, still more constantly than that of the muscular system, is to maintain *relations* and *communications* between the body and things external to it, and between its own different parts.

The ultimate formative elements of the nervous system are, 1, *nerve cells*, collected into *ganglia* or *nerve-centres;* and 2, *tubular nerve-filaments*, arranged in bundles called nerves and commissures. The centres are like the stations or offices of a telegraphic system; the nerves correspond with the wires. The simplest conceivable nervous system consists of one ganglion with two nerves; one of these conveying impressions from the surface to the ganglionic centre, and the other taking out impressions from that centre to a movable tissue. The first of these nerves is called an *afferent*, and the second an *efferent* nerve. Action produced by such a conveyance and *reflection* of an impression by a nerve centre is called *excito-motor* or *reflex* action.

Ganglia are said to receive, accumulate, generate, reflect, and radiate nerve-force. Most certain is the fact of their *reflection* of

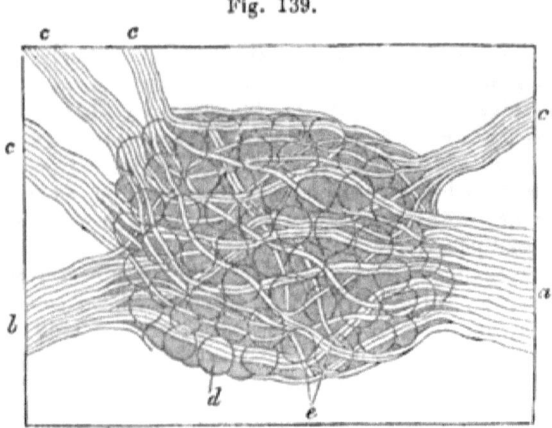

Fig. 139.

Diagram of a Ganglion. *a, b, c.* Nerves. *d, e.* Cells.

its impressions; the understanding of which is the key to the functions of nervous organs, from the lowest to the highest. *Radiation* of impressions appears to occur sometimes, as when disease in one tooth causes pain in the whole of that side of the face; or when inflammation of the hip-joint (coxalgia) brings on pain extending to the knee.

Nerves only transmit impressions. Their analogy to wires of the telegraph is close; we do not understand that anything "flows" along them, the term *current* being metaphorical; but, that a *wave* of *movement* of their particles is propagated along them, from end to end. Certain constant facts or laws of nervous transmission are important.

1. Every nerve transmits impressions *only in one direction;* either toward (afferent) or from (efferent) a centre. All afferent nerves are not sensory. Reflex action may occur without sensation. The latter only exists when an impression is conveyed to the *brain*, the seat of the perception of sensations.

There is clear proof that muscular movement may take place without sensibility of the parts concerned. Thus, in the experiment of Brown-Séquard and others, of dividing the spinal marrow of a frog in sunder; if a hind foot of the animal be then pinched, the limb will be jerked with force; although it is certain that no sensation can be conveyed to the brain. John Hunter saw a paraplegic patient, quite deprived of feeling in his feet; yet tickling the sole of one of them caused the limb to be withdrawn. This was *involuntary, automatic,* reflex or excito-motor action.

2. Each nerve can convey only *one kind* of impression. *Efferent* nerves may be nerves of *motion*, if they are distributed to muscles; or, the excitation they bear out will cause *secretion*, if they terminate in glandular organs. This, when it results from an impression reflected by a ganglion which receives it from an impressible (exterior or interior) surface, is *excito-secretory* action. So, the presence of food in the mouth excites the salivary glands to secretion; in the stomach, it draws out the secretion of the gastric juice; &c. Morbidly, we find the irritation of the gums in the dentition of weakly children inducing diarrhœa, from excessive excito-secretory action. This is parallel to the violent morbid *excito-motor* action, denominated *convulsions*, which incomplete dentition may cause, through irritation of the spinal marrow; and which may also be brought on, in like-manner, by the presence of indigestible food in the stomach or intestines, or accumulated feces in the rectum.

Nerves of *sensation* are capable, each, of only one kind of sensibility. Some are nerves of *touch*, one of *sight*, another of *hearing*, *taste*, or *smell;* no one of *more* than one of these. If the optic nerve is irritated, a flash of light, not pain, results. Irritation of the auditory nerve or its sensorial centre will cause a "subjective" impression of sound: as, for example, the "tinnitus aurium" produced by quinine in large doses. Substitution of the guidance of one sense by that of another may occur, as when a blind person walks by direction of his hearing and touch. But the actual transfer of one kind of special sensibility to a nerve possessed naturally of another function, is impossible.

3. Sensory nerves usually report, so to speak, their impressions as if coming from their terminations; even when it is the trunk of the nerve that is acted upon. What is commonly called the *crazy-bone* at the elbow is the ulnar nerve; when it is struck, the principal pain is not at the elbow, but in the last two fingers, to which its terminations are distributed. After the amputation of a limb, sometimes sensations in the stump seem to the patient to be in the missing toes. A flap of skin being brought over the nose to replace a deficiency of the latter, for a few days a fly lighting on the nose will produce an itching, referred in the individual's feeling to his forehead.

Nerve-filaments never inosculate (*i. e.*, actually join into one, as bloodvessels do), although they are often packed together in the same trunk, called "a nerve." When such a trunk is divided, intentionally or by accident, it is slow to unite again. In course of time it will do so, however; and then the function of its filaments may be, though it is not always, entirely restored. This is remarkable, since a trunk, containing a number of filaments, some motor, some sensory, and others ganglionic or sympathetic, must, when divided, suffer displacement of their corresponding ends. Each filament must, therefore, be restored to its proper connection, notwithstanding the displacement; as though function had a sort of control over nutrition.

Nothing in the appearance or structure of any nerve shows a reason for its character, *i. e.*, whether it be sensory or motor, etc. This would seem to be determined by the connections of its extremities. Dr. Beale believes that no nerves have free ends; but that each filament makes a complete circuit.

The nature of nerve-force has been the subject of much speculation. The favorite hypothesis with many has been, that it was identical with electricity. Galvanism will, undoubtedly, *stimulate* the muscles and other organs through the nerves of a living or recently dead animal. But so also will the point of a knife, high heat, a drop of nitric acid, etc. Electricity is only one of the stimulants which call out, as it were, nerve-force. Moreover, Matteucci has shown that no special electrical current passes through even the largest nerve of the limb of a horse, when the muscles are stimulated to action. If a nerve be cut across, and the ends be placed in contact, or with a copper wire between them, or if a ligature be drawn tightly around it, in either case electricity will traverse it freely, but nerve-impressions will not. Muscle is a decidedly better conductor of electricity than nerve; and copper wire is many million times a better conductor than either. Still, electrical currents are proved to be constantly present in different parts of the living body. Among the conversions or transmutations of force (modes of motion) taking place, it is quite probable that the generation of electricity may have a place, not as yet clearly defined. This is rendered the more probable by the fact that several species of fish, as the *torpedo*, *gymnotus*, and *silurus*, have special electrical organs or batteries, directly connected in each with the nervous system.

REFLEX ACTION.

This is the key to the whole physiology of the nervous system. Although before discerned by Unzer and other physiologists, to Marshall Hall belongs the credit of developing our knowledge especially of the reflex functions of the spinal marrow. Laycock, Carpenter, and others have extended the study of reflex action to the physiology of the brain and of mental action; Longet and Campbell, into that of the secreting organs. The simplest and most common of reflex actions, in all the subdivisions of the animal kingdom, is *excito-motor* action. Next, is *excito-secretory* action. We may add, after Carpenter, *sensori-motor* and *ideo-motor* actions. Besides, also, the *voluntary* motor actions, wherein will directs the choice between different possible actions of the reflex system, it would be equally correct to speak of *emoto-motor* actions; in which, not volition, but emotion gives character to the effective impulse.

Excito-motor actions are those in which an impression upon a receptive surface (interior or exterior) is, by an afferent nerve, conveyed to a motor nerve-centre, and, thence, reflected through an efferent (motor) nerve to a muscle, which it brings into action. Thus, in a complex animal, movements are effected by stimuli at a distance from the muscles, which could not, without inconvenience, be made to reach them directly. In this way the nerves and nerve-centres are *internuncial*, in the body itself.

Familiar acts give clear examples of such reflex agency. The pupil of the eye contracts when strong light falls upon the retina; this is because the impression is conveyed to a ganglionic centre, and thence reflected through a nerve to the circular muscular fibres of the iris. The impression of the want of oxygen in the blood is conveyed (chiefly from the lungs by the pneumogastric nerve) to the medulla oblongata; thence it is reflected, as a motor impulse, to the diaphragm and intercostal muscles, for inspiration. A morsel of food is introduced into the pharynx; the impression made by it upon the mucous membrane is carried by the glosso-pharyngeal nerve to a part of the medulla oblongata, whence it is reflected to the constrictor muscles of the pharynx which act in deglutition. A similar account might, of course, be given of defecation, parturition, &c.

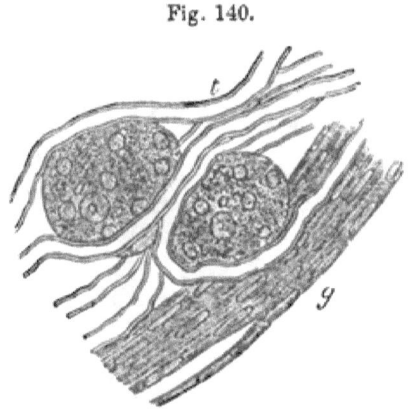

Fig. 140.

GANGLION-CELLS.

Excito-motor actions under *morbid* irritation are seen in convulsions, lock-jaw (trismus, tetanus), and, with less violence, in coughing, sneezing, vomiting. Such movements are usually involuntary, and often not controllable by the will. Extreme reflex excitability is met with especially in infancy (most of all during dentition), and in *hysteria* and *hydrophobia*.

GANGLIONIC NERVOUS APPARATUS.

In *Anatomy*, this is commonly described as "the Sympathetic Nerve." All the ganglia upon the spinal column and in the cavities and internal organs of the body, have two sets of communications: 1, with the spinal cord; 2, with the organs of circulation, digestion, assimilation, secretion, or reproduction.

Fig. 141.

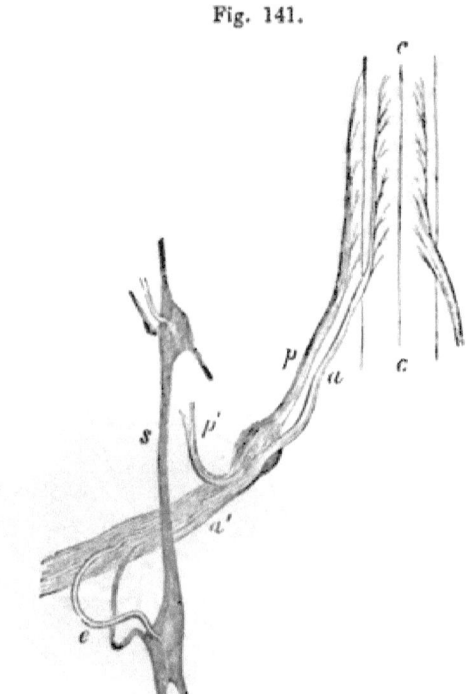

JUNCTION OF SPINAL AND SYMPATHETIC SYSTEMS.—*a*. Anterior root of spinal nerve. *p*. Posterior root. *c*. Columns of spinal cord. *s*. Sympathetic. *a'*. Spinal nerve. *e*. Connecting filaments.

Whether any motor or other power belongs to or originates in the ganglia, or whether all their energy is derived from the spinal marrow, is a question. At all events they exercise a modifying, regulating influence over the functions of the organs to which, in many parts, they alone send nerves. This is the case with the arteries, intestines, glands and their ducts, &c. Bernard and others have shown, for instance, that division of the sympathetic nerve in the neck causes passive dilatation of the bloodvessels, with redness and heat.

A question exists, also, whether in each of these instances, the "sympathetic" ganglia control and regulate the organic functions, only by their influence over the action and calibre of the *blood-vessels*, or by a more *direct* power exerted over secretion and nutrition. Much has been made recently (especially in the *neuropathy* of Dr. John Chapman) of the vaso-motor action of the ganglia, and the modifications of it produced by temperature, &c. Pflüger's experiments, however, clearly demonstrate that the nerves exert an influence over the secretory action of glands *apart from*, or over and above, that belonging to changes in the circulation of the blood. These changes, also, are, no doubt, important.

SPINAL MARROW.

Consisting of a central ganglionic mass, inclosed in columns of white nerve-substance, the functions of the spinal cord are complex. After Sir Charles Bell had demonstrated that different nerves going to the same parts may have different functions, Magendie

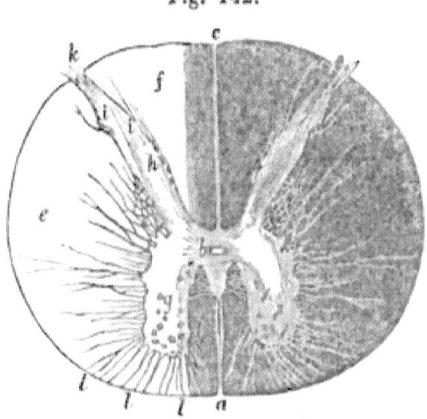

Fig. 142.

SECTION OF SPINAL CORD.—*a*. Anterior fissure. *b*. Gray central substance. *c*. Posterior fissure. *e*, *f*. White substance. *k*. Posterior nerve-root. *l, l, l*. Anterior filaments.

extended this discovery to the roots of the spinal nerves. He proved that the *posterior* roots are entirely *sensory*, or afferent, and the *anterior* roots exclusively *motor*. This is shown by the fact that, when a posterior root alone is divided in a living animal, irritation of the end left in connection with the spinal cord will produce signs of pain; while irritation of the distal end will cause no result. On the contrary, if an anterior root be cut across, irritation of the end next the spinal marrow will produce no effect; but excitation of the end connected with the muscles will throw them into action. Beyond their roots, the spinal nerves are mixed, containing both sensitive and motor filaments.

Part of the functional use of the spinal cord is, to transmit impressions, sensory and motor, to and from the brain. This is effected both by the white and the gray substance; principally, it may be supposed, by the former. A difference in function between the anterior and posterior *columns* of the cord has been proved. The *anterior* columns, under excitation, emit motor impulses through their motor nerves, producing action of muscles. The *posterior* columns, when irritated, display sensibility instead.

Crossing or *decussation*, however, occurs in the cord. The *anterior* columns decussate at the medulla oblongata, just within the skull. Therefore, disease of the brain (apoplexy, compression from fracture, &c.) on one side, commonly produces paralysis of motion on the other side of the body; as, in many cases of what is called *aphasia*,[1] disease of the left side of the brain is attended by right hemiplegia, *i. e.*, palsy of the right half of the body.

Brown-Séquard has inferred from his experiments that the *sensory* filaments of the spinal cord cross each other through the *whole length* of the cord. If, for instance, one lateral half of the spinal marrow of a dog be divided in the dorsal region, sensation will remain on that side of the body, but will be lost on the opposite side. The same physiologist asserts, that the sensory filaments of the spinal nerves, after entering the posterior columns, pass *through* them, and go up toward the brain in the central portion of the cord. Therefore, if the posterior columns be alone divided in any part of the cord, sensibility is not destroyed in all the nerves below that part, but only in those which enter near the place of division.

Besides thus acting as a medium of communication between the brain and all parts of the body except the head and face, the spinal marrow has more special functions of its own, as the seat of reflex actions. The drawing away of a limb when it is touched by anything very hot or very cold, or pinched, is often entirely involuntary. It may happen during sleep, or when, from paralytic disease, the part is incapable of sensation. The spinal cord seems, here, to be capable of maintaining action independently of the brain. This has, also, been abundantly proved in decapitated animals.

The habitual passive action of the muscles, though perhaps resulting in part from the essential nature of living muscle, is under the *influence* of the spinal marrow; as shown by an experiment of Wilson Philips. He found that, when a red-hot iron rod was suddenly thrust through the whole length of the spinal cord, all the muscles became relaxed at once.

The *sphincter* muscles, which guard the outlets of the rectum and bladder, are kept in due contraction under this influence of the

---

[1] Loss of language, from cerebral disorder.

spinal marrow. Its reflex agency is also, no doubt, concerned, differently, in the evacuation of the rectum and bladder; in order for which, the contraction of the sphincters yields to dilatation under expulsive effort. Ordinarily, the will can control and regulate these actions.

If the spinal marrow be seriously injured low down, loss of power over the sphincters results; and involuntary defecation, retention, and incontinence of urine follow. If high up, disturbance of the secreting organs occurs, from the influence of the spinal cord over the ganglia. Injury of the spine in the neck, as high as the third vertebra, is almost always fatal at once, by interruption of respiration; to which a sound state of the phrenic and intercostal nerves is essential.

ENCEPHALON OR BRAIN.

All the contents of the cranium, together, constitute the *encephalon;* the average weight of which to the whole body, in man,

Fig. 143.

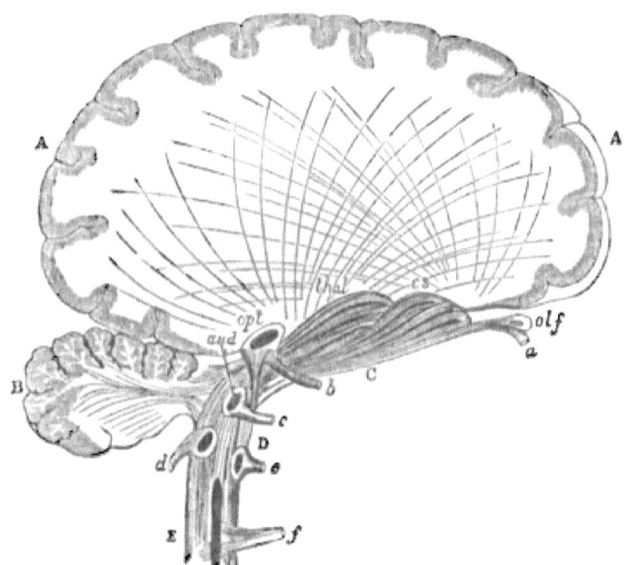

DIAGRAM OF ENCEPHALON.—*A*. Cerebrum. *B*. Cerebellum. *C*. Sensori-motor tract. *D*. Medulla oblongata. *E*. Spinal cord. *a*. Olfactory nerve. *b*. Optic. *c*. Auditory. *d*. Pneumogastric. *e*. Hypoglossal. *f*. Spinal.

bears the proportion of 1 to 36. In mammals (animals which suckle their young) as a class, the average proportion of brain to body is 1 to 186.

The *parts* of the encephalon are, the *medulla oblongata* and *pons Varolii*, the *cerebellum*, and the *cerebrum*. Further subdivision of these, of course, may be made, upon both anatomical and physiological grounds. If the whole encephalon were divided into 204 equal parts, the cerebral hemispheres would make of these 170 parts, the cerebellum 21, the medulla oblongata, corpora striata, and thalami 13 parts. On the same scale, the spinal cord would weigh 7 parts. The comparative size of these different portions of the nervous system is very different in different animals.

Medulla Oblongata.

Although several nerves of different functions begin or terminate about the place of union of the medulla oblongata with the base of the brain, the peculiar attributes of this centre are connected with respiration and deglutition. It is the seat of the reflex actions essential to those functions; the performance of which is, in its nature, spinal, while, for obvious reasons of convenience, they are both voluntary to a certain degree. For speech, we must have some control over the expiratory muscles; and so we must be able to manage those used in swallowing, on various occasions.

Because of its being the centre presiding over respiration, the functional activity of the medulla oblongata is, in man and all the

Fig. 144.

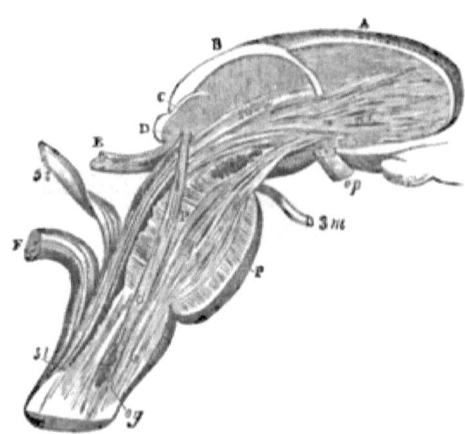

MEDULLA OBLONGATA.—*A*. Corpus striatum. *B*. Thalamus. *C, D*. Corpora quadrigemina. *E*. Commissure. *F*. Corpora restiformia. *P, P*. Pons Varolii.

higher animals, indispensable to life. Mechanical injury or disease affecting it considerably, or narcotism of it by large quantities of chloroform, &c., is fatal.

The *pons Varolii* or tuber annulare is, in its greater part, commissural; *i. e.*, connective, between the two halves of the cerebellum, and, by a small number of its filaments, passing also into the two hemispheres of the cerebrum. A mass of gray vesicular nervous matter, however, is imbedded within it; which Longet and others believe to be "the ganglion by which impressions, conveyed inward through the nerves, are first converted into conscious sensations; and in which the voluntary impulses originate, which stimulate the muscles to contraction."[1]

Cerebellum.

Gall, the founder of the system of phrenology, proposed the opinion, founded upon a few coincidences, that the cerebellum is

Fig. 145.

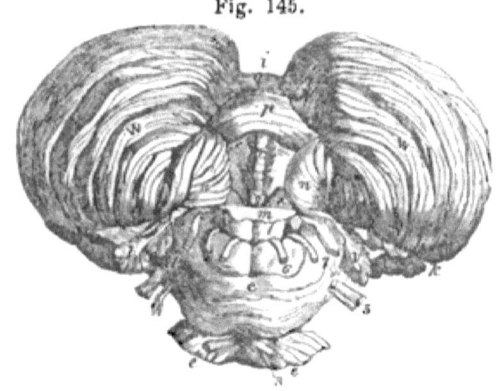

CEREBELLUM.—*m.* Medulla oblongata. *c.* Pons Varolii. *w.* Hemispheres of cerebellum. *i.* Middle notch. *p.* Pyramids. *e, e.* Crura. 3 to 7. Nerves.

the seat of "amativeness," or the sexual propensity. Investigation has not sustained this view. Upon any such question, three principal methods of inquiry are open: 1, comparison of structure and function in different animals; 2, the results of disease or injury; 3, experiments upon living animals. Flourens, Longet, and others have, on the basis of the experimental method, proposed the theory, that the cerebellum has the duty of co-ordinating, or harmonizing together, voluntary muscular movements. A bird or animal from which the cerebellum has been removed, loses the power to regulate its movements—like one intoxicated.

Disease of the cerebellum has not often been, after death, shown to have been connected with special symptoms; but some instances have occurred, which at least do not oppose the above theory.

Comparison of the structure of the brain, and the corresponding endowments of different animals, is, upon this as on other allied topics, the most instructive method. No relation appears between

[1] Dalton's Physiology, p. 423.

the size of the cerebellum and the sexual appetite. It is large, comparatively, in some animals (fishes) which do not copulate; and proportionately small in some, as the frog, in which the propensity is powerful. With locomotor activity, and especially *complexity* of movements, there does appear to be proportionate development of the cerebellum. So, in birds of rapid and varied flight, as the swallow and many birds of prey, it is large, while in the polygamous but heavily flying pheasant family, of which the common barn fowl is a member, it is small. The climbing ape has it quite large. The bear, which stands naturally on two feet, using the other two as hands, has a larger cerebellum than the dog, to which that position is unnatural. Lastly, removal of the testicles early in life does not, in the horse, at all lessen the growth of the cerebellum, if the animal is kept at work; that of the gelding, at full maturity of age, is often larger than that of the stallion or the mare. Altogether, the evidence is sufficient to make the theory of Flourens much the most probable. Perhaps the cerebellum may also be the seat of the muscular sense of Sir Charles Bell. Some recent experiments of Drs. B. W. Richardson and S. Weir Mitchell, in *congealing* different parts of the brain in animals by jets of ether spray, are very interesting, but their results are not yet sufficiently definite for positive conclusions.

Sensorial Ganglia.

Anatomically, sufficient distinction exists for naming, under this category, the *corpora striata, thalami*, and *tubercula quadrigemina*.

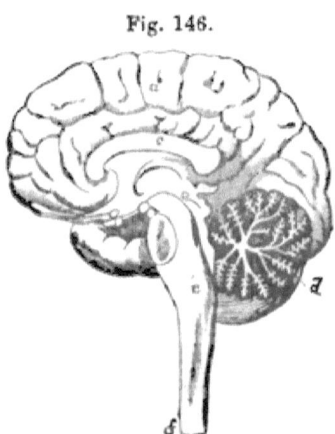

Fig. 146.

LONGITUDINAL SECTION OF THE BRAIN.
—*a*. Right hemisphere of cerebrum. *c*. Corpus callosum. *d*. Cerebellum, showing *arbor vitæ*. *e*. Medulla oblongata. *f*. Spinal cord.

Physiological reasons suggest the propriety of recognizing the association with these, of those other small ganglionic masses in which terminate the nerves of hearing, smell, and taste.

Experiments point to the *corpora striata* as the probable source from which emanate the *motor impulses* of muscular actions. Like evidence exists for the view that the thalami (formerly called *nervorum opticorum*) are the receptacles of impressions of *common sensation* or touch. Each *corpus striatum* is *anterior* to the corresponding *thalamus;* being thus continuous, the former with the anterior columns, and the latter with the posterior columns, of the spinal marrow.

The *tubercula quadrigemina* are the principal terminations of the *optic nerves;* the latter, however, sending filaments also to the thalami.

The reason for the close contiguity and connection between these different sensorial centres, and between them and the corpora striata, as well as between them and the spinal marrow below and the cerebral hemispheres above, is explained by the facts of what is called *sensori-motor guidance.* Every voluntary action is guided by a sensation; walking, speaking, writing, playing upon an instrument, working with tools. If one shuts his eyes and tries to walk a plank, he will almost certainly fail. Even the disturbance of vision which attends looking down from a great height may render walking dangerous, though otherwise safe. All confused or unusual impressions make actions uncertain. *Mutes* are commonly so because born *deaf.* Those who become deaf after learning to talk, retain speech. For musical performance one must have a good *ear;* for painting or sculpture, a good visual perception and discernment.

Some sensori-motor (reflex) actions are involuntary; as, winking, sneezing. Common acts, as walking, may become so, when habitual, though at first requiring attention and will. Blind persons exemplify the guidance of diverse sensations; instead of sight, they walk by hearing and touch. Sensori-motor actions under morbid causes often occur; as coughing, vomiting, hysterical convulsions. *Instincts* are, all of them, sensori-motor impulses and actions, more or less voluntary.

Cranial or Cephalic Nerves.

Anatomists still number these as nine on each side. Functionally, a very different classification might be made out. Comparing them, in plan of arrangement, to the spinal nerves, Dr. Dalton gives the following table:—

CRANIAL NERVES.

NERVES OF SPECIAL SENSE.

1. OLFACTORY. 2. OPTIC. 3. AUDITORY.

	Motor Nerves.	*Sensitive Nerves.*
1ST PAIR	Motor oculi communis Patheticus Motor oculi externus Small root of 5th pair Facial	Large root of 5th pair.
2D PAIR	Hypoglossal.	Glosso-pharyngeal.
3D PAIR	Spinal Accessory.	Pneumogastric.

Following the common enumeration, the functions of the different nerves may be briefly described:—

1st pair. Olfactory nerve.—Peculiar in having a bulb at its anterior extremity, which is apparently the true ganglionic portion.

2d pair. Optic nerve.—Remarkable for the *chiasm* or union of the two optic nerves, by which an entire unity of the visual image is produced, from the two actual images, one on each retina.

Fig. 147.

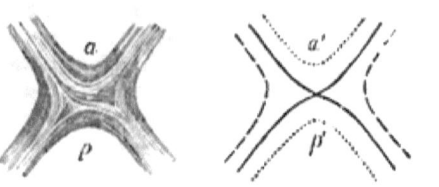

CHIASM OF OPTIC NERVES.

3d pair. Motor oculi communis.—All the muscles of the eyeball (four *recti* and two *obliqui*) are supplied by this nerve; also, the *levator palpebræ*, and the *iris*. Besides the simple ordinary movements of the eyeball, therefore, the opening of the eye and the contraction of the pupil are controlled by it.

4th pair. Patheticus.—Only the superior oblique muscle is supplied with motor influence by this nerve.

6th pair. Motor externus.—The external rectus muscle of the eyeball receives all the terminations of this. It is sometimes called *abducens oculi.*

7th pair. Portio dura ; Facial nerve.—This is the *motor* nerve of the superficial muscles of the face ; including the orbicularis oculi. When it is paralyzed, which happens sometimes from exposure to cold, the lower eyelid falls, and so does the corner of the mouth ; and all expression of that side of the face is lost.

The sensibility which belongs to some branches of this nerve over the face is derived entirely from the union in the same sheaths of filaments from the *fifth* nerve ; the seventh is altogether motor.

Portio mollis of 7th pair ; *Auditory nerve.*—To this belongs exclusively the conveyance of impressions of sound from the ear to the base of the brain.

9th pair. Hypoglossal.—Being distributed to the tongue, this is the *motor* nerve of that organ.

Having thus disposed of the simpler nerves of the series, the *complex* ones remain to be considered. They are the 5th and 8th.

5th pair. Trifacial, or Trigeminus.—Like the spinal nerves, the 5th has two roots; an anterior, smaller, motor, and a posterior larger, sensory root. The latter also has a ganglion upon it, the ganglion of Casser.

After the ganglion, the trunk of the nerve divides into three branches. Of these, the *ophthalmic* (supra-orbital, etc., see *Anatomy*) and *superior maxillary* are entirely sensory ; to them being due the sensibility of the upper parts of the face. The inferior maxillary receives all the filaments of the anterior *motor* root of the 5th, besides sensory filaments like the other branches. It supplies the *muscles* of *mastication ;* including the buccinator. One branch of the inferior maxillary (lingual nerve) is one of the two nerves of *taste ;* going to the tip of the tongue.

The 5th nerve is, perhaps, the most acutely sensitive in the body. Injuries of it produce an indirect disturbance of nutrition ; the exact rationale of which has not been entirely explained. *Reflex* symptoms sometimes follow irritation of its branches ; as, when a severe blow over the supra-orbital notch causes sympathetic blindness (Morgagni); or a mis-fitting artificial tooth brings on convulsions (Lederer). *Tic douloureux*, or neuralgia of the face, is an affection of the 5th pair.

8th pair.—Three really separate nerves are included under this; the *par vagum* or *pneumogastric*, the *glosso-pharyngeal*, and the *spinal accessory.*

Pneumogastric nerve.—Although mostly sensory, or at least *afferent*, some motor influence seems to belong to this nerve. Probably the latter is derived from filaments contributed by the spinal accessory ; as such enter into the *inferior laryngeal* branch of the par vagum, which supplies the vocal muscles. The *superior* laryngeal is a sensitive nerve.

The pneumogastric is the main afferent nerve of respiration ; by which the impression of the need of fresh air (*besoin de respirer* of the French) is transmitted from the lungs to the medulla oblongata. Probably, however, other nerves may contribute to the same end, by conveying to the same respiratory centre an impression of the need of air from all parts of the body.

Digestion is also to some extent under the influence of the same nerve. When one pneumogastric is divided in a living animal, near the stomach, its digestion is interrupted. After some days, however, it may recover from the injury, and digest as before. If, then, that of the other side be also cut, digestion is more seriously interfered with, and for a longer time; but, at length, before either of the nerves has reunited, the functional action of the stomach is restored. These facts show, that, while the pneumogastric nerve has an *influence* over the digestive process, it is not *essential* to it.

How the *heart* is affected by the normal functional action of the pneumogastric is a difficult question. Vivisection has introduced a problem concerning it. When, in a living animal, the chest is opened, and the pneumogastric divided, if a *powerful* galvanic current be made to act on the end of the nerve towards the heart, that organ ceases to beat almost at once. Two theories have been proposed to explain this. One, that of *inhibition ;* namely, that it is the function of the pneumogastric to *check* or *retard* the heart's action. The other, that, since it is only a very strong current of galvanic electricity that will arrest the heart's action, this result is produced by *exhausting* or overwhelming the heart by excessive irritation. The latter is probably true.

Glosso-pharyngeal nerve.—As indicated by its name, this is distributed to the tongue and pharynx. It is doubtful whether a motor function belongs to any of its fibres. It is the afferent nerve of deglutition, and the general sensory nerve of the pharynx. One branch of it is a nerve of *taste;* being distributed to the base and middle of the tongue, and to the soft palate.

Spinal accessory nerve.—The two branches of this, *external* and *internal*, are of separate origins. The *internal* branch comes from the medulla oblongata, and is distributed to the vocal muscles of the larynx; being their motor nerve. The *external* branch comes from the spinal cord, enters the cranium by the foramen magnum—and, going out with the internal branch, is distributed to the *sterno-cleido-mastoid* and *trapezius* muscles. It has only motor functions.

Cerebral Hemispheres.

A marked peculiarity of the brain, besides the proportionately large supply of blood which it receives, is, the *variation* in the flow of blood through it from time to time, according to the presence or absence of mental excitement. As, however, a certain degree of increase of *pressure* will arrest the functions of the encephalon, even fatally (as in compression of the brain from fracture of the skull), provision must be made to avoid that pressure. This is done, partly, by the adapted increase of rapidity in the escape of blood from the brain by the veins; but more effectually by the movements of the *cerebro-spinal fluid.* This is a serous liquid, contained beneath the arachnoid membrane of the brain and spinal marrow, and passing readily from the one to the other. If more blood enters the brain, a larger amount of sub-arachnoid fluid leaves it to go into the spinal cavity; and *vice versâ.*

Upon the large and difficult subject of the functions of the cerebral hemispheres, much might be written (see *Carpenter's Human Physiology*). For economy of space it is proper to attempt here a statement only of all the most positive, and some of the most probable conclusions as yet attainable in regard to it.

1. The *cerebral hemispheres* are the sole organs of *mind;* that is, of intellect and emotion, as well as of will. The *active* portions of the hemispheres are, on good ground, believed to be the superficial or peripheral *convolutions;* the white substance being commissural or communicative only.

2. The two hemispheres, under all ordinary conditions at least,

act *as one*. Probably the *corpus callosum* is the main medium of their intimate correspondence.

3. The brain is, as regards the mental faculties, a *multiple* organ. This is proved by many facts; as, the *partial* mental activity of dreaming and somnambulism; partial insanity or monomania; the limited effects, upon the mental and moral powers, of some injuries; and those peculiar, special gifts of mind, so different in different individuals, to which we give the name of genius.

4. The system of "phrenology," or cranioscopy, taught by Gall, Spurzheim, and Combe, is not sustained by a sufficient preponderance of evidence to enable it to take its place as a part of physiological science.

Lately, the most remarkable suggestion tending in a similar direction to that of Gall's inquiries, is that of Broca and others, growing out of the facts concerning *aphasia*. This name is applied to loss of the power of expression in language, spoken or written, from cerebral disease. *Post-mortem* examination in a number of cases has exhibited some lesion of the anterior portion of left hemisphere of the cerebrum. Broca and others have hence inferred that this part of the brain is the seat and organ of the "faculty of language." The principal objection to this view is, that it supposes an *unsymmetrical* location of function, in a part of the system, the cerebro-spinal axis, which, elsewhere, so far as is known, is entirely symmetrical.[1]

5. *Reflex action* affects the cerebrum, as it does all lower nerve-centres. Especially is excito-motor agency visible in *emotional* actions.

Dr. Carpenter's theory is, that *emotion* consists only in the attachment, in our consciousness, of either pleasure or pain to an *idea*. This is not, however, well sustained by familiar facts. Especially is it contradicted by the evident distinction that emotions, unlike ideas, are always *impulsive;* they form our *motives;* we are moved by them. Hence it is probable that different parts of the brain are the seats respectively of the intellectual faculties and of the emotions.

6. Not only is mental action often reflex, being induced by the *suggestion* of impressions external or internal—but it is also in many instances *automatic*. Will acts, not by direct compulsion of the faculties, but by the directing and selecting power of *attention*. We thus encourage and sustain preferred trains of thought; or, on the contrary, avert attention from those not approved. So, also, with emotions. Dwelling upon an object which naturally excites feeling, will still maintain this, even if the will struggles directly against the emotion itself.

---

[1] See an account in *American Journal of Medical Sciences*, July, 1868, p. 296, of a remarkable case of injury of the head, in which loss of language followed penetration of the *right* half of the brain.

Childhood exhibits the automatism of mental action most fully. Emotional impulses are stronger than mere instincts in youth; maturity brings the full development of the intellectual powers. Over all of these, the will ought to dominate; but some human beings remain, all their lives, automata, under the continued dominion of impulses.

Remarkable exemplification, however, of the spontaneity or automatic character of the action of the mind, is witnessed in the manner of performance of great and peculiar genius. Often, highly gifted men are notably deficient in power of will; as shown by their impulsive nature and want of self-control. Carpenter adduces, as good instances of this, Mozart, the musical composer, and Coleridge, the poet and philosopher. Zerah Colburn in calculation, and Morphy in chess, are examples of different kinds of extraordinary natural superiority in particular gifts. Not always are such powers attended by uncommon general intelligence. Of the absence of this, Blind Tom, the negro pianist, a musical prodigy, but otherwise almost an idiot, is said to be an example.

7. There is reason, even, to believe, that mental action or "cerebration" may sometimes be *unconscious*. Dr. Carpenter's arguments in favor of this view are interesting and convincing; but would occupy too much space here.

SLEEP.

Only the cerebrum, the organ of the mental faculties, among all the nerve-centres of the body, has prolonged periods of complete repose in the order of nature. Short intervals, every centre, every organ in the body has; in the *alternations* or rhythmic actions which all exhibit. Thus, between each two inspirations, and between each two beats of the heart, short rests occur. But the brain must have several hours of continuous repose, every night or day.

In the soundest sleep, total unconsciousness exists. Between this and sleep-walking, or somnambulism, all degrees may occur. The entire explanation of the physiology of sleep is not yet attained. The most interesting fact fully developed of late years is the observation of Durham and others, that during natural sleep less blood flows through the brain than at any other time; the cerebral hemispheres becoming, from diminished activity of their arterial circulation, comparatively anæmic.

In somnambulism, the sensori-motor powers appear to be, sometimes, as perfect as in the waking state; but acting automatically. Artificial somnambulism, which may be produced in some susceptible persons, has been exaggerated in description, into mesmerism or animal magnetism, &c.

ORGANS OF SENSATION.

For the perception of an impression made upon any part of the body, three things are required: an organ or surface to receive the impression, a nerve to convey it, and a sensorial ganglion in the brain to perceive it. It is also requisite that *attention* be given to the sensation; as many impressions, especially if they be slight, may be made upon our organs, while the mind is strongly preoccupied, without our taking any cognizance of them.

Touch.

This is the simplest and most extensively diffused of all the senses; being sometimes, therefore, called *common sensation*. All the superficial parts of the body, and the inlets and outlets of the cavities, are endowed with it, in different degrees. The deeply-seated organs are not possessed of sensibility, although uneasiness and pain result from diseased conditions affecting them.

Susceptibility to *pain* appears to be a different thing from the simple sense

Fig. 148.

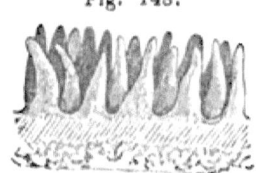

PAPILLÆ OF PALM.

Fig. 149.

Fig. 150.

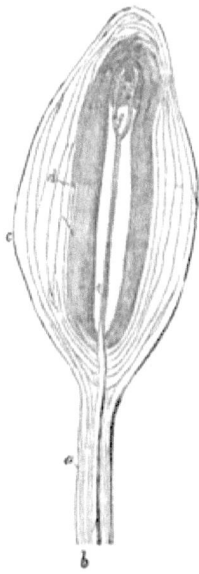

PACINIAN CORPUSCLES.—Portion of digital nerve.

PACINIAN CORPUSCLE.—*a*. Peduncle. *b*. Nerve-fibre. *e*. Axis-cylinder. *f*. Subdivision of same. *c, d*. Laminated sheath.

of touch. Anæsthetic inhalation (as of ether, nitrous oxide, or chloroform) may prevent, sometimes, the pain of an operation, while the patient feels every movement made.

The special organs of touch are the *papillæ tactus*. They are little irregularly conical projections, seen in rows, with wrinkles between them, especially upon the hands. Each papilla contains a (looped ?) termination of a nerve-filament, and a loop of capillary bloodvessels. Only on the digital nerves (of the fingers and toes), are the minute fibrous *Pacinian corpuscles* found.

By pressing the points of a pair of compasses, tipped with cork, upon the skin, it may be shown that it is only when they are at a certain distance from each other that they are felt as two instead of one object. As the distance needed for their separate perception is different on different parts of the body, the comparative sensitiveness of each may thus be measured. The following table exhibits the results of such experimentation (Weber, Valentin) :- -

Point of tongue,	½ line	Lower forehead,	10 lines.
Tip of forefinger,	1 "	Skin of occiput,	12 "
Red surface of lips,	2 lines	Back of hand,	14 "
Palm of hand,	5 "	Skin of back,	30 "
Skin of cheek,	5 "	Middle of arm,	30 "
End of great toe,	5 "	Outside of thigh,	30 "

The term *anæsthesia* is applied to any loss of sensibility, local or general, however produced. Not unfrequently it occurs as a form of *paralysis* or palsy.

Taste.

The tongue, and, to a less degree, the soft palate, are the seats of this sense. The nerves connected with it are, for the tip or

Fig. 151.

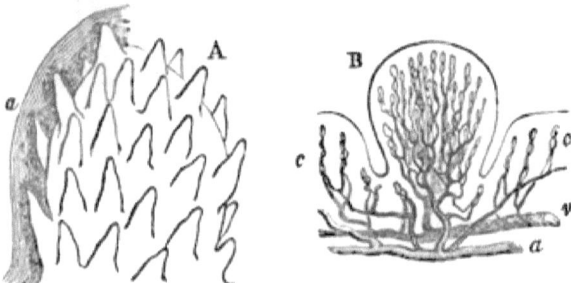

FUNGIFORM PAPILLÆ OF TONGUE.—A. Papilla, with secondary papillæ. *a.* Epithelium. B. Papilla with capillaries injected.

end of the tongue, the lingual branch of the inferior maxillary of the 5th pair, and the lingual branch of the glosso-pharyngeal (8th.) Over the tongue are scattered numerous papillæ (see *Anatomy*), simple, filiform, fungiform, and circumvallate. Each of these, like the papillæ tactûs, contains the end of a nerve and a loop of capillaries. For taste, it is necessary that particles of a sapid substance come in *contact* with the tongue in a state of *solution*. Unless the tongue is morbidly dry, the saliva will suffice to moisten sufficiently an otherwise dry material. A sensation of taste may be produced by a galvanic current; as in the familiar experiment of placing a silver coin under the tongue, and a copper one upon it, so that they touch beyond the end of the tongue.

Impressions of taste are considerably heightened or modified by the simultaneous perception of odor in the same substance. Taste may be cultivated, as in wine-tasters and tea-tasters, to a high degree of delicacy.

The *use* of the sense of taste evidently is, not only to give pleasure in eating and drinking, but to guide in the selection of food. Among *natural* products, it is a generally sufficient guide. Of course artificial substances may, and frequently do, combine pleasantness of taste with poisonous qualities.

Smell.

The seat of this sense is the upper part of the lining membrane of the nasal cavities; receiving the distribution of the terminations of the olfactory nerve. As in the case of taste, actual *contact* of odorous particles is necessary; but these particles may be extremely minute or subtle in diffusion. Thus a small piece of musk may give an odor to a very large apartment. A drop of creasote upon a table may leave a perceptible smell after repeated washings; and it seems sometimes almost impossible to wash the scent of the dissecting room from the hands. Most remarkable, however, is the power of certain animals to detect and pursue their prey by the scent, as the hound does the fox or hare, even while running at full speed. The delicacy of sensitiveness to impressions required for this is almost inconceivable.

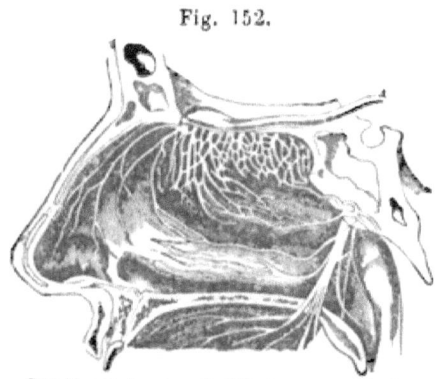

Fig. 152.

THE NASAL CAVITY.—1. Olfactory nerve. 2. Superior turbinated bone. 3. Middle bone. 4. Inferior bone. 5. Fifth nerve.

Among animals, to find their prey, or mates, is evidently a principal use of the organ of smell. We find it also capable of another service, in making known the presence of things which contaminate the air and make it insalubrious. All ordinary causes of atmospheric impurity, and consequently of ill health, are offensive to us.

Hearing.

Vibrations of sonorous bodies induce waves of air or of other media, which, impinging upon the ear, produce the impression

Fig. 153.

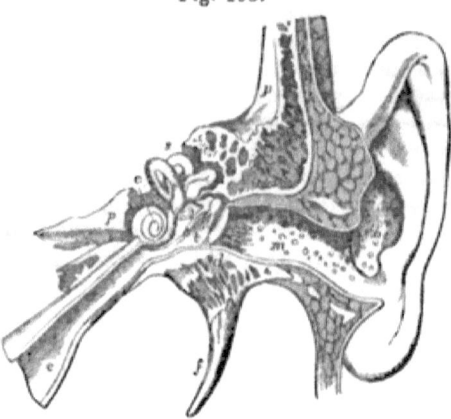

THE EAR.—*b*. Concha. *m*. External meatus. *d*. Membrana tympani. *t*. Cavity of tympanum. *e*. Eustachian tube. *f*. Mastoid cells. *s*. Semicircular canals. *c*. Cochlea. *p*. Petrous portion of temporal bone. *f*. Styloid process.

upon the auditory nerve which we call sound. Rapid undulations produce high notes, slow ones grave or low notes; whether they come from a stringed or wind instrument, a bell that is struck, or a human voice. In a receiver exhausted by the air-pump, the ringing of a bell will produce no sound; because there is no air to vibrate.

The essential part of the organ of hearing is the *internal ear*, where the terminations of the auditory nerve (portio mollis of seventh pair) are distributed. The *middle* and *external* ear, however, are important to the perfection of the organ as an instrument not only for the reception but also for the discrimination of sounds.

Referring the student to the Manual of Anatomy for the description of the structure of the ear, no more will be repeated here than is absolutely necessary to make our language intelligible.

External Ear.—Many animals have their ears movable in different directions, so as to receive sounds from various quarters more distinctly. Man, although possessing *rudimentary* (undeveloped) motor muscles of the ear, has not that power. Instead, the external ear is curiously curved and excavated; thus presenting a number of surfaces in different directions. The total result of these curves and channels is, the concentration or pouring of the aërial waves and reflections, as through a funnel, into the *external meatus*.

This passage ends in the *membrana tympani*. At its outer

orifice are a number of hairs. Its lining tegument secretes, by follicles, the *cerumen* or wax of the ear. The purpose of this wax, whose odor is unpleasant, seems to be, to exclude insects and animalcules. The meatus is endowed with delicate sensibility. Probably we may be aided in determining the *direction* of sounds, by the mechanical impression of air-waves, when sound is most intense, upon its surface. The loss of the external ear lessens distinctness of hearing, but does not cause deafness.

Middle Ear, or Tympanum.—This is a cavity in the temporal bone, bounded on the outside by the *membrana tympani*, and within by the vestibule and part of the cochlea of the internal ear. It communicates with the upper part of the pharynx by the Eustachian tube. Within it, surrounded by air, is the chain of *ossicles*, the *malleus, incus, orbiculare,* and *stapes ;* which connect the membrana tympani with the membrane of the foramen ovale (fenestra ovalis) of the vestibule. These little bones having a slight motion upon each other, three muscles affect their condition and the tension or relaxation of the tympanic membrane; the *tensor tympani, luxator tympani,* and *stapedius.*

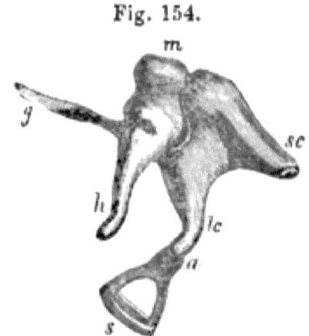

OSSICLES OF THE EAR.

The tympanum is, essentially, a drum. It is not air-tight, because, by the Eustachian tube, air is admitted into it. The use of this seems to be, that sonorous waves, impinging upon the membrana tympani from without, through the meatus, may be to a sufficient extent *balanced* within. Violent effects are thus provided against.

Internal Ear, or Labyrinth.— Again referring to *Anatomy* for description, we may call attention only to the facts, that the vestibule, cochlea, and semicircular canals contain fluid, and that within them are spread out the terminations of the auditory nerve. The vestibule communicates with the membrana tympani by the chain of ossicles. The cochlea has a foramen, the *foramen rotundum,* or *fenestra*

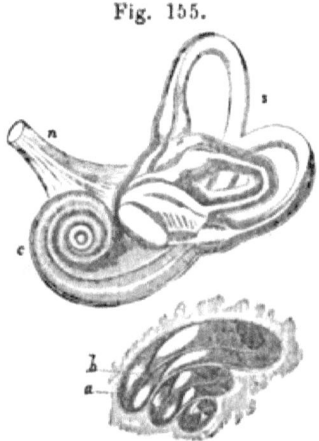

LABYRINTH OF THE EAR.—*n.* Auditory nerve. *s.* Semicircular canals. *c.* Cochlea. *a.* Osseous wall of the spiral tube of the cochlea. *b.* Spiral staircase.

25

rotunda, opening upon the cavity of the tympanum, but covered with membrane. Within the vestibule, also, are the *otoconia* or *otoliths*, minute calcareous crystals; whose whole use must be supposed to be, by their vibrations, to reinforce feeble impulses of sound and make them more appreciable.

The uses of the spiral shape and double "staircase" of the cochlea, and those of the three differently placed semicircular canals, are not yet known. The idea that the canals may serve to enable us to determine the *direction* of the origin of sounds (a not uncommon conjecture) is excluded by the fact that almost all sonorous vibrations enter by the *meatus externus*, and, after striking upon the *membrana tympani*, have thereafter the same direction, whatever their origin.

More probable is the theory that the spiral "ascending and descending" staircase of the cochlea (communicating by a minute aperture at the *humulus*) allows of the *roll* of every distinct wave

Fig. 156.

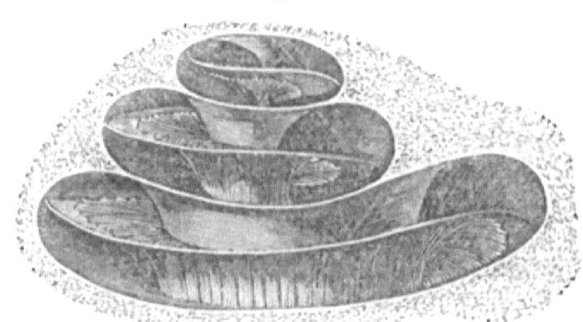

THE COCHLEA.

of sound, intensifying it by the condensation and rarefaction produced by the narrowing and subsequent expansion of the double spiral; no abrupt arrest or rebound of the wave being thus produced, but a gradual exhaustion of it. The rounded form and varied direction of the semicircular canals suggest an analogous purpose; to give place for the escape and dying out of undulations which have made their impression upon the auditory nerve, and which, if they returned or reverberated in the vestibule and tympanum, would confuse the hearing. This has been compared with the arrest or "absorption," by the black pigment layer of the choroid coat of the eye, of rays of light which have made their impression upon the retina. Probably the cochlea may also afford the means of determining the *pitch* of sounds.

Transmission of Sound.—Sonorous vibrations move through the *air* at the rate of somewhat more than 1100 (1118–1142) feet in a second; through *water*, 4900 feet in a second; through solid bodies, much more rapidly, according to their density. When such waves are transmitted by contact from one body to another, some loss of momentum and of rapidity in the undulations occurs; the note, if it be a continuous musical sound, is lowered. Savart proved that sounds may be audible which result from the succession of 24,000 impulses in a second; and, at the other extreme, of lowness, he found them audible when produced by seven or eight impulses, or fourteen or sixteen half-vibrations in a second.

If sound is conveyed by contact from air to water, a considerable loss is produced. But if a solid body, such as a metallic plate or an animal membrane, be placed upon the surface of the liquid, intervening between it and the vibrating air, there is less diminution.

Precisely this arrangement exists in the ear. Sonorous waves are conveyed to the extremities of the auditory nerve by undulations (reciprocal and resonant) of *solids*, *liquid* and *air*. First, as we ordinarily hear, the waves of air enter the meatus and strike upon the membrana tympani. Thence the undulations are carried inward, by 1st, the *walls* of the tympanum; 2d, the chain of bones; 3d, the air within the tympanum. Thus they reach the *fenestra ovalis* by the foot of the stirrup-bone (stapes); the *fenestra rotunda* through the air; and the general surface of the labyrinth by the walls of the tympanum. Then the *liquid* (perilymph, endolymph) of the vestibule, cochlea, and canals receives the vibrations, and transmits them to the auditory nerve.

Defect of hearing may be produced by several causes. Slight degrees of it may attend the accumulation of hard wax in the ear; thickening, from catarrhal inflammation, of the membrana tympani; or obstruction of the Eustachian tube. More serious deafness follows perforation or destruction of the membrana tympani, or partial destruction of the chain of bones in the tympanum. *Total* loss of hearing only occurs when the auditory nerve, or its sensorial ganglionic centre, is paralyzed. A simple test will show whether deafness, in any case, be nerve-paralytic, or connected only with the mechanical apparatus of the ear. If it be the latter, the ticking of a watch may be heard if it be placed between the teeth; as the solid bones of the face and head, and the air entering the Eustachian tube, conduct the sound. This will not be the case, and no sound will be heard, if the auditory nerve has lost its receptive or transmitting power.

Vision.

An expanded surface sensitive to light, an optic nerve to convey impressions made by it, and a sensorial ganglion to receive and *perceive* them; these are the parts essential to sight. Beyond these, the eye is a visual *apparatus*, compared often to a *camera obscura*, or a photographer's instrument.

Light is the subject of an abstruse mathematical science, Optics. Some account of it is usually given in connection with chemical physics (see *Chemistry*). Brief allusion to a few facts concerning it is unavoidable here.

Adopting the *undulatory* theory, we conceive light to consist of extremely rapid vibrations, conveyed from luminous bodies to the eye through the *æther* which pervades space. This universal æthereal fluid penetrates even solid bodies, affecting and being affected by the state and movement of their particles. Heat, electricity, galvanism, and magnetism, no doubt, involve its wave-movements, as well as those of ponderable substances. Light travels through space at the rate of from 186,000 to 192,000 miles[1] in a second.

Certain bodies *transmit* light; they are transparent. *Reflection* of light is a universal phenomenon or property of solids and liquids, although in very different degrees. The *direction* of reflected rays is governed by their line of incidence, and the reflecting surface; according to the law of mathematical physics, that *the angle of incidence and the angle of reflection are equal.* The colors of bodies depend upon the colors of the rays of light which they reflect. A red body reflects only red rays; a green one, only green rays, &c. A black body, however, does not (theoretically) reflect any light at all. A white one reflects the whole of the light falling upon it.

Passing *through* transparent media, as water, glass, &c., rays of light are often made to change their course. This is called *refraction.* Thus a marble, at the bottom of a cup, not visible to an eye at a certain distance beyond its edge, when the cup is empty, will seem to *rise up* and be seen over the edge when water is poured into it.

Emerging from water into air, a ray of light is bent in one direction; going from air into water, in another. Referring both, for comparison, to a line drawn perpendicularly to the surface entered or left, the law is thus expressed: a ray of light, in passing from a denser to a rarer medium, is refracted *from* the perpendicular; in passing from a rarer to a denser medium, *towards* the perpendicular.

But refraction does not occur to the same extent or in just the same way with all substances, or with bodies of all forms. *Double refraction* and *polarization* of light are complex optical phenomena which cannot be entered upon here. By the action of a simple prism of glass, a ray of light in its transmission is *decomposed,* or separated into the seven rays of the *spectrum,* identical with those of the rainbow. Three *primary* rays are admitted, red, yellow, and blue; the others being combinations of these in different proportions. That white light is really made up of a union of all the

---

[1] According to different authorities.

colored rays, may be demonstrated by a physiological experiment. If a circular disk or flat piece of pasteboard be divided into seven equal parts, by lines radiating from the centre to the circumference, and the seven colors of the spectrum be painted upon these, when the disk is made to revolve rapidly it will appear to be white. This happens because the impression of any visible object upon the retina *remains* for a moment after looking at it. The quick succession of impressions, therefore, in the revolution of the disk, blends all the colors into one, and the result is whiteness.

Upon modifications in the reflection and refraction of light caused by different materials, forms, positions, and other conditions of bodies, depend the uses of the parts of the eye as an organ. For the full description of these parts we must refer to *Anatomy*.

The transparent media of the eye are, the *cornea, aqueous humor, crystalline lens,* and *vitreous humor.* The receptive surface upon

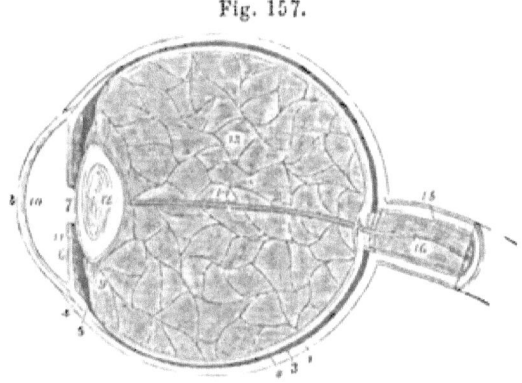

Fig. 157.

SECTION OF EYEBALL.

which, within the chamber of the eye, the picture or image is thrown (as upon the sensitive prepared plate of the photographer's apparatus) is the *retina.* Of this, the most important layers, physiologically, are, 1, the layer of rods and cones, which suggest analogy to the *papillæ* of touch; 2, the layer of ganglionic nerve-cells; 3, the radiating expansion of the terminations of the optic nerve.

It is very remarkable that the point of entrance of this nerve into the retina, which is just within (*i. e.*, nearer to the nose than) the centre of its surface, is devoid of sensibility to light, as shown by careful experiments. At the centre of the retina, the *yellow spot* of Sœmmering is the point of most distinct vision. Evidently, the image of any object upon the retina must be *inverted;* because, the rays from its upper and lower extremities crossing each other as they enter the pupil, those rays from the top of the object must pass through to the bottom of the retina, and those from the bottom must go to the top of the same receptive surface. How, then, do we see everything right side up? Several theories have been proposed to explain this, which is certainly not a mere result of experience or education. Perhaps the crossing of the fibres of optic nerves, from the top and bottom of the retina, respectively, may *reverse*, in their conveyance of impressions to the tubercula quadrigemina, the picture itself. More reasonable, perhaps, it is, to suppose that the mind really sees, not *the retina with the image* upon it, as some imagine—but, *from* the retina; that is, there receives and perceives waves of light from the reflecting visible *object*. So, as we refer a blow upon the hand or body to one or another source by our minds tracing backwards the direction of its impression upon us, we in like manner refer a ray of light to an object by tracing backwards its direction. When we look at a mirror, for example, and see an image of anything in it, the image *seems* to be *beyond* the glass; and it requires an effort of thought, or some experience, to correct this impression.

The crossing of the filaments of the two optic nerves in all possible directions, from side to side, undoubtedly explains the *unity* of human vision with two eyes, upon each of which a distinct impression is formed. If the eyes are placed out of correspondence, as in strabismus, double vision results. In habitual or permanent strabismus, the mind is apt to accustom itself to give attention only to one of the two images seen. Double vision is sometimes *subjective; i. e.*, connected as a symptom with disorder of the brain.

The cornea, aqueous humor, crystalline lens, and vitreous humor act upon light transmitted through them as a series of lenses. The anterior surface of the cornea is convex; the crystalline lens is doubly convex, the posterior surface being the most convex of the two. Applying to convex lenses the law of refraction already mentioned, that, in passing from a rarer to a denser medium, rays of light are bent or refracted *towards* the perpendicular, and remembering that at any part of the surface of a sphere a perpendicular to it is continuous with a straight line going to its centre, it will be obvious that parallel rays (or those nearly parallel) must, by such a lens, be made to *converge*, and, finally, to meet. The place at which the rays from any body thus unite is called the *focus* of the lens. Rays not parallel but divergent, as from a small and near body, cannot be brought so soon to a focus, unless the lens is modified so as to increase its refraction.

Adjustment of the eye to different distances, then, needs explanation. The theory now adopted is, that, as the eye *reposes* in distant vision, effort is required only for the sight of near objects; and that this is effected by pressure upon the lens by the *ciliary muscle;* which is described by Wharton Jones, H. Müller, and

Rouget as consisting, like the iris, of circular and radiating fibres. Perhaps the vascular processes of the *ciliary body* may also, as an erectile structure, assist in compressing the lens when distended with blood.

The *iris* contracts when we look at near objects, and thus excludes the most divergent rays. The general purpose of the iris is to regulate the amount of light entering the eye through the pupil. By reflex action, its circular fibres contract under the stimulus of light. When that stimulation is withdrawn, the radiating fibres dilate the pupil. The contraction of the pupil may be artificially produced by the action of the Calabar bean, or by opium in poisonous doses. Dilatation of the pupil follows the local or systemic action of stramonium, belladonna, hyoscyamus, or their active principles. The *afferent* or sensory nerve of the iris is the 5th; the motor nerve in contraction of the pupil is the 3d; dilatation is under spinal and sympathetic influence.[1]

Optical instruments are liable especially to two defects: *chromatic* and *spherical aberration.* The first consists in change of color in objects seen through lenses, owing to the differently-colored rays being unequally refracted in their transmission. The second is the deviation and confusion of rays impinging upon different parts of the surface of a convex lens. Those rays which pass from the centre of an object straight through the centre of the lens, are not *refracted* at all. Those striking at a small distance from the centre are made to *converge* somewhat; those entering near the margin or circumference, converge very much. So, they do not all meet in one focus, and an indistinct image results.

The correction of *chromatic* or color aberration is, by opticians, effected by using several lenses, of different materials, refracting the rays of light differently; so that one corrects (by reversing it) the inequality produced by another. Thus, if the arrangement be perfect, colorless rays, or rays without change of color, are transmitted. In the construction of the eye, this principle is illustrated by a combination of several refracting media; the cornea, aqueous humor, crystalline lens, and vitreous humor.

Spherical aberration is, in the eye, corrected in two ways. For one, the *iris* serves to exclude those rays which would strike the margin of the convexity of the lens; and thus their tendency to confuse the image is prevented. Also, the *structure* of the crystalline *lens* admirably preserves the eye against this deviation of light. The lens is *most dense at the centre*, and lessens gradually in density towards the circumference. Thus those rays which strike it *near* the centre are made to converge more, in proportion, than those striking it towards the margin; more, that is, than the form

---

[1] The pupil contracts, under strong light, after death; according to Brown-Séquard, in eels and frogs, sixteen days after death.

of the lens would cause them to do, if it were homogeneous; but just the same as the others, by this correction; so that a clear image is made.

Our ideas of the *distance* of objects are obtained through *experience*, with the comparison of other senses than sight, and the use of our powers of locomotion. To a person cured of congenital cataract, as in a case reported by Cheselden, or to one like Caspar Hauser, brought out into the light after years of imprisonment in darkness, everything seems at first *upon* or very near to the eye.

Size, also, is only very imperfectly determined by sight *alone*. The *visual angle* gives us some estimate of it; that is, the angle made by the rays coming from the top and bottom, or the right and left side, of an object, and meeting upon the retina. Also, we are aided in measuring the size of an object seen, by the *muscular sense* of the muscles which move the eye up and down, or from side to side, in looking at it all over.

Our knowledge of the size of any object will facilitate our making an estimate of its distance; and so, if we know its distance, we can judge better of its size. Hence, on a foggy day, many things look larger than they actually are, because haziness gives an impression of distance; while, on the contrary, in an extremely clear atmosphere, things of unknown size may appear to be very near but small.

The *thaumatrope* is a simple toy, which well illustrates the fact that impressions made upon the eye remain for a little while. It consists of a piece of card, on each side of which is drawn or painted a different picture, the two bearing some relation to each other; as, on one side a bird, on the other a cage; or, on one side a man and on the other a horse. When the card is made to revolve rapidly so as to present the two sides alternately, the images are combined into one; and the bird may be seen in the cage, or the man on the horse.

The *stereoscope* of Wheatstone is an instrument composed of two lenses, so placed as to cause two pictures of the same object, one representing the right side of it and the other the left (Fig. 158), to appear as one. The impression of *solidity* and perspective, which, in actually solid objects, we gain by the use of both eyes together, is thus perfectly conveyed by flat pictures; as photographs, for example.

Imperfections of sight, in various degrees, are common and important. Their study and treatment belong to the special department of ophthalmic surgery. A few of them may be here briefly mentioned.

Near-sightedness, or *myopia*, is the most frequent of all. It is owing to too great a *convexity* of the crystalline lens, or else to an excess of length of the eyeball antero-posteriorly; the rays falling upon the eye from a distant object being brought to a focus *too soon*, that is, in front of the retina. The correction of this is obtained by the use of *concave* glasses, which by their refraction move back the image to the retina.

Fig. 158.

STEREOSCOPIC VISION.

Long-sightedness, or the incapacity to see objects at short distances, is due to causes the reverse of the above; the lens being *too flat*, or the eye too short from before backwards. The rays then converge too little to make an image upon the retina; their focus falling *behind* it. This (very common in old people) is corrected by *convex* glasses, which bring the rays to a focus sooner. Because of the frequency with which old persons experience this defect of vision, it is commonly called *presbyopia*. But age is apt to bring with it other kinds of failure of sight also; and it is oftenest a loss of power to *accommodate* the eye to the distance of objects, the *range* of vision becoming abnormally restricted.

Near point at	10th year	. . .	$2\frac{3}{4}$ inch from cornea.
" "	20th "	. . .	$3\frac{5}{8}$ " " "
" "	30th "	. . .	$4\frac{1}{8}$ " " "
" "	40th "	. . .	$6\frac{8}{9}$ " " "
" "	50th "	. . .	12 " " "
" "	60th "	. . .	24 " " "
" "	70th "	. . .	144 inches.

(Donders, Fellenberg, &c.)

Terms now much used by opticians are, *emmetropic*, *myopic*, (hypometropic) and *hypermetropic* vision; also, *paralysis of accommodation*, and *astigmatism*. *Emmetropic* is *normal* sight, with the full natural range of accommodation. Many persons can see to read, for example, with ease, letters at a distance of seven inches from the eyes, or at thirty-six inches, or at any distance between. That is their range of accommodation. *Myopic* persons sometimes can only read with the book two or three inches from the eye. *Hypermetropia* is deficiency in the refracting power of the eye, so that only convergent rays make distinctness of image, and thus only objects at some distance can be well seen. It is what is most generally called presbyopia. *Astigmatism* is indistinctness of vision from differences in refraction by the eye according to the part of it upon which rays fall; a want of harmony in refraction. It is tested by looking at the two lines of a cross, or a large letter T; the lines in different directions not being seen with equal clearness. It may be remedied by employing cylindrical lenses (see *Laurence and Moon's Ophthalmic Surgery*).

Amblyopia is indistinctness of vision. *Asthenopia* is *feebleness* of visual power, shown by the eyes becoming soon exhausted when used, as in reading, and images then becoming confused. This is a symptom of *fatigue* of the retina, optic nerve, or tubercula quadrigemina; calling for *repose* of the organs of sight.

Amaurosis (a term repudiated now by some ophthalmologists) is nerve-blindness, or paralysis of the optic nerve.

Cataract is opacity either of the crystalline lens or of its capsule. Though sometimes congenital, it occurs in much the largest number of instances in advanced life. Occasionally it is observed in connection with *diabetes mellitus;* and, experimentally, it has been produced in animals, by injecting sugar into the tissues. The theory of endosmosis has been brought forward to aid in explaining some of these facts.

Accessory Apparatus of the Eye.—The *lachrymal gland* is very important, on account of the protection its constantly flowing moisture affords in clearing fine particles from the surface of the eyeball, and keeping it from shrinking with dryness. The *Meibomian* glands of the eyelids furnish a sebaceous material which maintains the flexibility of the eyelashes. The *lashes* themselves (cilia) are especially protective in their use, acting somewhat like cow-catchers in front of a locomotive, their convexities being turned towards each other. *Winking* is a (generally automatic and often unconscious) motion of the lids, by which the lachrymal secretion is brought frequently over the ball, and, when needful, the eye is closed against the entrance of foreign bodies. The *eyebrows* (supercilia) are protective overhanging ridges, covered with hairs, which aid in keeping blows or falls from affecting the eyes, and also avert from them drops of perspiration of the forehead. For an account of all these auxiliary structures see *Anatomy*.

CHAPTER VIII.
THE VOICE.

THE larynx is the organ of voice in man. Referring to Anatomy for the particular description of its parts, we may consider what is its nature as an instrument.

Fig. 159.

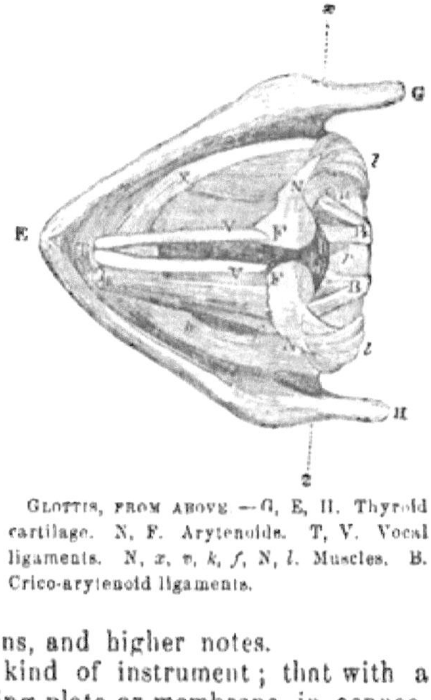

Among musical instruments, two kinds are most decidedly different: *wind* and *stringed* instruments. Of the former, the trumpet, horn, trombone, flute, and fife are examples. Of the latter, the violin, guitar, &c. In wind instruments, the notes vary with the length (and width) of the *column of air*, affected by the closing and opening of the holes, and the manner of impelling air from the mouth. In stringed instruments, the *length, thickness, and degree of tension* of the strings determine the notes principally. In both, the larger vibrating mass, moving in slower waves, gives low notes; the shorter column or cord produces quicker vibrations, and higher notes.

GLOTTIS, FROM ABOVE.—G, E, H. Thyroid cartilage. N, F. Arytenoids. T, V. Vocal ligaments. N, x, v, k, f, N, l. Muscles. B. Crico-arytenoid ligaments.

But there is an intermediate kind of instrument; that with a *tongue* or *reed*, that is, a vibrating plate or membrane, in connection with a tube. The clarionet, bassoon, hautboy, and reed stops of an organ are of this kind. The human larynx is a reed instrument, neither precisely a wind nor a stringed one. The reeds are the *vocal ligaments;* the *epiglottis* is also of that nature.

Except in the production of very high notes, only the *inferior* vocal cords are shown, by experiment and by observations with the laryngoscope, to be essential to voice. The *anterior* part of the glottis, also, is the vocal part. The changes which occur in vocalization are, the *widening* and *narrowing* of the *glottis*, and the increase and diminution of the *tension* of the *ligaments*. These changes are effected by muscles. The *posterior crico-arytenoid* muscles draw back the arytenoid cartilages, and make tense the ligaments; the *crico-thyroid* muscles aid in the same effect. The *posterior arytenoid* muscles, *transverse* and *oblique*, narrow the orifice of the glottis. Both of these movements cause elevation of the pitch of sounds. The *thyro-arytenoid* muscles relax the ligaments by drawing the arytenoid cartilages forward. The *lateral crico-arytenoid* muscles widen the aperture of the glottis. Thus grave or low notes are produced. The lowest human tone has 80 vibrations per second; the highest, 1024 vibrations.

The compass of the voice extends in singers to two or three octaves. The lowest note of the female voice is about an octave higher than the lowest of the male voice, and the highest of the female voice an octave above the highest of the male. This depends chiefly on the greater length of the ligaments in the male larynx. The larynx in boys is like that of the adult female. At puberty it changes, with other sexual developments.

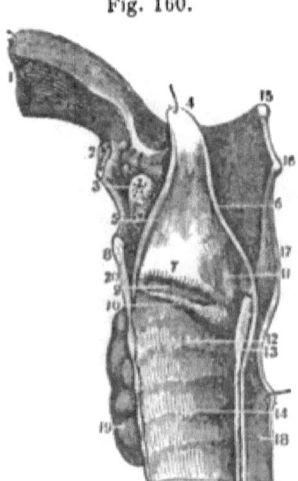

Fig. 160.

THE LARYNX OPENED.—7. Superior vocal ligaments. 8. Thyroid cartilage. 9. Ventricle of Galen. 10. Inferior vocal ligament. 11. Arytenoid cartilages. 12. Cricoid cartilage.

"There are two kinds of male voice, the bass and tenor, and two kinds of female voice, the contr'alto and soprano. The bass voice usually reaches lower than the tenor, and its strength lies in the low notes; while the tenor extends higher than the bass. The contr'alto has generally lower notes than the soprano, and is strongest in the lower notes of the female voice, while the soprano voice reaches higher in the scale. But the difference of compass, and of power in different parts of the scale, are not the essential distinctions between them. The important difference consists in their tone or 'timbre,' which distinguishes them even when they are singing the same note. The qualities of the barytone and mezzo-soprano voices are less marked, the barytone being intermediate between the bass and tenor; the mezzo-soprano between the contr'alto and soprano." (*Kirkes*.)

In each octave there are twelve semitones. Within each semitone at least six different notes can be produced; 120 notes or more can, then, be made distinctly by an ordinary singer. Yet the vocal ligaments vary in length but ⅛ of an inch. Each note requires, therefore, a variation in the state of the ligaments of but $\frac{1}{600}$ of an inch. In falsetto notes, it would appear that only the edges of the vocal cords approach each other.

The lungs are the *wind-box* of the vocal organ; the manner in which air is impelled through the larynx making much difference in vocalization. The solid and flexible structures of the mouth and nostrils also affect and modify it considerably.

For speech, *articulation* as well as voice is necessary. In this, the tongue, palate, lips, and teeth are employed. Besides the *vowel* sounds, some of which may be continuously pronounced—as ah, au, e, eh, ih, uh, oo, while others are compound, as a, i, o, oi, ow, yu, the *consonants* are by some writers divided into *linguals*, *dentals*, *labials*, &c. These distinctions are not very accurate. The tongue is used in c, hard and soft, d, g, hard and soft, h, j, k, l, n, q, r, s, t, x, y, z. The lips especially in b, p, f, m, v, w. The teeth are expressly touched by the tongue in s, z, and th. Reverberation in the nasal cavities occurs in the *prolonged* sound of b, d, g, l, m, n, r, v, w, y, z, and also in certain compound sounds, as ng. In other languages than the English, many sounds exist which no combination of our letters is capable of expressing.

Stammering is owing to a want of control of the will over the muscles of articulation. It is to be treated and cured, therefore, by systematic and well devised *vocal gymnastics;* instruments afford only very slight palliation for it.

CHAPTER IX.
DEVELOPMENT.

CONCEPTION.

ACTUAL contact of the spermatozoa and ovule is necessary for impregnation; more than this, according to recent observers, not only contact but penetration of the ovule by the spermatozoa. Barry, Nelson, Keber, and others have seen spermatozoa inside the ovule in various animals. In most, if not all cases, there appears to be a penetration by many of them. Spermatozoa, at the same time, pass through the coats of the vitellus, and, massing around the germ cell, lose their motor power, become disintegrated and disappear, after having awakened the germ to its wonderful life.

THE EMBRYO.

The first change in the ovule (henceforth known as the ovum) is cleavage or segmentation of the yelk. The embryo cell elongates, becomes somewhat fiddle shaped, divides into two cells by the ordinary process of cell division. At the same time the yelk divides into two parts, so that each cell is contained in its own separate yelk mass. By repetition of this process we have 4—8—16 cells and yelk masses formed, and so on until segmentation is completed, and out of the germ cell and yelk is formed a homogeneous mass of cells—the *germ mass*. During this, there has been an evident increase of cells at the expense of the yelk.

After the formation of the germ mass the cells nearest the surface immediately under the zona pellucida or vitellary membrane become aggregated to form a membrane, the so-called *germinal membrane*, which soon divides itself into the external or *serous layer* and the internal or *mucous layer*, between which the other germinal cells are collected forming the *vascular layer*. The first of these gives origin to the vertebral column, skull, extremities, nerves, general skeleton, brain, and spinal cord. In the second the glands and mucous structures are formed, whilst the vascular system, including the heart, originates in the intervening layer—the vascular. Soon there is an evident thickening at one end of this fourfold ball, as it were; a heaping together of cells, and the dense area thus formed receives the name of *area germinativa*, which is first roundish, then oval, then pyriform. In the centre of this, which is of course made up of three layers, the cells of the serous and mucous layers become comparatively few in number, and a semitransparent spot is thus caused, the so-called *area pellucida*, which is bounded laterally by an accumulation of the cells of the vascular layer, the *area vasculosa*. In the serous lamina, in the centre of the *area pellucida*, appears soon the *primitive trace*, a transparent groove, the first sign of the fœtus. It is surrounded above by two ridges, the *laminæ dorsales;* which, during development, arch over the groove and finally unite above, enclosing the groove in which the cerebro-spinal nerve-centres are afterwards formed, whilst the laminæ are transformed into the spinal column and skull.

In the same way two ridges grow out below the primitive trace, arch over (or rather under) and unite, thus forming the abdominal cavity of the fœtus. These are the *laminæ ventrales*. It is readily seen that if either the dorsal or ventral laminæ fail to unite, there will result a monstrous fœtus; if it be the former which do not join, the spinal column will be incomplete, and *spina bifida* will occur; if the latter, the anterior part of the body will be incomplete, and *hare-lip, cleft palate*, or other malformation dependent on insufficient development of the walls of the chest or abdomen will be present. In the area vasculosa, which is inclosed in the laminæ ventrales, the bloodvessels are formed by the union of rows of cells, and obliteration of their separating walls; their nuclei (according to Dr. Carpenter) forming the blood disks. In like manner the heart is formed after the vessels; the first flow of blood is towards the heart, the so called *punctum saliens* (Fig. 163.)

Amnion.

The *amnion*, the essential envelope of the ovum, appears very early in the formation of the embryo. A little beyond the extremities of the ovum, on the outside of the area pellucida, the serous lamina projects in two hollow processes, which arch over the dorsum of the embryo and unite to inclose it. At first these two processes are widely separated in their central portion; but soon they spread also over the ventral surface of the embryo, and ultimately surround the elements of the umbilical cord: of the two layers, which thus form the amnion, the outer becomes adherent to the external or maternal membrane, the other constitutes the especial envelope of the fœtus.

Fig. 161.

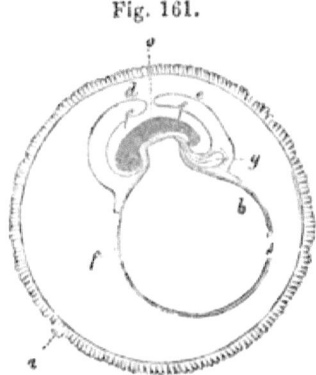

Fig. 162.

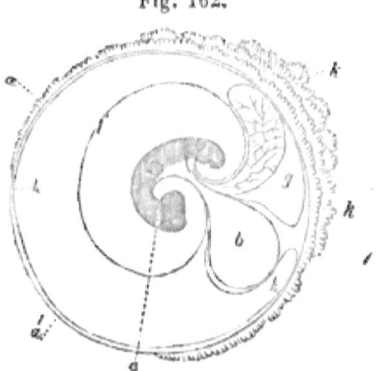

THE AMNION AND CHORION.—*a.* Chorion. *b.* Umbilical vesicle, surrounded by the serous and vascular laminæ. *c.* Embryo. *d, e,* and *f.* External and internal folds of the serous layer, forming the amnion. *g.* Incipient allantois.

DIAGRAM REPRESENTING A HUMAN OVUM IN SECOND MONTH.—*a* 1. Smooth portion of chorion. *a* 2. Villous portion of chorion. *k, k.* Elongated villi, beginning to collect into placenta. *b.* Yelk sac, or umbilical vesicle. *c.* Embryo. *f.* Amnion (inner layer). *g.* Allantois. *h.* Outer layer of amnion, coalescing with chorion.

The membrane thus formed embraces the embryo very closely at an early period, and is continuous with the common integument of the fœtus, at the open abdominal parietes. At a later period it is distended with fluid, and so separated from the fœtus, and after being reflected upon the funis, of which it forms the outer coat, it terminates at the umbilicus. It is thin and transparent, but of a firm texture, resisting laceration much more than the other membranes. Its external surface is somewhat flocculent, but internally it is quite smooth, like serous membrane, and like it secretes a bland fluid. This fluid resembles dilute serum, and is called *liquor amnii*. It varies in amount from half a pint to several quarts, the average quantity being about half a pound. It subserves several useful ends. It probably serves as nutriment to the fœtus during the early months; it preserves an equable temperature for it while remaining in utero; it protects it from the effects of sudden blows, shocks, etc. It is also useful in dilating the os uteri, by protruding the membranes in the commencement of labor.

It will be remembered that the mucous layer is the most internal, being that which immediately surrounds and contains the yelk. The first change in it is the appearance of a constriction caused by the growing from it, towards the centre of the yelk, of a pair of processes, which increase in size until they finally meet and divide the cavity of the mucous layer into two; in the smaller of these, the one next to the embryo, the mucous membranes, glands, etc., are developed, whilst the larger consists largely of the yelk and is the so-called *umbilical vesicle*. Connecting these cavities is a small canal, the *vitelline duct*. The embryo, at this period, chiefly subsists on the yelk contained in the umbilical vesicle; which reaches it not merely through the duct but by means of vessels, the *vasa omphalo-mesenterica*, developed on the walls of the part of the mucous layer composing the duct. These vessels terminate in the superior mesenteric artery and vein. When, finally, the store of yelk is exhausted, the umbilical vesicle shrinks up and leaves a sort of scar, the white spot, *vesicula alba*.

Allantois.

The *allantois* is formed at the lower and anterior part of the embryo, rather from a mass of cells than by a reduplication of one of the primary layers. It is at first a delicate membranous elongated sac, but, like the cavity of the mucous layer, it is divided into two by constriction. The smaller of these divisions becomes the urinary bladder, the other, the allantois proper. The *urachus* of the perfected animal (the cord leading from the bladder to the umbilicus) is the remains of the duct connecting the fœtal bladder with the allantois. The allantois extends itself until at last it comes in contact with the uterine surface. Its further history will be traced directly.

The Decidua.

While this development has been going on, certain changes have taken place in the uterus, which it is now necessary to notice. The earliest change in this organ after impregnation consists in an enlargement of the tubular glands of the mucous membrane, and of the capillaries between them; the glands becoming visible

Fig. 163.

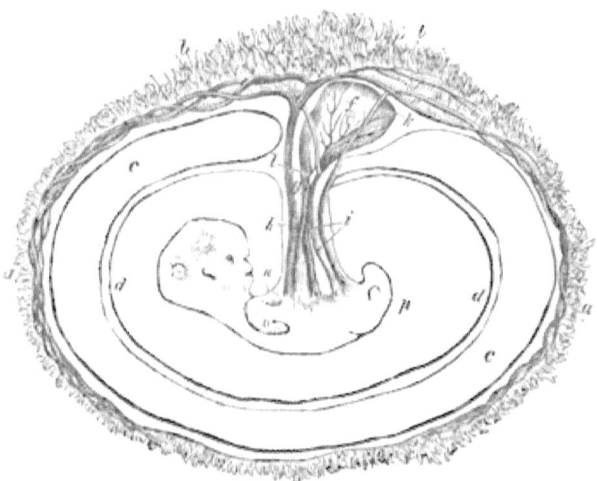

DIAGRAM OF THE FŒTUS AND MEMBRANES ABOUT THE SIXTH WEEK. FROM CARUS.—*a*. Chorion. *b*. The larger absorbent extremities, the site of the placenta. *c*. Aliantois. *d*. Amnion. *e*. Urachus. *e*. Bladder. *f*. Vesicula umbilicalis. *g*. Communicating canal between the vesicula umbilicalis and intestine. *h*. Vena umbilicalis. *i, i*. Arteria umbilicalis. *l*. Vena omphalo-mesenterica. *k*. Arteria omphalo-mesenterica. *n*. Heart. *o*. Rudiment of superior extremity. *p*. Rudiment of lower extremity.

to the naked eye, and pouring out profusely a peculiar secretion of albuminous matter and cells. These together, *i. e.*, the entire altered mucous membrane of the uterus, constitute the *decidua vera* which lines the cavity of the uterus. According to Coste, the ovum on entering the uterus is imbedded in the soft altered mucous membrane, and as it were sinks into it, the latter throwing out granulations which arch over the ovum and form a membrane covering it, *decidua reflexa*. Other authorities have believed that this membrane arises as follows: by the time the ovum reaches the uterine orifice of the Fallopian tube, the uterine mucous membrane has become so swollen as completely to occlude it, so that the ovum necessarily in entering the uterus pushes before a part of the mucous membrane which constitutes the membrana reflexa. It will be seen that there is therefore a double membrane; one part of which lines the uterus, the decidua vera; the other enwraps the fœtus, the decidua reflexa; the whole of which constitutes the *membrana decidua*, so called, because cast off at the birth of the child.

Chorion.

The *chorion* is another envelope of the fœtus obtained from the mother and partially formed before the ovum arrives at the uterus. When the ovum leaves the ovary it is enveloped in a portion of the proliferous disk, which adheres to it and constitutes the first appearance of the chorion. In the passage through the Fallopian tube, new gelatinous material is secreted; until, by the time the ovum arrives at the uterus, it is inclosed in a thick membrane, which, at its first appearance, was smooth, but now is rough and shaggy from the enlargement of numerous villi.

This new formation is one of great importance, as it is through it that the whole subsequent nutrition of the embryo is derived; this is accomplished at first by means of the villous processes which proceed from the whole surface of the chorion and give it a rough, shaggy appearance. These villous processes serve as absorbing radicles, drawing in the fluids supplied by the mother, until a more perfect communication is afforded by the placenta. As the ovum advances in age, these villi diminish in number, assume a vesicular appearance, and finally disappear altogether; *except* at that part of the chorion which is in contact with the uterus, and where the *placenta* is subsequently formed. In some animals, this original connection between the villous coat of the chorion and the uterine surface is the only one that exists; hence they are called *non-placental*.

Placenta.

It will be remembered that we left the allantois passing out of the umbilicus towards the maternal organs. By the time it reaches the decidua reflexa, this, as well as the decidua vera, has become much thickened at the spot which it touches, and the villi of the chorion have there become remarkably enlarged. The placenta is thus formed, by the fusion of the chorion and decidua; the former being connected by the allantois with the ovum. The villi of the placenta are the villi of the chorion capped by the decidua. The human placenta, at full term, is of a soft spongy texture, round or slightly oval, from six to eight inches in diameter and from one to two inches thick; its fœtal or inner surface covered by the amnion is smooth, and generally has the umbilical cord inserted near its centre; sometimes to one side, when it is known as the *battledore* placenta. The external or maternal face of the cast-off placenta is seen to be divided somewhat irregularly into lobes, and covered by the hypertrophic mucous membrane of the uterus, into which the tufts of the chorion have penetrated. The placenta is destitute of nerves and lymphatics, and is in its structure essentially vascular. It will be seen that there are two parts to it; the fœtal and maternal.

The *fœtal* portion of the placenta consists of the branches of the umbilical vessels, which divide minutely where they enter the organ, and constitute by their ramifications a large portion of its substance; each subdivision terminating in a villus. Each villus contains a capillary vessel, which forms a series of loops, communicating with an artery on one side and with a vein on the other. The vessels of the villi are covered by a layer of cells inclosed in basement membrane. The *maternal* portion may be considered as a large sac, consisting of a prolongation of the internal coat of the great uterine vessels. Against the fœtal surface of this sac the placental tufts push themselves, dipping down into it and carrying before them a portion of its thin wall, so as to constitute a sheath to each tuft. The blood is conveyed into the cavity of the placenta by the "curling arteries," so named from their tortuous course, which proceed from the arteries of the uterus, and the blood is returned through large uterine veins called *sinuses*. It must not, however, be understood that there is any direct communication between the vessels of the fœtus and those of the mother, the fœtal tufts being merely *bathed* in the maternal blood and drawing nourishment from it by cells, which have the power of selecting, and of elaborating their own materials. The placenta performs the threefold office of an absorbing, respiratory, and excreting organ. It begins to be formed about the end of the second month; acquires its peculiar character during the third, and goes on increasing in proportion to the development of the ovum.

Umbilical Cord.

The *umbilical cord* is formed from a portion of the allantois; it connects the fœtus to the placenta and affords a passage to the blood from one to the other. It contains two umbilical arteries, and one umbilical vein, the duct of the umbilical vesicle, the omphalo-mesenteric vessels, the urachus, and sometimes more or less of the intestinal canal, the whole imbedded in the *Whartonian jelly*, and invested by a reflection from the amnion. Its length varies much; the average, however, is about eighteen inches. Sometimes it is so short as seriously to impede the progress of labor.

FŒTAL GROWTH.

The head is one of the first portions to be developed. At the beginning of the second month, it constitutes almost half of the entire mass of the fœtus. The upper extremities appear in the early part of the fifth week, as two blunt processes; and, immediately afterwards, the lower limbs. In both cases, the distal parts, *i. e.*, the hands and feet, are the first to be formed, so that they appear, as it were, sessile upon or fixed directly to the trunk; then the parts next to these, *i. e.*, the forearms and legs, and, finally, the arms and thighs. The motions of the child are rarely perceptible before the third month, and the beating of the heart is generally first heard during the fifth month. The average length of a full-grown child at birth, is about eighteen inches; the weight varies from four to eighteen pounds; the average probably being seven or eight pounds—a male weighing about one pound more than a female.

Circulation of the Fœtus.—There are some peculiarities in this. The chief are owing to the presence of the foramen ovale and ductus arteriosus in the thorax, umbilical artery and vein and the ductus venosus in the abdomem.

The *foramen ovale* is an opening between the two auricles of the heart, allowing the blood to pass freely from one to the other. The *ductus arteriosus* arises from the pulmonary artery near its origin,

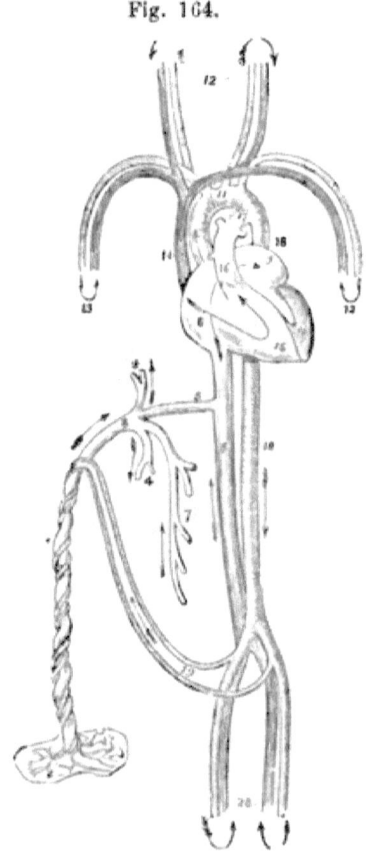

Fig. 164.

THE FŒTAL CIRCULATION.—1. The umbilical cord, consisting of the umbilical vein and two umbilical arteries, proceeding from the placenta (2). 3. The umbilical vein dividing into three branches; two (4, 4) to be distributed to the liver; and one (5) the ductus venosus, which enters the inferior vena cava (6). 7. The portal vein, returning the blood from the intestines, and uniting with the right hepatic branch. 8. The right auricle: the course of the blood is denoted by the arrow, proceeding from 8 to 9, the left auricle. 10. The left ventricle: the blood following the arrow to the arch of the aorta (11), to be distributed through the branches given off by the arch to the head and upper extremities. The arrows 12 and 13 represent the return of the blood from the head and upper extremities through the jugular and subclavian veins, to the superior vena cava (14), to the right auricle (8), and in the course of the arrow through the right ventricle (15) to the pulmonary artery (16). 17. The ductus arteriosus, which appears to be a proper continuation of the pulmonary artery—the offsets at each side are the right and left pulmonary artery cut off; these are of extremely small size as compared with the ductus arteriosus. The ductus arteriosus joins the descending aorta (18, 18) which divides into the common iliacs, and these into the internal iliacs, which become the umbilical arteries (19), and return the blood along the umbilical cord to the placenta; while the other divisions the external iliacs (20) are continued into the lower extremities. The arrows at the termination of these vessels mark the return of the venous blood by the veins to the inferior cava.

and conducts the blood from the right auricle to the aorta, into which it empties just below the arch. The *ductus venosus* passes from the umbilical vein to the vena cava ascendens along the thick edge of the liver.

The following is the course of the blood : It is conveyed from the placenta along the umbilical vein, partly, at once, to the vena cava ascendens by means of the ductus venosus, partly into the liver, whence it reaches the vena cava by the hepatic vein. Having passed through the two great fœtal depurating organs (liver and placenta) it is in the state of arterial blood. It enters the right auricle; but, being directed through the foramen ovale by the Eustachian valve, arrives at the left auricle, thence goes to the left ventricle, and is thrown out into the aorta. Pure arterial blood is, therefore, sent to the head and upper extremities. At the contraction of the ventricles, the right one throws its venous blood into the pulmonary artery; but a small part of it passes to the lungs, the greater portion going through the ductus arteriosus into the descending aorta, where it mingles with the descending arterial current. Thus the blood supplied to the trunk and lower extremities is mixed arterial and venous. Of the descending current just spoken of, a small part goes to supply the trunk and lower extremities with nutriment, but the chief portion passes directly to the placenta through the umbilical arteries; whence it is returned by means of the umbilical vein, after having been aerated and nourished by endosmotic action between it and the maternal blood.

CHANGES AT BIRTH.

After birth and the expansion of the lungs, this course of the blood is entirely changed. The ductus venosus and ductus arteriosus become impermeable, and, finally, consolidated into fibrous cords; the foramen ovale closes, by the apposition and union of its valves; the pulmonary artery and vein enlarge in accordance with the increase of their function, and the regular round of the human circulation commences. If for any reason these changes do not take place, serious organic disease is the result. Thus, when the foramen ovale remains patulous, there is in consequence a constant mixture of arterial and venous blood, resulting in the production of the well-known phenomenon of the so-called *cyanosis* or "*blue disease*."[1]

INFANCY AND ADOLESCENCE.

A new-born child has a very large head, upper extremities and abdomen, in proportion to its pelvis and lower limbs. The umbi-

---

[1] Produced also by insufficient enlargement of the pulmonary artery.

licus is nearly at the middle of its body. Gradually all the parts approach to the perfect symmetry of adolescence and maturity.

Breathing by the lungs is less active, and that by the skin more so, in infancy than later in life. The power of generating animal heat is less, and the need of protection or communicated heat

Fig. 165.

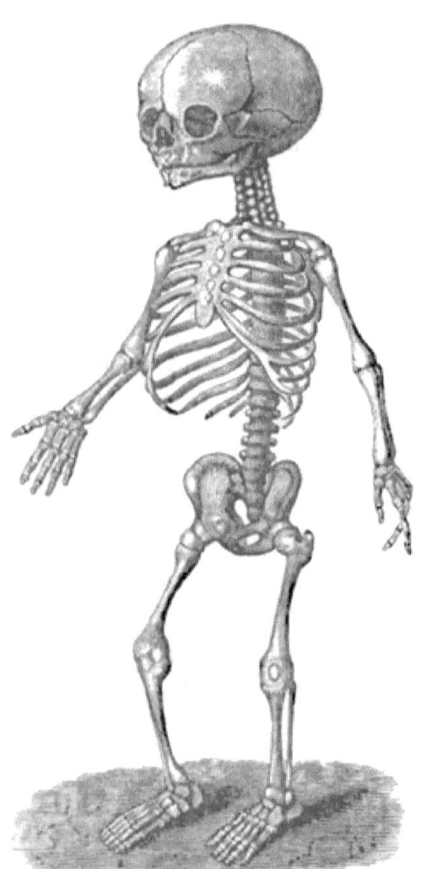

FŒTAL SKELETON.

much greater at that time than afterwards. The skin and alimentary canal are extremely delicate and impressible. Kölliker states that all the hairs of the head and body are renewed in the first year.

Of the nervous system, the organic or ganglionic (sympathetic) and spinal reflex apparatus are most nearly mature at birth. Excito-

INFANCY AND ADOLESCENCE.

motor involuntary actions are predominant over those of purpose and will, throughout childhood. Convulsions are more readily produced by irritating causes in infants and children than in adults.

At birth, the twenty first teeth are partly formed in both jaws, but beneath the gums. The following is the usual order of their protrusion or "cutting." Two central incisors, in each jaw, in the seventh month; two lateral incisors, eighth month; two anterior molars (first jaw teeth), end of twelfth month; two canines ("stomach teeth" of lower jaw, "eye-teeth" of upper), eighteenth month; two second molars or jaw teeth, twentieth to twenty-fourth month. Those of the lower jaw are generally first.

At about seven years of age, the second teeth begin to replace the others. The first to appear, commonly, is the first permanent molar tooth, at six years and a half. Next, at seven years, the permanent middle incisors. At eight, the lateral incisors follow. The first and second bicuspids before ten years of age. The four canines, upper and lower, before twelve. At thirteen, the second permanent molars. The third and last molars ("wisdom teeth") between seventeen and twenty-one; making in all thirty-two teeth of the second set; in each jaw, four incisors, two canines, four bicuspids, and six molars.

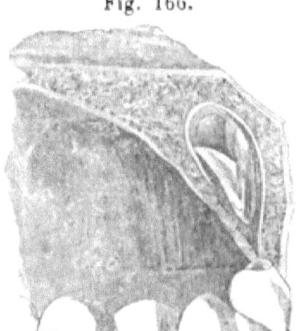

Fig. 166.

DEVELOPMENT OF TEETH.

The age of *puberty*, from fourteen to sixteen in temperate climates, is marked by important changes. Sex is now manifested, by menstruation and the development of the breasts in the female, and by the change of voice and growth of beard in the male. Emotional differences also appear.

The muscular system and the bony skeleton develop progressively from infancy to maturity. This is seldom reached before the twenty-fifth year. The osseous union of the ends and shafts, epiphyses and diaphyses of the long bones, and that of the different bones of the head, may sometimes be incomplete even to a later period; and, according to some anatomists, the brain may grow until forty, and the heart still longer.

The following table, from Dalton, exhibits the comparative proportion of different organs to the whole body, at birth and maturity:—

	Fœtus at term.	Adult.
Weight of entire body	1000.00	1000.00
" Thymus gland	3.00	0.00
" Thyroid "	0.60	0.51
" Renal capsules	1.63	0.13
" Kidneys	6.00	4.00
" Heart	7.77	4.17
" Liver	37.00	29.00
" Brain	148.00	23.00

Old age is shown by general *atrophy* or failure of nutrition as well as of power; frequently attended by various forms of degeneration of structure and defective functional action, tending towards death.

www.ingramcontent.com/pod-product-compliance
Lightning Source LLC
Chambersburg PA
CBHW022057230426
43672CB00008B/1201